Adsorption Technology for Water and Wastewater Treatments

Adsorption Technology for Water and Wastewater Treatments

Editor

Hai Nguyen Tran

Basel • Beijing • Wuhan • Barcelona • Belgrade • Novi Sad • Cluj • Manchester

Editor
Hai Nguyen Tran
Duy Tan University
Vietnam

Editorial Office
MDPI
St. Alban-Anlage 66
4052 Basel, Switzerland

This is a reprint of articles from the Special Issue published online in the open access journal *Water* (ISSN 2073-4441) (available at: https://www.mdpi.com/journal/water/special_issues/320NWX363A).

For citation purposes, cite each article independently as indicated on the article page online and as indicated below:

Lastname, A.A.; Lastname, B.B. Article Title. *Journal Name* **Year**, *Volume Number*, Page Range.

ISBN 978-3-0365-8585-7 (Hbk)
ISBN 978-3-0365-8584-0 (PDF)
doi.org/10.3390/books978-3-0365-8584-0

© 2023 by the authors. Articles in this book are Open Access and distributed under the Creative Commons Attribution (CC BY) license. The book as a whole is distributed by MDPI under the terms and conditions of the Creative Commons Attribution-NonCommercial-NoDerivs (CC BY-NC-ND) license.

Contents

About the Editor vii

Preface ix

Hai Nguyen Tran
Adsorption Technology for Water and Wastewater Treatments
Reprinted from: *Water* **2023**, *15*, 2857, doi:10.3390/w15152857 1

Yu Zou, Bozhi Ren, Zhendong He and Xinping Deng
Enhanced Removal of Sb (III) by Hydroxy-Iron/Acid–Base-Modified Sepiolite: Surface Structure and Adsorption Mechanism
Reprinted from: *Water* **2022**, *14*, 3806, doi:10.3390/w14233806 9

Miguel Pereira de Oliveira, Carlos Schnorr, Theodoro da Rosa Salles, Franciele da Silva Bruckmann, Luiza Baumann, Edson Irineu Muller, et al.
Efficient Uptake of Angiotensin-Converting Enzyme II Inhibitor Employing Graphene Oxide-Based Magnetic Nanoadsorbents
Reprinted from: *Water* **2023**, *15*, 293, doi:10.3390/w15020293 23

Meriem Zamouche, Mouchira Chermat, Zohra Kermiche, Hichem Tahraoui, Mohamed Kebir, Jean-Claude Bollinger, et al.
Predictive Model Based on K-Nearest Neighbor Coupled with the Gray Wolf Optimizer Algorithm (KNN_GWO) for Estimating the Amount of Phenol Adsorption on Powdered Activated Carbon
Reprinted from: *Water* **2023**, *15*, 493, doi:10.3390/w15030493 43

Yubo Zhao, Kexun Li, Bangsong Sheng, Feiyong Chen and Yang Song
Nitrogen-Doped Core-Shell Mesoporous Carbonaceous Nanospheres for Effective Removal of Fluorine in Capacitive Deionization
Reprinted from: *Water* **2023**, *15*, 608, doi:10.3390/w15030608 67

Marcela Paredes-Laverde, Diego F. Montaño and Ricardo A. Torres-Palma
Montmorillonite-Based Natural Adsorbent from Colombia for the Removal of Organic Pollutants from Water: Isotherms, Kinetics, Nature of Pollutants, and Matrix Effects
Reprinted from: *Water* **2023**, *15*, 1046, doi:10.3390/w15061046 81

Lotfi Khezami, Abueliz Modwi, Kamal K. Taha, Mohamed Bououdina, Naoufel Ben Hamadi and Aymen Amine Assadi
Mesoporous Zr-G-C_3N_4 Sorbent as an Exceptional Cu (II) Ion Adsorbent in Aquatic Solution: Equilibrium, Kinetics, and Mechanisms Study
Reprinted from: *Water* **2023**, *15*, 1202, doi:10.3390/w15061202 103

Hai Nguyen Tran
Applying Linear Forms of Pseudo-Second-Order Kinetic Model for Feasibly Identifying Errors in the Initial Periods of Time-Dependent Adsorption Datasets
Reprinted from: *Water* **2023**, *15*, 1231, doi:10.3390/w15061231 123

Fatima Elayadi, Mounia Achak, Wafaa Boumya, Sabah Elamraoui, Noureddine Barka, Edvina Lamy, et al.
Factorial Design Statistical Analysis and Optimization of the Adsorptive Removal of COD from Olive Mill Wastewater Using Sugarcane Bagasse as a Low-Cost Adsorbent
Reprinted from: *Water* **2023**, *15*, 1630, doi:10.3390/w15081630 137

Younes Dehmani, Dison S. P. Franco, Jordana Georgin, Taibi Lamhasni, Younes Brahmi, Rachid Oukhrib, et al.
Comparison of Phenol Adsorption Property and Mechanism onto Different Moroccan Clays
Reprinted from: *Water* **2023**, *15*, 1881, doi:10.3390/w15101881 . . . . **155**

Fahad A. Alharthi, Riyadh H. Alshammari and Imran Hasan
In Situ Polyaniline Immobilized ZnO Nanorods for Efficient Adsorptive Detoxification of Cr (VI) from Aquatic System
Reprinted from: *Water* **2023**, *15*, 1949, doi:10.3390/w15101949 . . . . **179**

Naila Farid, Amin Ullah, Sara Khan, Sadia Butt, Amir Zeb Khan, Zobia Afsheen, et al.
Algae and Hydrophytes as Potential Plants for Bioremediation of Heavy Metals from Industrial Wastewater
Reprinted from: *Water* **2023**, *15*, 2142, doi:10.3390/w15122142 . . . . **199**

Xu Liu, Huilai Liu, Kangping Cui, Zhengliang Dai, Bei Wang, Rohan Weerasooriya and Xing Chen
Adsorption-Reduction of Cr(VI) with Magnetic Fe-C-N Composites
Reprinted from: *Water* **2023**, *15*, 2290, doi:10.3390/w15122290 . . . . **213**

About the Editor

Hai Nguyen Tran

Dr. Hai Nguyen Tran is currently the director of the Center for Energy and Environmental Materials at Duy Tan University, Vietnam. He earned a Master's degree (Soil Science) from Can Tho University, Vietnam, and a Ph.D. degree (Environmental Engineering) from Chung Yuan Christian University, Taiwan. His interesting research includes adsorption science and technology, physical and chemical materials, catalysis, advanced oxidation processes, and waste management.

Dr Tran (ORCID: 0000-0001-8361-2616) has published more than 100 papers in peer-reviewed journals and some book chapters. He is a reviewer for many reputable journals (>100) in different publishers. His Scopus author identifier is 57052152100 (current H-index: 43). Based on the Scopus database, Stanford University annually published a ranking of the World's Top 2% Scientists. Dr. Tran was ranked in this list in 2022 (ranking: 13,713 and self-citation: 6.24%), 2021 (14,704 and 8.85%), and 2020 (25,844 and 13.56%). Moreover, he was ranked #2 in Vietnam and #4687 in theWorld among Best Scientists in the field of Environmental Sciences for 2023 (according to Research.com that is a leading academic platform for researchers).

He has served as an editor (*Water Science and Technology*), associate editor (*Environment, Development, and Sustainability; Frontiers in Environmental Chemistry; Vietnam Journal of Science, Technology, and Engineering*), academic editor (*Adsorption Science & Technology*), advisory editorial board member (*Journal of Chemical Technology and Biotechnology*), and editorial board member (*Chemosphere; Science of the Total Environment; Journal of Hazardous Materials Advances; Biochar; Carbon Research; Separation & Purification Reviews; Green Processing and Synthesis; Nanotechnology for Environmental; Current Pollution Reports; Experimental Results; Air, Soil, and Water Research*) of several prestigious journals.

Preface

The current levels of emerging and traditional pollutants in water are remarkably increasing, which is an enormous concern. Adsorption is acknowledged as a potential method for treating various kinds of pollutants in water and wastewater. This method has an operational cost, a fast kinetic removal ability, a minimal generation of secondary pollutants (i.e., sludge), and a high efficiency rate to remove different pollutants even under low initial concentrations of pollutants. However, the adsorption efficiency and affinity of pollutants are strongly dependent on the properties of adsorbents, the nature of adsorbates, and adsorption conditions. These can be reflected through adsorption mechanisms. Natural adsorbents often exhibit a high adsorption capacity to target pollutants. For example, activated carbon is an excellent material used for adsorbing organic pollutants (i.e., dye) in water. Some synthetic materials (suitable surfactant-modified zeolite) can serve as dual-electronic and amphiphilic adsorbents used for removing potentially toxic metals (cationic and oxyanionic ions) and organic compounds with low and high water solubility.

This reprint is expected to provide innovative and outstanding reference materials for water decontamination and bring more updated and advanced theories on science and technology related to adsorption for internal and national scientific communities. The Guest Editor would like to thank all authors and colleagues for their great contributions

Hai Nguyen Tran
Editor

Editorial

Adsorption Technology for Water and Wastewater Treatments

Hai Nguyen Tran [1,2]

[1] Center for Energy and Environmental Materials, Institute of Fundamental and Applied Sciences, Duy Tan University, Ho Chi Minh 700000, Vietnam; trannguyenhai@duytan.edu.vn or trannguyenhai2512@gmail.com
[2] Faculty of Environmental and Chemical Engineering, Duy Tan University, Da Nang 550000, Vietnam

Abstract: This Special Issue includes 12 research papers on the development of various materials for adsorbing different contaminants in water, such as Sb, Cr(VI), Cu(II), Zn(II), fluorine, phenol, dyes (indigo carmine, Congo red, methylene blue, and crystal violet), and drugs (dlevofloxacin, captopril, and diclofenac, and paracetamol). The commercial, natural, and synthetic materials used as adsorbents comprise commercial activated carbon, natural clay and montmorillonite, biosorbent based on sugarcane bagasse or algal, graphene oxide, graphene oxide-based magnetic nanomaterial, mesoporous Zr-G-C_3N_4 nanomaterial, nitrogen-doped core–shell mesoporous carbonaceous nano-sphere, magnetic Fe-C-N composite, polyaniline-immobilized ZnO nanorod, and hydroxy-iron/acid–base-modified sepiolite composite. Various operational conditions are evaluated under batch adsorption experiments, such as pH, NaCl, solid/liquid ratio, stirring speed, contact time, solution temperature, initial adsorbate concentration. The re-usability of laden materials is evaluated through adsorption–desorption cycles. Adsorption kinetics, isotherm, thermodynamics, and mechanisms are studied and discussed. Machine learning processes and statistical physics models are also applied in the field of adsorption science and technology.

Keywords: adsorption; mechanism; modelling; artificial intelligence; emerging pollutant; machine learning; statistical physics; water treatment

Citation: Tran, H.N. Adsorption Technology for Water and Wastewater Treatments. *Water* **2023**, *15*, 2857. https://doi.org/10.3390/w15152857

Received: 1 August 2023
Accepted: 2 August 2023
Published: 7 August 2023

Copyright: © 2023 by the author. Licensee MDPI, Basel, Switzerland. This article is an open access article distributed under the terms and conditions of the Creative Commons Attribution (CC BY) license (https://creativecommons.org/licenses/by/4.0/).

1. Introduction

The presence of pollutants (i.e., emerging contaminants, radionuclides, potential toxic metals, and dyes) in water environments has a negative impact on the environment and presents a potential health risk for inhabitants. Among the existing technologies (advanced oxidation process, membrane filtration, biodegradation, coagulation, and flocculation, etc.) for treating pollutants in water, adsorption has garnered great interest due to its low operational cost and fast removal. In particular, this method can effectively remove various pollutants at their low concentrations (trace levels) from water, if adsorbent materials exhibit excellent adsorption capacity and high adsorption affinity to pollutants. Unlike coagulation and flocculation processes, adsorption techniques do not generate by-products (i.e., sewage sludge). There is a remarkable and increasing amount of sewage sludge being discharged into the environment, which represents an enormous concern with regard to waste management aspects.

In recent times, the development of advanced materials and their application as potential adsorbents for water treatments have caught the attention of researchers. This Special Issue (SI) aims to establish the state of the art on adsorption and biosorption technologies for water and wastewater treatments.

Apart from traditional methods (i.e., response surface methodology), some machine learning processes are currently applied for estimating the adsorption capacity of the adsorbent (q_e) or the removal efficiency of the adsorption process (%R). Some typical examples of the machine learning are (1) the Gaussian process regression (GPR) coupled with the particle swarm optimization (PSO) model and (2) the nearest neighbor coupled with the gray wolf optimizer algorithm. The papers relating to the application of artificial intelligence in anticipating and optimizing operational conditions (strongly affecting the q_e or %R value) are given a major priority in this SI.

The adsorption isotherm plays an essential role in this field. In the literature, many isotherm models have been used to fit adsorption equilibrium datasets. They include the Langmuir and Freundlich models (the most widely used model), Lui model, Redlich–Peterson model, Hill model, Radke–Prausnitz model, Khan model, Toth model, Sips model, and Koble–Corrigan model. Unlike these models, several statistical physics models can help to well explain the adsorption phenomena and mechanisms. This SI also gives a priority for collecting this topic.

This SI belongs to the section "Wastewater Treatment and Reuse" of the journal (*Water*; SSN 2073-4441). After the Editor-in-Chief of *Water* (Dr. Jean-Luc PROBST) had approved this SI, "Adsorption Technology for Water and Wastewater Treatments", it was announced online on the website of the journal and opened for submissions from 20 September 2022 to 30 June 2023. The acceptance rate of this SI is approximately 63%. The impact factor (Web of Science), CiteScore (Scopus), and H-index (Scimago) values for this journal are 3.4, 5.5, and 85, respectively.

This SI successfully compiled 12 unique papers contributed by 75 authors from 12 countries (France, United Kingdom, Vietnam, China, Brazil, Colombia, Algeria, Saudi Arabia, Sudan, Morocco, Pakistan, and Sri Lanka). All of them [1–12] are of the "Article" type. In particular, one published paper was recommended as a "feature paper" by two academic editors (Andrea G. Capodaglio and Antonio Panico) of this journal. A Feature Paper is a substantial original article that presents "the most advanced research with significant potential for high impact in the field".

Various adsorbates that were investigated and published in this SI included antimony (Sb), hexavalent chromium, copper, zinc, fluorine, phenol, organic dye (indigo carmine, Congo red, methylene blue, and crystal violet), and pharmaceutical adsorbates (levofloxacin, captopril, diclofenac, and paracetamol). To provide considerable insights into the essential aspects of this SI, the Editor highlights some brief summaries on all the works published [1–12] in the next section.

2. Summary of Papers Published in This Special Issue

2.1. Enhanced Removal of Sb (III) by Hydroxy-Iron/Acid–Base-Modified Sepiolite: Surface Structure and Adsorption Mechanism

Zou and co-workers [1] developed a new composite derived from a sepiolite mineral (magnesium-enriched silicate clay) through two steps: treating sepiolite with acid–base (HCl and NaOH) and then impregnating it with $FeCl_3$.

The composite and the sepiolite were applied for adsorbing antimony (Sb) in an aqueous solution. The pKa of antimony is around 3. Antimony exists in water as $Sb(OH)_3$ (within a solution pH from 1.0 to 6.0) and as $Sb(OH)_6^-$ (pH from 1.0 to 14). At pH = 3, the percentage of $Sb(OH)_3$ and $Sb(OH)_6$—species was 50% and 50%. The batch adsorption experiments were conducted under different conditions: solid/liquid (m/V) ratio (0.5–3.0 g/L), solution pH (2–12), contact time (3–240 min), initial Sb(III) concentration (C_o = 10–150 mg/L), and temperatures (298–318 K).

The powder X-ray diffraction (PXRD) data indicated that the primary iron form present in the composite was iron(III) oxide-hydroxide FeO(OH). The BET (Brunauer–Emmett–Teller) specific surface area (S_{BET}) of the composite (152.4 m^2/g) was slightly lower than that of the sepiolite (160.1 m^2/g).

The maximum adsorption capacity (Q_{max}) calculated based on the Langmuir model of the composite (at 25 °C, 35 °C, and 45 °C) was 25.73, 26.89, and 32.45 mg/g, values that were higher than the corresponding values of the sepiolite (14.66, 14.87, and 16.42 mg/g, respectively). The adsorption mechanisms were evaluated via the X-ray photoelectron spectroscopy (XPS) technique.

2.2. Efficient Uptake of Angiotensin-Converting Enzyme II Inhibitor Employing Graphene Oxide-Based Magnetic Nanoadsorbents

Oliveira and colleagues [2] used graphite to generate a magnetic graphene oxide (M-GO) and applied M-GO as an adsorbent to eliminate captopril from water. The magnetic graphene oxide was prepared through the co-precipitation method of graphene oxide with $FeCl_2$ at pH 9.0.

The magnetization saturation (M_S) of M-GO was 45 emu/g. The results of Raman spectroscopy indicated that the I_D/I_G ratio of GO (0.96) increased to 1.22 after the magnetization procedure (M-GO).

Different conditions were applied for the batch adsorption processes such as solid/liquid (m/V) ratio (0.125–1.0 g/L), solution pH (2–9), NaCl (0.01–1 M), contact time (5–180 min), initial captopril concentration (C_o = 10–200 mg/L), and temperatures (20–40 °C). Regeneration and reuse studies were applied by using NaOH (0.25 mol/L) as a desorbing agent.

The result indicated that under the same adsorption conditions, M-GO (99.4 mg/g) exhibited a slightly higher adsorption capacity of captopril than GO (91.8 mg/g). The Q_{max} values of M-GO towards captopril at 20 °C, 30 °C, and 40 °C were 101.9, 101.5, and 99.3 mg/g, respectively. The standard enthalpy change ($\Delta H°$) of this adsorption process was −64.19 kJ/mol (exothermic in nature). The laden M-GO can be well reused after five adsorption–desorption cycles.

2.3. Predictive Model Based on K-Nearest Neighbor Coupled with the Gray Wolf Optimizer Algorithm (KNN_GWO) for Estimating the Amount of Phenol Adsorption on Powdered Activated Carbon

The application of machine learning in optimizing and predicting the amount of phenol (diameter: 0.56 nm) adsorbed by commercially available powdered activated carbon (PAC) under aqueous solution was examined by Zamouche et al. [3]. This paper has been recommended as a "feature paper".

The S_{BET} value and total pore volume (V_{Total}) of PAC were 463.4 m^2/g and 0.13 cm^3/g. Its strong acid carboxylic group and hydroxyl and phenol groups. The effect of several adsorption conditions on the phenol adsorption processes was explored, such as solid/liquid ratio (0.4–1.6 g/L), solution pH (2.1–12), stirring speed (200–600 rpm) contact time (5–60 min), initial phenol concentration (C_o = 10–200 mg/L), and temperature (20 °C–50 °C).

The Q_{max} values of PAC towards phenol were 156.3 mg/g at 20 °C, 150.0 mg/g at 30 °C, and 150.8 mg/g at 40 °C. The $\Delta H°$ of this process was −19.23 kJ/mol (an exothermic process).

The result of applying K-nearest neighbor (common method in machine learning) coupled with the gray wolf optimizer algorithm for predicting the phenol amount adsorbed by PAC demonstrated an advantage in providing highly precise values because of very low statistical errors (closing to zero) and high statistical coefficients (reaching unity).

2.4. Nitrogen-Doped Core–Shell Mesoporous Carbonaceous Nanospheres for Effective Removal of Fluorine in Capacitive Deionization

Zhao and coworkers [4] synthesized a mesoporous carbonaceous nanospheres (MCS) doped with nitrogen (NMCS) using a popular sol–gel method. The SEM images showed that MCS and NMCS exhibited a rough surface morphology with spherical shapes. Meanwhile, their core–shell structure was confirmed via transmission electron microscope (TEM). The Raman data provided a higher I_D/I_G ratio of NMCS (0.999) than MCS (0.982). The S_{BET} and V_{Total} of MCS (1058 m^2/g and 1.89 cm^2/g) decreased after doping with nitrogen NMCS (1049 m^2/g and 1.78 cm^2/g).

The presence of nitrogen-containing functional groups in NMCS was confirmed through a new band at around 3430 cm^{-1} (i.e., N–H) in its FTIR spectrum. The XPS data provided the information on the nitrogen content of MCS (0.31%; %atomic) and NMCS (4.30%; %atomic) as well as the fraction of nitrogen (%) in two samples. The high-resolution

N 1s spectrum of MCS and NMCS suggested their nitrogen-containing functional groups were graphitic-N (401 eV), pyrrolic-N (400 eV), and pyridinic-N (398 eV).

NMCS was applied for removing fluorine (F^- anions) from water. Its maximum electrosorption capacity to F^- anions (C_o = 1000 mg/L at 12 V) was 13.34 mg/g. The result of adsorption–regeneration cycles showed that NMCS exhibited an excellent repeatability, with its electrosorption capacity slightly decreasing after 10 cycles.

2.5. Montmorillonite-Based Natural Adsorbent from Colombia for the Removal of Organic Pollutants from Water: Isotherms, Kinetics, Nature of Pollutants, and Matrix Effects

The study of utilizing a natural clay (montmorillonite) as adsorbent for treating various organic pollutants in water was undertaken by Paredes-Laverde and colleagues [5]. The target adsorbates included traditional pollutants (Congo red (CR), indigo carmine (IC), crystal violet (CV), and methylene blue (MB) dye) and emerging pollutants (diclofenac and levofloxacin drugs).

The montmorillonite was characterized through thermogravimetric analysis (TGA), Fourier transform infrared spectroscopy (FTIR), scanning electronic microscopy (SEM), XRD, and BET analyzer. Its S_{BET} and V_{Total} were 85.5 m^2/g and 0.004 cm^3/g.

The montmorillonite exhibited a relatively low Q_{max} value for dyes: CV (18.2 mg/g), CR (16.2 mg/g), MB (14.1 mg/g), and IC (1.07 mg/g) under aqueous solutions. The montmorillonite can remove 99% CV dye under an aqueous solution and 85% CV dye under real textile wastewater. The result of adsorbing pharmaceuticals in urine indicated that the montmorillonite can remove approximately 90% of levofloxacin and 8% of diclofenac after a 60 min contact.

2.6. Mesoporous Zr-G-C3N4 Sorbent as an Exceptional Cu(II) Ion Adsorbent in Aquatic Solution: Equilibrium, Kinetics, and Mechanisms Study

Khezami et al. [6] slightly modified an one-step ultrasonication method to synthesize a mesoporous nanomaterial (Zr-G-C$_3$N$_4$) that was graphitic carbon nitride (g-C$_3$N$_4$) doped with ZrO$_2$.

The XRD data confirmed the existence of both g-C$_3$N$_4$ and ZrO$_2$ peaks. The textural parameters of Zr-G-C$_3$N$_4$ were 95.7 m^2/g (S_{BET}) and 0.216 cm^3/g (V_{Total}).

Adsorption kinetic of Cu(II) by Zr-G-C$_3$N$_4$ reached an equilibrium at around 500 min. The studies on the effect of pH solution (pH = 1.0–8.0) indicated that the optimal pH value for the Cu(II) adsorption process was established at 5.0. The adsorption isotherm (C_o = 5–200 mg/L) provided the Q_{max} value of Zr-G-C$_3$N$_4$ was 144.1 mg/g. After four continuous cycles of the Cu(II) adsorption–desorption processes using 0.1 M NaOH as a desorbing agent, the removal efficiency was still higher than 80%. The XPS data confirmed that the primary adsorption mechanism was surface complexation.

2.7. Applying Linear Forms of Pseudo-Second-Order Kinetic Model for Feasibly Identifying Errors in the Initial Periods of Time-Dependent Adsorption Datasets

Tran [7] studied the kinetic process of paracetamol adsorption onto commercial activated carbon (its S_{BET} = 1275 m^2/g, and V_{Total} = 0.670 cm^3/g) under experimental conditions: C_o = 3.45 mmol/L; pH = 7.0; 25 °C; m/V = 1 g/L. The contact time was 1, 5, 15, 30, 45, 60, 90, 120, 180, 240, 300, 360, 720, 1440, 2880, and 4320 min. Some adsorption kinetic models were applied for modelling time-dependent adsorption datasets, such as the pseudo-n^{th}-order, pseudo-first-order, pseudo-second-order (PSO), Elovich, and Avrami models. The best-fitting model was evaluated through the highest value of determination coefficient (R^2) and adjusted determination coefficient (adj-R^2) and the lowest value of chi-squared (χ^2), reduced chi-square (red-χ^2), and Bayesian information criterion (BIC).

Among those models, the PSO model is the most commonly used in this field. Six linear forms (Types 1–6) and one non-linear form of the PSO model were used for exploring the error points under the initial periods (potential outliers). The results indicated that although the use of a non-linear optimization method is often recommended to minimize error functions, it cannot help to identify errors in the initial periods of time-dependent adsorption datasets. Among six forms of the PSO model, those error experimental points (i.e., very high q_t values at 1 and 5 min) can be observed though Types 2–5.

In the literature, the Type 1 is often used for fitting the experimental data of time-dependent adsorption because of very adj-R^2 (i.e., 0.9999). However, in many cases, this high adj-R^2 or r^2 indicates a spurious correlation. Therefore, the estimation of the rate constant k_2 of time-dependent adsorption processes based on the Type 1 should be avoided. After removing the outliers, the experiential data were well described by the following models: Avrami > PNO > PSO > PFO > Elovich.

2.8. Factorial Design Statistical Analysis and Optimization of the Adsorptive Removal of COD from Olive Mill Wastewater Using Sugarcane Bagasse as a Low-Cost Adsorbent

Sugarcane bagasse that was utilized as a biosorbent for treating COD (chemical oxygen demand) from real olive mill wastewater was published by Elayadi and colleagues [8]. The authors applied the 2^{5-1} fractional factorial design (using Minitab 18) for optimizing the effects of various operational conditions on the COD removal efficiency.

The experimental variables (their low and high levels) included biosorbent dosage (10 and 60 g/L), stirring speed (80 and 300 rpm), adsorption time (1 and 24 h), pH (2 and 12), and solution temperature (20 °C and 60 °C).

The result showed that the optimal conditions reached at 10 g/L, 80 rpm, 1 h, pH 12, and 60 °C, with the COD removal efficiency being 55.1%. The authors reported that the Q_{max} value of the sugarcane bagasse obtained from the adsorption isotherms was 331.9 mg/g.

2.9. Comparison of Phenol Adsorption Property and Mechanism onto Different Moroccan Clays

Dehmani and coworkers [9] investigated the adsorption capacity and mechanisms to phenol in aqueous solutions using two natural clay materials (RCA and RCG). Those clays were characterized via X-ray Fluorescence (XRF), XRD, FTIR, TGA, and SEM–EDS. The pH_{PZC} and CEC values of RCA were 8.60 and 10.6 meq/100 g, and those of RCG were 8.60 and 16.6 meq/100 g, respectively. The clay samples (RCA and RCG) exhibited low values of S_{BET} (51.41 and 25.31 m2/g) and V_{Total} (0.13 and 0.03 cm/g).

The adsorption process occurred fast and reached equilibrium after 30 min of contact. Adsorption isotherms were conducted at 30 °C, 40 °C, and 50 °C. Isotherm modeling was evaluated through physical–statistical approaches such as the monolayer model with a single energy site, monolayer model with two energy sites, double-layer model with one energy site, double-layer model with two energy sites, and multilayer model. The best fitting of models was evaluated based on R^2, adj-R^2, minimum squared error, BIC, and average relative error.

The result showed that the adsorption equilibrium data were satisfactorily described by the monolayer model with two different energy sites. The primary adsorption mechanism of phenol onto RCA was van der Waals, and those onto RCG were van der Waals and hydrogen bond.

2.10. In Situ Polyaniline Immobilized ZnO Nanorods for Efficient Adsorptive Detoxification of Cr(VI) from Aquatic System

Alharthi and colleagues [10] applied an in situ free radical polymerization method to synthesize a PAni-ZnO composite that was zinc oxide nanorod (ZnO-nanorod) immobilized with polyaniline. Some techniques that were used for characterizing the materials (ZnO-nanorod, polyaniline, and PAni-ZnO composite) included XRD, TGA, FTIR, SEM–EDS, and TEM.

The S_{BET} value and the average pore width of PAni-ZnO composite were 113.5 m^2/g and 7 nm. The analysis result of zeta potentials of the composite indicated its isoelectric point (IEP) at pH = 4.32 (its pH_{IEP} = 4.32).

The adsorption process of Cr(VI) by the PAni-ZnO composite was conducted at different operational conditions such as pH, contact time, adsorbent dose, initial Cr(VI) concentration, and solution temperature. The effect of the nanorod sizes (10.5, 19.0, 32.6, and 52.2 nm) on the Cr(VI) adsorption behavior of the composite was also evaluated.

The result of the absorption isotherm demonstrated that the Q_m values of the composite increased (142.3, 219.2, and 310.4 mg/g) within an increase in solution temperatures (25 °C, 35 °C, and 45 °C, respectively). The value $\Delta H°$ of this adsorption process was 32.01 kJ/mol, suggesting dominantly physical adsorption.

2.11. Algae and Hydrophytes as Potential Plants for Bioremediation of Heavy Metals from Industrial Wastewater

Biosorption of potential toxic metals from real industrial wastewater using algae (*Zygnema pectiantum* and *Spyrogyra* species) and hydrophytes (*Typha latifolia* and *Eicchornia crassipes*) for 40 days was studied by Farid and coworkers [11]. The industrial wastewater (IWW) discharged from various industries (steel, textile, marble, chemical, ghee, and soap and oil).

The physicochemical characteristics of IWW were pH (6.85), EC (electrical conductivity; 6.35 mS/cm), COD (chemical oxygen demand; 236 mg/L), BOD (biological oxygen demand; 143 mg/L), DO (dissolved oxygen; 114 mg/L), TDS (total dissolved solids; 559.67 mg/L), TSS (total suspended solids; 89.30 mg/L), Cu (7.13 mg/L), and Zn (5.73 mg/L). The authors concluded that the concentrations of COD, BOD, Cu, and Zn of the raw IWW were higher than the corresponding values currently published by the Pakistan Environmental Protection Agency (maximum permissible limits).

The results demonstrated that the removal efficiencies of both hydrophytes and algae towards various selected parameters were EC (74.01–91.18%), DO (13.15–62.20%), BOD (21.67–73.42%), COD (25.84–73.30%), TDS (14.02–95.93%), and TSS (9.18–67.99%). In particular, they can effectively remove approximately 85% of Cu ions and 82% of Zn ions from IWW. Namely, the removal efficiency of Cu and Zn was 85.13% and 71.55% (using *Typha latifolia*), 80.78% and 62.12% (*Eicchornia crassipes*), 81.20% and 82.19% (*Zygnema pectiantum*), and 81.90% and 60.90% (*Spyrogyra* species).

2.12. Adsorption–Reduction of Cr(VI) with Magnetic Fe-C-N Composites

Liu et al. [12] reused an iron-enriched coagulant sludge collected from a dye chemical wastewater plant to develop magnetic Fe–C–N composites through a direct calcination process at different temperatures (200, 400, 500, and 600 °C). The developed material was then used for studying the adsorption of Cr(VI) anions.

The results of the materials' properties showed that the highest magnetization saturation (M_S = 4.3 emu/g) was recorded for the FCN-500 sample (prepared by calcinating the sludge at 500 °C). A high M_S value resulted from the existence of Fe_3O_4 nano-particles in this material that was confirmed via XRD analysis. FCN-500 can be classified as a non-porous material because of its low S_{BET} value (47.77 m^2/g). The XPS survey data indicated that FCN-500 contained mainly carbon (41.24%; %atomic), oxygen (34.74%), iron (12.8%), and nitrogen (11.22%). The XRF data demonstrated that FCN-500 and the sludge had a high percentage of iron (74.01% and 68.57%, respectively).

The Langmuir maximum adsorption capacity of FCN-500 towards Cr(VI) anions in aqueous solutions (adsorption isotherm conditions: 35 °C, pH = 2.0, 120 min, m/V = 0.8 g/L, and C_o = 10–50 mg/L) was 52.63 mg/g. The study of consecutive adsorption–desorption cycles (using 1 M NaOH as a desorbing agent) confirmed that FCN-500 exhibited an excellent reusability, with its adsorption capacity decreasing by approximately 20% after the five cycles.

The adsorption mechanism was considered at pH 2.0. The Cr 2p XPS spectrum of FCN-500 after adsorbing Cr(VI) indicated the dominant species of Cr(III) in the laden FCN-500, suggesting adsorption-coupled reduction mechanisms. Cr(VI) anions in solution were reduced to Cr(III) ions by contacting the nitrogen groups on the surface of FCN-500. The reduced Cr(III) ions were adsorbed in FCN-500 by mainly chelating with aprotic nitrogen atoms in its surface. Another mechanism was surface precipitation that was confirmed by the appearance of a peak at around 42° (involved in the crystal plane of Cr–N–Fe species) in the laden FCN-500.

Notably, under real electroplating wastewater conditions (C_o = 117 mg/L, 25 °C, m/V = 2.4 g/L, and pH = 3), FCN-500 can effectively treat Cr(VI) in this wastewater sample, with its removal efficiency reaching 99.93% after a 120 min contact.

3. Conclusions

Different materials were synthesized and used as potential adsorbents for removing a variety of adsorbates: inorganic and inorganic pollutants (especially some emerging pollutants) from aqueous solution, synthetic wastewater, and real wastewater samples. Various operational parameters affecting adsorption processes were instigated. Adsorption mechanisms were discussed and proposed.

However, all adsorption experiments [1–12] were conducted via a batch technique. With regard to real applications, adsorption phenomena under column conditions need to be continuously investigated. Current data for a leaching test and waste management after adsorption processes are missed. Further studies should be carried out to estimate the cost preparation of materials. Techno-economic analyses for treating pollutants from water via adsorption-based techniques are also suggested.

Acknowledgments: As the guest editor of this Special Issue, I gratefully acknowledge the following: (1) all colleagues and authors for their submissions, patience, and great contributions; (2) the staff members of the journal and publisher for their considerable assistance; and (3) expert reviewers for their valuable time and constructive comments. Finally, I would like to take this great opportunity to thank my dearest wife (Nguyen Phuong Van) and daughter (Tran Hai An) for their encouragement and support in completing this SI and this book.

Conflicts of Interest: The author declares no conflict of interest.

References

1. Zou, Y.; Ren, B.; He, Z.; Deng, X. Enhanced Removal of Sb (III) by Hydroxy-Iron/Acid–Base-Modified Sepiolite: Surface Structure and Adsorption Mechanism. *Water* **2022**, *14*, 3806. [CrossRef]
2. de Oliveira, M.P.; Schnorr, C.; da Rosa Salles, T.; da Silva Bruckmann, F.; Baumann, L.; Muller, E.I.; da Silva Garcia, W.J.; de Oliveira, A.H.; Silva, L.F.O.; Rhoden, C.R.B. Efficient Uptake of Angiotensin-Converting Enzyme II Inhibitor Employing Graphene Oxide-Based Magnetic Nanoadsorbents. *Water* **2023**, *15*, 293. [CrossRef]
3. Zamouche, M.; Chermat, M.; Kermiche, Z.; Tahraoui, H.; Kebir, M.; Bollinger, J.-C.; Amrane, A.; Mouni, L. Predictive Model Based on K-Nearest Neighbor Coupled with the Gray Wolf Optimizer Algorithm (KNN_GWO) for Estimating the Amount of Phenol Adsorption on Powdered Activated Carbon. *Water* **2023**, *15*, 493. [CrossRef]
4. Zhao, Y.; Li, K.; Sheng, B.; Chen, F.; Song, Y. Nitrogen-Doped Core-Shell Mesoporous Carbonaceous Nanospheres for Effective Removal of Fluorine in Capacitive Deionization. *Water* **2023**, *15*, 608. [CrossRef]
5. Paredes-Laverde, M.; Montaño, D.F.; Torres-Palma, R.A. Montmorillonite-Based Natural Adsorbent from Colombia for the Removal of Organic Pollutants from Water: Isotherms, Kinetics, Nature of Pollutants, and Matrix Effects. *Water* **2023**, *15*, 1046. [CrossRef]
6. Khezami, L.; Modwi, A.; Taha, K.K.; Bououdina, M.; Ben Hamadi, N.; Assadi, A.A. Mesoporous Zr-G-C3N4 Sorbent as an Exceptional Cu (II) Ion Adsorbent in Aquatic Solution: Equilibrium, Kinetics, and Mechanisms Study. *Water* **2023**, *15*, 1202. [CrossRef]
7. Tran, H.N. Applying Linear Forms of Pseudo-Second-Order Kinetic Model for Feasibly Identifying Errors in the Initial Periods of Time-Dependent Adsorption Datasets. *Water* **2023**, *15*, 1231. [CrossRef]
8. Elayadi, F.; Achak, M.; Boumya, W.; Elamraoui, S.; Barka, N.; Lamy, E.; Beniich, N.; El Adlouni, C. Factorial Design Statistical Analysis and Optimization of the Adsorptive Removal of COD from Olive Mill Wastewater Using Sugarcane Bagasse as a Low-Cost Adsorbent. *Water* **2023**, *15*, 1630. [CrossRef]
9. Dehmani, Y.; Franco, D.S.P.; Georgin, J.; Lamhasni, T.; Brahmi, Y.; Oukhrib, R.; Mustapha, B.; Moussout, H.; Ouallal, H.; Sadik, A. Comparison of Phenol Adsorption Property and Mechanism onto Different Moroccan Clays. *Water* **2023**, *15*, 1881. [CrossRef]
10. Alharthi, F.A.; Alshammari, R.H.; Hasan, I. In Situ Polyaniline Immobilized ZnO Nanorods for Efficient Adsorptive Detoxification of Cr(VI) from Aquatic System. *Water* **2023**, *15*, 1949. [CrossRef]

11. Farid, N.; Ullah, A.; Khan, S.; Butt, S.; Khan, A.Z.; Afsheen, Z.; El-Serehy, H.A.; Yasmin, H.; Ayaz, T.; Ali, Q. Algae and Hydrophytes as Potential Plants for Bioremediation of Heavy Metals from Industrial Wastewater. *Water* **2023**, *15*, 2142. [CrossRef]
12. Liu, X.; Liu, H.; Cui, K.; Dai, Z.; Wang, B.; Weerasooriya, R.; Chen, X. Adsorption-Reduction of Cr(VI) with Magnetic Fe-C-N Composites. *Water* **2023**, *15*, 2290. [CrossRef]

Disclaimer/Publisher's Note: The statements, opinions and data contained in all publications are solely those of the individual author(s) and contributor(s) and not of MDPI and/or the editor(s). MDPI and/or the editor(s) disclaim responsibility for any injury to people or property resulting from any ideas, methods, instructions or products referred to in the content.

Article

Enhanced Removal of Sb (III) by Hydroxy-Iron/Acid–Base-Modified Sepiolite: Surface Structure and Adsorption Mechanism

Yu Zou [1,2], Bozhi Ren [1,2,*], Zhendong He [1,2] and Xinping Deng [3]

[1] Hunan Provincial Key Laboratory of Shale Gas Resource Exploitation, Xiangtan 411201, China
[2] School of Civil Engineering, Hunan University of Science and Technology, Xiangtan 411201, China
[3] 402 Geological Prospecting Party, Xiangtan 411201, China
* Correspondence: bozhiren@126.com

Abstract: To improve the removal of antimony (Sb) from contaminated water, sepiolite (Sep) was chosen as the feedstock, modified with an acid–base and a ferric ion to yield a hydroxy-iron/acid–base-modified sepiolite composite (HI/ABsep). The surface structure of the HI/ABsep and the removal effect of the HI/ABsep on Sb (III) were investigated using potassium tartrate of antimony as the source of antimony and HI/ABsep as the adsorbent. The structural features of the HI/ABsep were analyzed by SEM, FTIR, PXRD, BET, and XPS methods. Static adsorption experiments were performed to investigate the effects of adsorption time, temperature, adsorbent dosage, and pH on the Sb (III) adsorbed by HI/ABsep. This demonstrates that sepiolite has a well-developed pore structure and is an excellent scaffold for the formation of hydroxy-iron. HI/ABsep adsorption of Sb (III) showed the best fit to the pseudo-second-order model and the Freundlich model. The maximum saturated adsorption capacity of the HI/ABsep regarding Sb (III) from Langmuir's model is 25.67 mg/g at 298 K. Based on the research results, the HI/ABsep has the advantages of easy synthesis and good adsorption performance and has the potential to become a remediation for wastewater contaminated with the heavy metal Sb (III).

Keywords: antimony; hydroxy-iron; sepiolite; adsorption

1. Introduction

The heavy metal pollution of water supplies and sewers has become a common problem in the production of food safety and human health [1]. Antimony and its compounds, as a crucial strategic nonrenewable metal, have been extensively used in flame retardants, alloys, ceramics, pigments, and synthetic fibers [2,3]. However, antimony is a widely distributed toxic element. Antimony (Sb) is classified as one of the main pollutants by the US Environmental Protection Agency and the European Environmental Protection Agency due to its proven and potential carcinogenicity, immunotoxicity, genotoxicity, and toxicity to the reproductive system [4–6]. Antimony in water exists primarily in two forms, Sb (V) and Sb (III), among which the toxicity of Sb (III) is approximately ten times higher than that of Sb (V) [7]. China also has the largest reserve of antimony and is the world's largest coal miner. Extensive antimony mining makes antimony pollution more serious in the southwest [8]. Therefore, it is crucial to find a rapid and efficient way to remove antimony.

In the past few decades, standard methods for removing Sb mainly include adsorption [9], ion exchange [10], coagulation [11], and microbial treatment [12]. The adsorption method is still one of the most widely used of these methods due to its advantages: simple operation, low cost, and greenness of the materials. The adsorbents include modified nonmetallic minerals, natural or synthetic metal oxides, and carbon materials, etc. The current research results show that iron oxyhydroxide has a good removal effect on antimony in water [13,14]. However, ferric hydroxide in water is often in a loose, amorphous flocculation state, with the disadvantage of not being easy to disband nor easy to separate from water. These shortcomings increase the cost of removing antimony from iron oxides and limit their practical application. The key approach to solving this problem is to fix iron oxide in coarse, porous granular carriers with a large specific surface area. At present, iron modification is widely applied in clay minerals. Bahabadi et al. [15] compared the adsorption effects of several naturally occurring clay minerals and iron-modified clay minerals on heavy metal adsorption and found that the modification of Fe^{3+} ions can be localized in the channel or that oxygen- or hydrogen–oxygen complexes are formed at the mineral surface to enhance the active sites of the mineral. Xu et al. [16] prepared a composite material of iron oxide and quartz sand from iron oxide-filled quartz sand and used it to adsorb Sb (III) into the water.

The mineral sepiolite (Sep) is a naturally occurring magnesium-rich silicate clay mineral with a standard crystalline chemical formula of $Mg_8Si_{12}O_{30}(OH)_4(OH_2)_4 \cdot 8H_2O$. China's sepiolite resources are mainly distributed in Hunan Province, with proven reserves exceeding 21.4 million tons. Therefore, it is an excellent choice to study iron modification with sepiolite from local materials. In the unique channel of the sepiolite nanostructure, the presence of "zeolite water" and some Mg^{2+} and Ca^{2+} exchangeable ions cause it to have strong adsorption properties and ion exchange capacities, so its potential application in the heavy metal adsorbing field has been widely investigated [17,18]. Sepiolite is cheap, readily available, and has a large surface area. It has great potential in the treatment of heavy metal wastewater. Natural sepiolite, however, has shortcomings, such as low surface acidity, a small channel, and low metal binding constant, so it needs proper modification processing [19].

Hence, this study is mainly conducted from the following three aspects: (I) the modified two-step sepiolite synthesis of antimony adsorption material, and SEM-EDS, FTIR, PXRD, BET, and XPS technology enhanced for analyzing the adsorbing process. (ii) The surface structure and adsorption conditions of HI/ABsep are investigated, including contact time, adsorbent dosage, pH, and initial concentration of Sb (III). (iii) The adsorption behavior and adsorption type of HI/Absep-adsorbed Sb (III) were determined by quasi-primary and quasi-secondary kinetic models, and Langmuir and Freundlich isothermal models.

2. Materials and Methods

2.1. Preparation of HI/ABsep

Sepiolite purity >80%, procured from Xiangtan Sepiolite Technology Co., Ltd. (Xiangtan, China). All reagents are of analytical grade and procured from Sinopharm Chemical Reagent Co., Ltd. (Shanghai, China).

Sepiolite (10 g) is dispersed in 2.5 mol/L HCl solution (250 mL) at room temperature, acid activation is performed at 250 r/min agitation speed for 24 h, then ammonia water is added, and added dropwise under magnetic stirring in the solution until the pH of the solution reaches 10.0. The supernatant was stirred for 2 h and then removed by centrifugation, followed by several items of deionized water and absolute ethanol. The precipitate is finally collected, dried in a drying oven at a constant temperature of 80 °C, crushed, and sieved to obtain a combined acid–base-modified sepiolite (abbreviated as ABsep). Disperse ABsep (10 g) in 1 mol/L $FeCl_3$ solution (10 mL) and sonicate for 10 min. Then, 1 mol/L NaOH (10 mL) was added, stirred with a magnetic stirrer at 250 r/min for 2 h, washed several times with deionized water, centrifuged, dried at 80 °C, then ground, continued to be cleaned until nearly neutral, and then after drying, passed through a 200-mesh sieve, and, finally, HI/ABsep is prepared.

2.2. Characterization Methods

Before the option of Sb (III) in simulated wastewater, sepiolite and HI/ABsep derived from sepiolite were characterized. Fourier-transform infrared spectroscopy (FTIR) (ALPHA, Bruker) was used to analyze the functional groups in the samples and their changes during the reaction. The surface morphology and structural characteristics of the sample were observed by scanning electron microscope (SEM) (JEOL JSM-6700F) with energy spectrum;

X-ray powder diffraction (PXRD) was obtained by X-ray diffractometer (D/Max 2500, Rigaku). The material pore structure was measured by the specific surface area and porosity analyzer (ASAP2020M). The composition and distribution information of chemical elements was obtained by X-ray photoelectron spectroscopy (XPS) (ESCALAB 250Xi, Thermo fisher, Waltham, MA, USA).

2.3. Batch Adsorption Experiments

The powdered potassium tartrate of antimony dissolved in diluted hydrochloric acid was prepared for 500 mg/L, mother liquor pH 6, and diluted by mother liquor to obtain simulated wastewater of different antimony concentrations. The separated or modified sepiolite powders were added to the simulated antimony wastewater and the vibrating adsorption at 1500 r/min under different experimental conditions and requirements. This was filtered through a 0.45 μm syringe diaphragm filter. An atomic absorption spectrometer (TAS-990) was used to determine the equilibrium concentration of Sb (III) in the supernatant. The equilibrium adsorption capacity (q_e) and adsorption efficiency η (%) of Sb (III) were calculated by the following formula:

$$q_e = \frac{(C_0 - C_e) \times V}{C_0} \tag{1}$$

$$\eta = \frac{C_0 - C_e}{C_0} \times 100\% \tag{2}$$

where C_0 and C_e are the initial and equilibrium (mg/L) concentrations of Sb (III), respectively, V is the volume of solution (L), and m is the mass of the adsorbent (g).

2.3.1. Adsorption Kinetics Experiment

Sep (1.0 g) and HI/ABsep (1.0 g) were added to the 20 mg/L Sb (III) solution (500 mL). At this time, the initial pH of the solution was not adjusted (in this case, pH = 6). The beaker with the mixture was transferred to a magnetic stirrer and continuously stirred at 150 r/min speed at room temperature. At each sampling time (3–240 min), samples were collected and immediately filtered through a 0.45 μm syringe membrane filter to determine the equilibrium concentration of Sb (III) in the supernatant.

The pseudo-first-order Equation (3) and pseudo-second-order Equation (4) are used for nonlinear fitting. The formula are as follows:

$$q_t = q_e \left(1 - e^{-K_1 t}\right) \tag{3}$$

$$q_t = \frac{K_2 q_e^2 t}{1 + K_2 q_e t} \tag{4}$$

where K_1 is the pseudo-first-order kinetic equation rate constant (min^{-1}); K_2 is the pseudo-second-order kinetic equation rate constant (g·mg^{-1}·min^{-1}); t is the adsorption time (min); q_t and we are the Sb (III) adsorption amount (mg·g^{-1}) at time t and adsorption equilibrium, respectively.

2.3.2. Effect of Adsorbent Dosage and pH Value

Experimental conditions stipulate a temperature of 35 °C, a volume of solution 50 mL, an initial concentration of Sb (III) 10 mg/L, an initial pH of the system 6, an adsorption time of constant temperature oscillation 240 min, a dosage of HI/ABsep (0.5–3 g/L), respectively, for the effect on adsorbing heavy metal Sb (III).

In the same way, HCl (0.1–1 mol/L) and NaOH (0.1–1 mol/L) were used to adjust the initial pH value of the adsorption system (2–12), the dosage of the adsorbent is 2 g/L, and the other conditions are the same as mentioned above. Multiple adsorption tests were carried out on heavy metal Sb (III), and the effect of pH on the adsorption of Sb (III) at HI/ABsep was obtained.

2.3.3. Adsorption Thermodynamics

Sep and HI/ABsep (0.1 g) were added to a series of 50 mL solutions of Sb (III) (10, 20, 50, 80, 100, and 150 mg/L), respectively. The solution was placed in a box with constant-temperature gas bath agitation and stirred at 150 r/min for 240 min at a temperature of 298 K, 308 K, and 318 K, respectively. Sb (III) was measured by filtering the supernatant through a 0.45 μm membrane syringe filter.

The experimental data were analyzed using two adsorption models. Langmuir Equation (5) and Freundlich Equation (6) perform non-linear fits of the adsorption isotherms at different temperatures (289 K, 308 K, and 318 K), respectively.

$$q_e = \frac{K_L Q_m C_e}{1 + K_L C_e} \tag{5}$$

$$q_e = K_F C_e^{1/n} \tag{6}$$

where q_e is the equilibrium adsorption amount (mg/g); C_e is the equilibrium concentration (mg/L); Q_m is the saturation adsorption amount (mg/g); K_L (L·mg^{-1}) and K_F [mg·g^{-1}·(L·mg^{-1})$^{1/n}$] are Langmuir adsorption constant and Freundlich adsorption constant, respectively; $1/n$ is a constant for adsorption strength, which varies with the heterogeneity of the material.

2.3.4. Thermodynamic Parameters

In order to further explain the spontaneous characteristics and energy changes in the preparation of HI/ABsep when adsorbing Sb (III) in the solid–liquid system, the thermodynamic parameters under three different temperatures (standard free energy $\Delta G°$ standard enthalpy $\Delta H°$, standard entropy $\Delta S°$) were calculated by using Equations (7) [20–22], (8), and (9).

$$K°_{Eq} = \frac{K_T \times C°_{Adsorbate}}{r_{Adsorbate}} \tag{7}$$

$$\Delta G° = -RT \ln K°_{Eq} \tag{8}$$

$$\ln K°_{Eq} = \frac{\Delta S°}{R} - \frac{\Delta H°}{RT} \tag{9}$$

where $K°_{Eq}$ (dimensionless) is the standard thermodynamic equilibrium constant of adsorption based on molar concentrations; K_T is the distribution coefficient obtained from the Langmuir isotherm (L/mol); $C°_{Adsorbate}$ (mol/L) is the standard concentration of adsorbate; and $r_{Adsorbate}$ (dimensionless) is the activity coefficient of the adsorbate.

Using $\ln K°_{Eq}$ and $1/T$ as the plot, the intercept and slope obtained by linear regression analysis can be used to calculate $\Delta S°$ and $\Delta H°$, respectively, and $\Delta G°$ can be directly calculated by Equation (8).

3. Results and Discussion

3.1. Characterization of Sep and HI/ABsep

3.1.1. N$_2$ Adsorption–Desorption Isotherm and Pore Size Distribution

Before compounding hydroxy iron into sepiolite, the sepiolite needs to be modified with acid and alkali. The N$_2$ adsorption–desorption isotherm at 77 K and pore size distribution of sepiolite, ABsep, and HI/ABsep are shown in Figure 1a,b. Both of these two materials are of type IV adsorption. Their pores are concentrated in size distribution at 2–50 nm, indicating that there are a large number of mesopores in the sample, and antimony can penetrate the internal sepiolite surface, which is beneficial for adsorption [23]. The mean pore diameters of the ABsep composite and Sep materials calculated by BJH are 1.4 nm and 3.7 nm, respectively. The BET formula gives a significantly higher average specific surface area for the ABsep composite material (260.5 m^2/g) than Sep (160.1 m^2/g); it can be seen that the surface performance after the acid–base modification has been greatly improved compared with that before modification, which provides a good foundation for the composite of hydroxyl iron. HI/ABsep surface area and average pore diameter were 152.4 m^2/g and 1.7 nm, respectively, both smaller than the surface area and average pore diameter of ABsep. The reason for this may be that, during the introduction of iron and supported iron oxide, some channels are blocked, forming a thick membrane on the surface that prevents N$_2$ from adsorption and desorption [24].

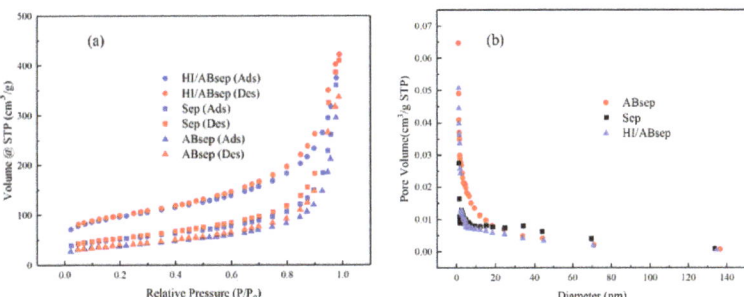

Figure 1. N$_2$ adsorption–desorption isotherm at 77 K (**a**) and BJH pore size distribution (**b**) of Sep, ABsep, and HI/ABsep.

3.1.2. SEM Analysis

As shown in Figure 2a, the original structure of the sepiolite material is a thin straight rod-like fiber structure. In Figure 2b, the SEM image of the ABsep material obtained by treatment of sepiolite with hydrochloric acid and ammonia is shown. As can be seen, the surface of the composite is non-uniform, and the nanofibers are fragmented in comparison to the SEM image of the original sepiolite structure. With the massive structure, the dispersibility is reduced, the agglomeration appears, and the pores increase. The result is consistent with Figure 1. When compared to the surface without iron modification, the iron-modified HI/ABsep showed irregular ripples and a highly roughened surface (Figure 2c), which could be affected by hydroxyl iron.

Figure 2. SEM image of Sep (**a**) and ABsep (**b**) and SEM-EDS image of HI/ABsep (**c**).

3.1.3. PXRD and FTIR Analysis

The PXRD spectrum of the sample is shown in Figure 3b. The Sep spectrum showed apparent diffraction of 7.2, 11.8, 19.7, 23.7, 26.7, 33.5, and 40.2 degrees. The normalized card number ((JCPDS: 14-0001) indicates that (110), (130), (300), (340), (270), (510), and (490) are known to correspond to the sepiolite orders. After modification, the sepiolite peak disappears, and the characteristic diffraction peaks 2θ of HI/ABsep are 20.9°, 26.6°, and 50.1°, which are following SiO$_2$ (JCPDS:46-1045) [25], indicating that the main component of HI/ABsep used in this study is SiO$_2$ [26,27]. In addition, the characteristic peaks of HI/ABsep, (130), (021), (221), and (151), correspond to the angles 2θ = 33.2°, 34.7°, 53.2°, and 59.0°, which are consistent with the standard card number (JCPDS: 81-0464). It shows that the main form of iron supported by modified sepiolite is iron oxyhydroxide FeOOH.

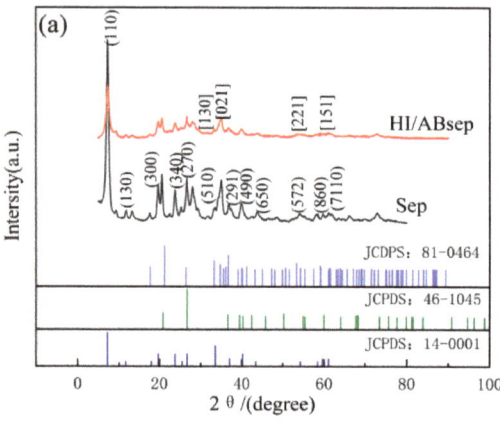 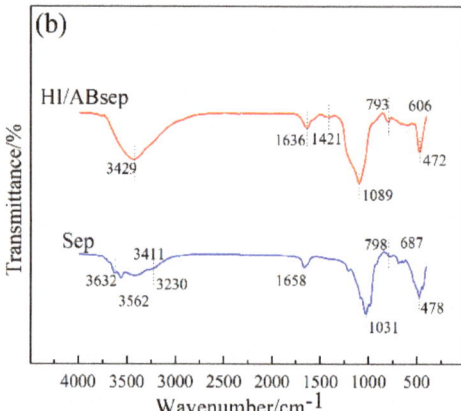

Figure 3. PXRD (**a**) and FTIR spectrum (**b**) of Sep and HI/ABsep. The red line in Figure 3a represents the XRD image of HI/ABsep, the black line is the XRD image of Sep, and the light blue refers to the data image corresponding to JCPDS:81-0464; Green is the data image corresponding to JCPDS:46-1045; The dark blue is the data image corresponding to JCPDS:14-0001. The red line in Figure 3b is the image produced by the infrared spectral data of HI/ABsep, and the black line is the infrared spectrum of Sep.

Figure 3a shows the FTIR spectra obtained by Sep and HI/ABsep. In the wavenumber range of 1300–900 cm^{-1}, it is a broad and strong stretching vibration band produced by Si-O-Si [28]; the strong absorption band near 3462 cm^{-1} belongs to the structure of the sepiolite ore fine octahedral coordination. The stretching vibration of hydroxyl water (Mg-OH) and coordination water (-OH$_2$), and the absorption peak near 687 cm^{-1} are lattice vibrations of MgO, Al$_2$O$_3$, and Ca$_2$O$_3$ oxides; the band of 479 cm^{-1} corresponds to vibrations on the Si-O-Si plane [29]; the 1658 cm^{-1} belongs to the peak of C=O stretching vibration; moreover, there are some differences in the peaks of the HI/ABsep spectrum. The hydroxyl peak (-OH) at 3429 cm^{-1} is enhanced. The -OH could be the hydroxyl peak of water. The absorption band characteristic at 1421 cm^{-1} can be caused by Cl-vibration. Especially at 900–450 cm^{-1}, which is the characteristic absorption peak of metal oxide, the change before and after modification is related to the appearance of FeOOH oxyhydroxide [30,31].

3.2. Adsorption of Sep and HI/ABsep on Sb (III)

3.2.1. Adsorption Kinetics

However, as shown in Figure 4, the amount of Sb (III) adsorbed by Sep and HI/ABsep increased rapidly and then stabilized. At 30 min contact time, the rate of Sb (III) removal by HI/ABsep had reached 67.28%. Within 30–120 min, the absorption rate of the material gradually decreases and tends to equilibrium, reaching an adsorption equilibrium at about 180 min. Sb (III) has a maximum adsorption capacity of 6.14 mg/g and a removal rate of 92%%, which is significantly higher than the original sepiolite (2.65 mg/g); the reason is that hydroxy-iron increases the number of active adsorption sites on the sepiolite surface and the adsorptive capacity is enhanced. The initial rapid adsorption may be due to the external diffusion of Sb (III), which is attached to the external surface of the material. Its main functions are ion exchange and physical adsorption. However, as time increases, on the one hand, the surface of the adsorbent can be used for the adsorption of Sb (III). However, as the concentration gradient of Sb (III) on the solution surface and adsorbent is reduced, the driving force for the concentration gradient decreases so that the resistance of the Sb (III) to adsorbate in solution increases, and the rate of adsorption slowly decreases [32,33].

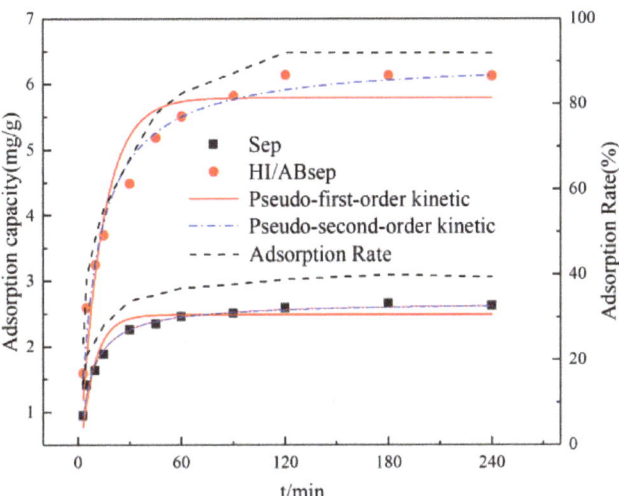

Figure 4. Adsorption kinetics of Sep and HI/ABsep.

In Table 1, the R^2 value of the quasi-second-order kinetics for the adsorption of Sb (III) by HI/ABsep is 0.9865, which is greater than that of the fitting quasi-first-order kinetics, suggesting that the quasi-second-order kinetics can more accurately describe the adsorption process of Sb (III) by HI/ABeep. Based on the assumption of the quasi-second-order kinetic model, combined with the related literature, the HI/ABsep adsorption process of Sb (III) can be known to be primarily chemical adsorption. The adsorbent and Sb (III) may form chemical bonds by sharing or exchanging electrons. Adsorption occurs through complexation, coordination, or ion exchange [6,9,13,34–37].

Table 1. Kinetic parameters of adsorption on Sb (III).

Adsorbent	Pseudo-First-Order Kinetic Parameters			Pseudo-Second-Order Kinetic Parameters		
	q_e (mg/g)	K_1 (min^{-1})	R^2	q_e (mg/g)	K_2 (min^{-1})	R^2
Sep	2.4875	0.1246	0.9078	2.6803	0.0672	0.9865
HI/ABsep	5.7879	0.0769	0.9158	6.3861	0.0163	0.9825

3.2.2. Effect of HI/ABsep Dosage and pH on the Removal of Sb (III)

It has obvious practical significance for the industrial application of sorbent to study the effect of the amount of sorbent and pH on sorbent adsorption performance and the maximum adsorption effect of the sorbent on different heavy metal ions. From Figure 5a, it can be seen that, as the dosage of HI/ABsep is increased, the rate of elimination gradually increases. When the dosage is 2 g/L, the removal rate reaches 95.46%. Further analysis shows that, when the dosage is low, the number of adsorption sites that can be provided is correspondingly more negligible. The adsorption site of the adsorbent material has a positive correlation with the adsorption efficiency. As a result, the rate of Sb (III) elimination by HI/ABsep at low doses is lower. When the dosage of the adsorbent gradually increases, on the one hand, the adsorption site of the solution also increases rapidly. Conversely, the probability that the reaction between the adsorbed substance and the adsorbate collide with each other also increases as a result; thus, the removal rate of Sb (III) is rapidly increased. Taking into account the economic benefits, we use a dose of 2 g/L in other single-factor experiments.

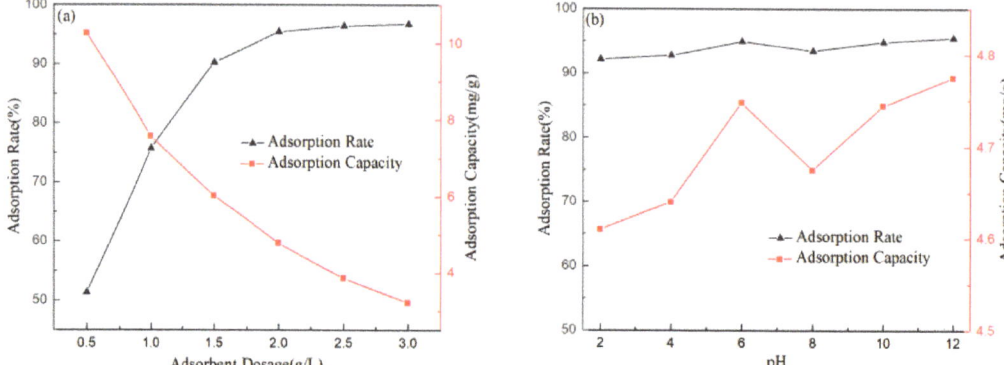

Figure 5. Effect of different dosages of adsorbent and pH change on Sb (III) adsorption. (**a**). Effect of different doses of adsorbents on the adsorption of Sb (III); (**b**). Effect of pH on the adsorption of Sb (III) by adsorbents.

From Figure 5b, it can be seen that the pH of the solution does not significantly affect the adsorption efficiency of Sb (III) by HI/ABsep. When pH is 2–6 and pH = 8–12, the removal rate of Sb (III) increased slowly. When pH is 6–8, the removal rate of Sb (III) decreased slowly. On the whole, the removal rate of Sb (III) remained at a high level (92–95.5%). When the pH of the solution is 2~10, the main form of Sb (III) in the solution is the neutral complex Sb (OH)$_3$. Under the strong acid condition, Sb (III) mainly exists as SbO$^+$; all of them may be adsorbed on the surface of goethite with a positive charge as inner sphere complexes [38]. Under the strong alkali condition, its preferential form is SbO$_2$ [35,39–41]. Sb (OH)$_3$ easily reacts with Fe ions to form X ≡ Fe-Sb (OH)$_2$ precipitate [41], and the reaction process is little affected by pH, so the pH has little effect on the adsorption and removal of Sb (III) by HI/ABsep.

3.2.3. Adsorption Isotherm

As can be seen from the nonlinear fitting results in Figure 6 and related parameters in Table 2, the Freundlich model has a higher temperature than the Langmuir model (R^2 = 0.9595, 0.9062 and 0.9545) at 298 K, 308 K, and 318 K temperatures. The values of the correlation coefficient are 0.9671, 0.9854, and 0.9660. This result demonstrates that the Freundlich model can describe the process of Sb (III) adsorption onto HI/ABsep well. Furthermore, at different temperatures, the Langmuir and Freundlich isotherm models can fit well. This shows that the intermolecular forces play a major role in the magnitude of the intermolecular forces: The higher the temperature, (1) the stronger the adsorption and (2) The greater the concentration of the solution. Langmuir isotherms are appropriate for single-layer adsorption where the adsorption sites are uniformly distributed over the adsorbent surface and there is no interaction between adjacent sites and the particles [42]. The Freundlich adsorption isotherm model is an empirical formula for multi-layer adsorption [43]. It can be deduced that Sb (III) is dominated by the uniform adsorption of multiple molecular layers under a variety of temperature conditions. Furthermore, the range of values of 1/n of the adsorption models of the Langmuir isotherm and Freundlich isotherm is from 0 to 1, indicating that the prepared HI/ABsep has a high affinity to Sb (III), and adsorption may spontaneously develop in the direction of adsorption [44]. According to the Langmuir isotherm (HI/ABsep) adsorption model, the maximum adsorption capacity of Sb (III) is found to be 25.73 mg/ at 298 K.

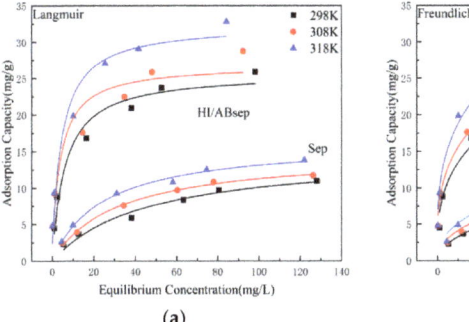
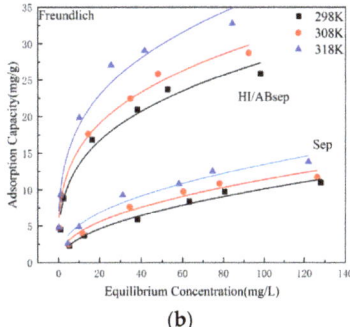

Figure 6. Isothermal adsorption line of HI/ABsep to Sb (III). (**a**). the Langermuir isotherm model for HI/ABsep adsorbed Sb (III), 298K, 308K, 318K; (**b**). the Freundlich isotherm model for HI/ABsep adsorbed Sb (III), 298K, 308K, 318K.

Table 2. Two adsorption isotherm-related parameters.

Adsorbent	T/K	Langmuir			Freundlich		
		K_L/L·mg^{-1}	Q_m/mg·g^{-1}	R^2	K_F/(L·mg^{-1})$^{1/n}$	n	R^2
Sep	298	0.0222	14.66	0.9654	1.0643	2.0454	0.9807
	308	0.0318	14.87	0.9952	1.4890	2.2601	0.9560
	318	0.0406	16.42	0.9922	1.9511	2.3808	0.9573
HI/ABsep	298	0.1673	25.73	0.9595	6.5908	3.2202	0.9671
	308	0.2617	26.89	0.9062	7.8975	3.4044	0.9854
	318	0.2245	32.45	0.9545	9.5873	3.4328	0.9660

The actual maximum adsorption capacity of the adsorbent HI/ABsep prepared in this study under optimal experimental conditions is 25.73 mg/g, which has adsorption potential compared with other Sb (III) adsorption materials, as shown in Table 3 below. At the same time, considering that the adsorption material in this study has the advantages of simple preparation process and easy physical separation from the treatment system, it has great application potential for the treatment of Sb (III)-contaminated wastewater

Table 3. Comparison of Sb (III) adsorption by various adsorbents.

Sorbents	Q_m (mg/g)	Experimental Conditions	Reference
PVA-Fe^0	6.99	pH = 7; T = 298 K	[45]
α-Fe$_2$O$_3$	23.23	pH = 4; T = 298 K	[14]
γ-FeOOH	33.08	pH = 4; T = 298 K	[14]
Graphene	10.919	pH = 11; T = 298 K	[37]
γ-Fe$_2$O$_3$ mesoporousmicrospheres	47.48	pH = 7; T = 298 K	[46]
HI/ABsep	25.73	pH = 6; T = 298 K	This work

3.2.4. Thermodynamic Parameters

As can be seen from Table 4, as the temperature is increased from 298 K to 318 K, the adsorption of Sb (III) on HI/ABsep has $\Delta G° < 0$ and $\Delta H° > 0$, indicating that the adsorption is a spontaneous endothermic reaction. If the absolute value of ΔH is between 0 and 20 kJ/mol, the adsorption reaction is related to physical adsorption, and if the absolute value of ΔH is between 40 and 80 kJ/mol, then the adsorption reaction is related to chemical adsorption [47]. The value of $\Delta H°$ used in this study is 160.18 kJ/mol at 298–318 K, which indicates that the adsorption of Sb (III) by HI/ABsep is chemical adsorption [48]. The mechanism of adsorption may be an electrostatic force and the filling of pores. This binding increases with an increasing temperature, indicating that increasing temperature is beneficial for the progress of adsorption.

Table 4. Thermodynamic parameters for Sb (III) adsorption by HI/ABsep.

Temperature (K)	$\Delta G°$ (kJ/mol)	$\Delta H°$ (kJ/mol)	$\Delta S°$ (J/mol × K)
298	−1.3062		
308	−6.4172	160.18	0.5409
318	−11.8262		

3.2.5. The Mechanism of HI/ABsep Adsorption on Sb (III)

In order to gain insight into the HI/ABsep adsorption mechanism of Sb (III), XPS is used to analyze the existence of elements on HI/ABsep and their interaction during the process of adsorption. The high-resolution spectra of HI/ABsep O and Sb before and after adsorption are shown in Figure 7a. The characteristic peaks of 531.84 eV binding energy and 532.53 eV binding energy in the pre-adsorption O1s high-resolution spectrum correspond mainly to Fe-O-OH and Si=O [49] (SiO_2), respectively. After adsorption, the two characteristic peaks at the binding energy of 530.65 eV and the binding energy of 540.28 eV correspond to the positions of the standard binding energy peaks of the two orbitals of Sb $3d_{5/2}$ and Sb $3d_{3/2}$, respectively. Among the valence states are Sb (III) and Sb (V), which indicates that the redox reaction took place during the adsorption process [50]. The binding energies for Fe 2p3/2 and Fe 2p1/2 of HI/ABsep are about 710.96 eV and 724.26 eV (Figure 7b). The photoelectron peak at 710.96 eV, which is characteristic of iron oxide species in HI/ABsep, shifts to the position of 710.72 eV in HI/ABsep-Sb, which indicates the interaction between adsorbed Sb (III) and iron oxide in HI/ABsep [51]. The removal mechanism of HI/ABsep-strengthened removal of Sb (III) is shown in Figure 8.

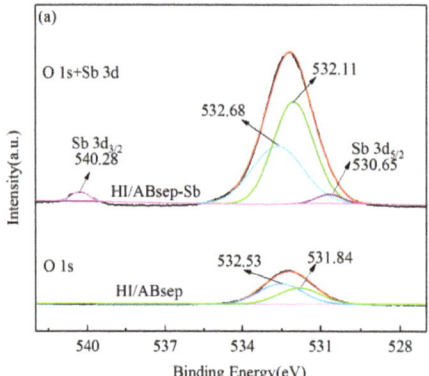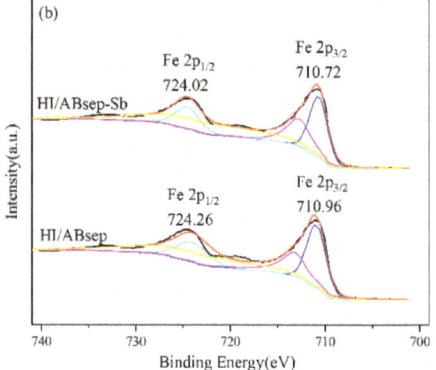

Figure 7. XPS spectra before and after HI/ABsep adsorption on Sb (III). (**a**) is the fine spectrum of O1s before and after the reaction, and (**b**) is the fine spectrum of Fe2p before and after the reaction). The red line in Figure 7a indicates the peak fitted to the data, indicating that it is composed of the peaks in the green, cyan, and purple lines; the green and cyan lines represent the two characteristic peaks of O1S corresponding to FeOOH and Si=O; and the purple line represents the characteristic peak of Sb (III); The XPS spectra before and after the adsorption of Sb (III) by HI/ABsep are shown in Figure 7b, respectively. The red line indicates the peak made by fitting the data, and the fuchsia line, cyan line and blue line are made by splitting the peak from the red line. The fuchsia line and the blue line indicate the two satellite peaks of the Fe2p3/2 orbit. The cyan line indicates the characteristic peak of Fe2p1/2 orbit.

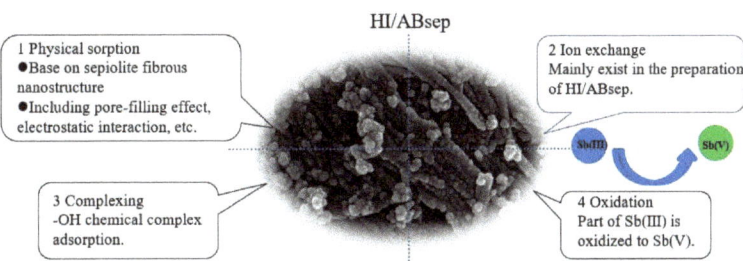

Figure 8. Mechanism of Sb (III) Removal by HI/ABsep.

Briefly, the adsorption of HI/ABsep Sb (III) is mainly chemical and physical adsorption. HI/ABsep adsorption onto Sb (III) is a very complex process, which can be combined through a variety of adsorption mechanisms such as redox, coordination complexation, and physical adsorption. A schematic diagram of hydroxyl iron formation and the mechanism of enhanced Sb (III) removal by HI/ABsep can be seen in Figure 8. At pH = 2–10, the main forms of antimony in water are Sb $(OH)_3$ and Sb $(OH)_6^-$. It is speculated that the surface complexation between antimony ions in the simulated water and the hydroxyl iron on the surface of HI/ABsep is shown in the following formula:

$$\equiv MOH + Sb(OH)_3 = MOSbO_2 + 2H_2O \tag{10}$$

$$2 \equiv MOH + Sb(OH)_3 = MOSb(OH)OM + 2H_2O \tag{11}$$

$$Sb(OH)_3 + 3H_2O - 2e^- = Sb(OH)_6^- + 3H^+ \tag{12}$$

$$\equiv MOH + Sb(OH)_6^- = MOSb(OH)_5^- + H_2O \tag{13}$$

$$2 \equiv MOH + Sb(OH)_6^- = MOSb(OH)_4^- OM + 2H_2O \tag{14}$$

where M stands for iron, silicon, magnesium, etc.

4. Conclusions

HI/ABsep modified based on sepiolite can be used as a promising material for adsorbing heavy metal Sb (III). XRD and XPS show that the Fe species supported on sepiolite are mainly FeOOH. Furthermore, the dispersion of FeOOH on sepiolite led to the exposure of a few new adsorption locations. The kinetics of adsorption follow a pseudo-second-order kinetic model. The Langmuir and Freundlich isotherm models were used to fit the composite adsorbent kinetics successfully, and the maximum adsorption capacity is 25.73 mg/g at 298 K, 26.89 mg/g at 308 K, and 32.45 mg/g at 318 K.

HI/ABsep adsorption of Sb (III) is barely affected by pH but is proportional to the content of the adsorbent. Future research will focus on the practical implementation of the live demonstration.

Author Contributions: Conceptualization, B.R. and Y.Z.; Methodology, Y.Z. and Z.H.; Software, Z.H.; validation, Y.Z., Z.H. and X.D.; formal analysis, Y.Z.; investigation, X.D.; Resources, X.D.; Data curation, Y.Z. and Z.H.; Writing—original draft, Y.Z. and Z.H.; Writing—review & editing, B.R. and Y.Z. All authors have read and agreed to the published version of the manuscript.

Funding: This work was financially supported by the National Natural Science Foundation of China (Nos. 41973078) and the Hunan Provincial Natural Science Foundation of China (2022SK2073).

Institutional Review Board Statement: The study did not deal with ethics.

Informed Consent Statement: The study did not involve humans.

Data Availability Statement: The study did not report any data.

Acknowledgments: We sincerely thank the National Natural Science Foundation of China and the Hunan Provincial Natural Science Foundation of China for their support in experiments and data analysis.

Conflicts of Interest: The authors declare no conflict of interest.

References

1. Nishad, P.A.; Bhaskarapillai, A. Antimony, a pollutant of emerging concern: A review on industrial sources and remediation technologies. *Chemosphere* **2021**, *277*, 130252. [CrossRef] [PubMed]
2. Fang, L.; Zhou, A.; Li, X.; Zhou, J.; Pan, G.; He, N. Response of antimony and arsenic in karst aquifers and groundwater geochemistry to the influence of mine activities at the world's largest antimony mine, central China. *J. Hydrol.* **2021**, *603*, 127131. [CrossRef]
3. Boreiko, C.J.; Rossman, T.G. Antimony and its compounds: Health impacts related to pulmonary toxicity, cancer, and genotoxicity. *Toxicol. Appl. Pharmacol.* **2020**, *403*, 115156. [CrossRef] [PubMed]
4. Chu, J.; Hu, X.; Kong, L.; Wang, N.; Zhang, S.; He, M.; Ouyang, W.; Liu, X.; Lin, C. Dynamic flow and pollution of antimony from polyethylene terephthalate (PET) fibers in China. *Sci. Total Environ.* **2021**, *771*, 144643. [CrossRef]
5. Li, J.; Zheng, B.; He, Y.; Zhou, Y.; Chen, X.; Ruan, S.; Yang, Y.; Dai, C.; Tang, L. Antimony contamination, consequences and removal techniques: A review. *Ecotoxicol. Environ. Saf.* **2018**, *156*, 125–134. [CrossRef]
6. Xiong, N.; Wan, P.; Zhu, G.; Xie, F.; Xu, S.; Zhu, C.; Hursthouse, A.S. Sb (III) removal from aqueous solution by a novel nano-modified chitosan (NMCS). *Sep. Purif. Technol.* **2020**, *236*, 116266. [CrossRef]
7. Wei, D.; Li, B.; Luo, L.; Zheng, Y.; Huang, L.; Zhang, J.; Yang, Y.; Huang, H. Simultaneous adsorption and oxidation of antimonite onto nano zero-valent iron sludge-based biochar: Indispensable role of reactive oxygen species and redox-active moieties. *J. Hazard. Mater.* **2020**, *391*, 122057. [CrossRef]
8. Jiang, F.; Ren, B.; Hursthouse, A.; Deng, R.; Wang, Z. Distribution, source identification, and ecological-health risks of potentially toxic elements (PTEs) in soil of thallium mine area (southwestern Guizhou, China). *Environ. Sci. Pollut. Res.* **2019**, *26*, 16556–16567. [CrossRef]
9. Bagherifam, S.; Brown, T.C.; Fellows, C.M.; Naidu, R.; Komarneni, S. Highly efficient removal of antimonite (Sb (III)) from aqueous solutions by organoclay and organozeolite: Kinetics and Isotherms. *Appl. Clay Sci.* **2021**, *203*, 106004. [CrossRef]
10. Zeeshan, M. Expropriation of Cd(II) from wastewater by ion-selective membrane electrode based on Polypyrrole-Antimony(III)iodophosphate nanocomposite. *Mater. Today Proc.* **2020**, *29*, 352–362. [CrossRef]
11. Ran, Z.; Yao, M.; He, W.; Wang, G. Efficiency analysis of enhanced Sb (V) removal via dynamic preloaded floc in coordination with ultrafiltration. *Sep. Purif. Technol.* **2020**, *249*, 117115. [CrossRef]
12. Long, X.; Wang, X.; Guo, X.; He, M. A review of removal technology for antimony in aqueous solution. *J. Environ. Sci.* **2020**, *90*, 189–204. [CrossRef] [PubMed]
13. Song, K.; Zhang, C.; Shan, J.; Wang, W.; Liu, H.; He, M. Adsorption behavior and surface complexation modeling of oxygen anion Sb (V) adsorption on goethite. *Sci. Total Environ.* **2022**, *833*, 155284. [CrossRef] [PubMed]
14. Guo, X.; Wu, Z.; He, M.; Meng, X.; Jin, X.; Qiu, N.; Zhang, J. Adsorption of antimony onto iron oxyhydroxides: Adsorption behavior and surface structure. *J. Hazard. Mater.* **2014**, *276*, 339–345. [CrossRef]
15. Bahabadi, F.N.; Farpoor, M.H.; Mehrizi, M.H. Removal of Cd, Cu and Zn ions from aqueous solutions using natural and Fe modified sepiolite, zeolite and palygorskite clay minerals. *Water Sci. Technol.* **2017**, *75*, 340–349. [CrossRef]
16. Xu, G.M.; Shi, Z.; Deng, J. Removal of antimony from water by iron-oxide coated sand (in Chinese). *Environmental Chemistry* **2006**, *25*, 481.
17. Wang, Z.; Liao, L.; Hursthouse, A.; Song, N.; Ren, B. Sepiolite-Based Adsorbents for the Removal of Potentially Toxic Elements from Water: A Strategic Review for the Case of Environmental Contamination in Hunan, China. *Int. J. Environ. Res. Public Health* **2018**, *15*, 1653. [CrossRef]
18. Alshameri, A.; He, H.; Zhu, J.; Xi, Y.; Zhu, R.; Ma, L.; Tao, Q. Adsorption of ammonium by different natural clay minerals: Characterization, kinetics and adsorption isotherms. *Appl. Clay Sci.* **2018**, *159*, 83–93. [CrossRef]
19. Wu, J.; Wang, Y.; Wu, Z.; Gao, Y.; Li, X. Adsorption properties and mechanism of sepiolite modified by anionic and cationic surfactants on oxytetracycline from aqueous solutions. *Sci. Total Environ.* **2019**, *708*, 134409. [CrossRef]
20. Tran, H.N.; Lima, E.C.; Juang, R.-S.; Bollinger, J.-C.; Chao, H.-P. Thermodynamic parameters of liquid–phase adsorption process calculated from different equilibrium constants related to adsorption isotherms: A comparison study. *J. Environ. Chem. Eng.* **2021**, *9*, 106674. [CrossRef]
21. Lima, E.C.; Sher, F.; Saeb, M.R.; Abatal, M.; Seliem, M.K. Reasonable calculation of the thermodynamic parameters from adsorption equilibrium constant. *J. Mol. Liq.* **2021**, *322*, 114980.
22. Lima, E.C.; Hosseini-Bandegharaei, A.; Moreno-Piraján, J.C.; Anastopoulos, I. A critical review of the estimation of the thermodynamic parameters on adsorption equilibria. Wrong use of equilibrium constant in the Van't Hoof equation for calculation of thermodynamic parameters of adsorption. *J. Mol. Liq.* **2019**, *273*, 425–434. [CrossRef]
23. Addy, M.; Losey, B.; Mohseni, R.; Zlotnikov, E.; Vasiliev, A. Adsorption of heavy metal ions on mesoporous silica-modified montmorillonite containing a grafted chelate ligand. *Appl. Clay Sci.* **2012**, *59–60*, 115–120. [CrossRef]

24. Xie, S.; Wang, L.; Xu, Y.; Lin, D.; Sun, Y.; Zheng, S. Performance and mechanisms of immobilization remediation for Cd contaminated water and soil by hydroxy ferric combined acid-base modified sepiolite (HyFe/ABsep). *Sci. Total Environ.* **2020**, *740*, 140009. [CrossRef]
25. Dou, Q.-S.; Zhang, H.; Wu, J.-B.; Yang, D. Synthesis and Characterization of Fe_2O_3 and FeOOH Nanostructures Prepared by Ethylene Glycol Assisted Hydrothermal Process. *J. Inorg. Mater.* **2007**, *22*, 213–218.
26. Chen, L.; Han, Y.; Li, W.; Zhan, X.; Wang, H.; Shi, C.; Sun, Y.; Shi, H. Removal of Sb (V) from wastewater via siliceous ferrihydrite: Interactions among ferrihydrite, coprecipitated Si, and adsorbed Sb (V). *Chemosphere* **2022**, *291*, 133043. [CrossRef]
27. Teo, S.H.; Islam, A.; Chan, E.S.; Choong, S.T.; Alharthi, N.H.; Taufiq-Yap, Y.H.; Awual, R. Efficient biodiesel production from Jatropha curcus using $CaSO_4/Fe_2O_3$-SiO_2 core-shell magnetic nanoparticles. *J. Clean. Prod.* **2019**, *208*, 816–826. [CrossRef]
28. Ji, Q.; Kamiya, S.; Jung, J.-H.; Shimizu, T. Self-assembly of glycolipids on silica nanotube templates yielding hybrid nanotubes with concentric organic and inorganic layers. *J. Mater. Chem.* **2004**, *15*, 743–748. [CrossRef]
29. Eren, E.; Gumus, H.; Ozbay, N. Equilibrium and thermodynamic studies of Cu(II) removal by iron oxide modified sepiolite. *Desalination* **2010**, *262*, 43–49. [CrossRef]
30. Lin, Z.; Huan, Z.; Zhang, J.; Li, J.; Li, Z.; Guo, P.; Zhu, Y.; Zhang, T. CTAB-functionalized δ-FeOOH for the simultaneous removal of arsenate and phenylarsonic acid in phenylarsenic chemical warfare. *Chemosphere* **2022**, *292*, 133373. [CrossRef]
31. Yao, S.; Zhu, X.; Wang, Y.; Zhang, D.; Wang, S.; Jia, Y. Simultaneous oxidation and removal of Sb (III) from water by using synthesized $CTAB/MnFe_2O_4/MnO_2$ composite. *Chemosphere* **2020**, *245*, 125601. [CrossRef] [PubMed]
32. Foo, K.Y.; Hameed, B.H. Insights into the modeling of adsorption isotherm systems. *Chem. Eng. J.* **2010**, *156*, 2–10. [CrossRef]
33. Liu, Y.; Li, C.; Lou, Z.; Zhou, C.; Yang, K.; Xu, X. Antimony removal from textile wastewater by combining PFS&PAC coagulation: Enhanced Sb (V) removal with presence of dispersive dye. *Sep. Purif. Technol.* **2021**, *275*, 119037. [CrossRef]
34. Tu, Y.-J.; Wang, S.-L.; Lu, Y.-R.; Chan, T.-S.; Johnston, C.T. New insight in adsorption of Sb (III)/Sb (V) from waters using magnetic nanoferrites: X-ray absorption spectroscopy investigation. *J. Mol. Liq.* **2021**, *330*, 1156911. [CrossRef]
35. Hossain, M.D.F.; Akther, N.; Zhou, Y. Recent advancements in graphene adsorbents for wastewater treatment: Current status and challenges. *Chin. Chem. Lett.* **2020**, *31*, 2525–2538. [CrossRef]
36. Huang, L.; Huang, X.; Yan, J.; Liu, Y.; Jiang, H.; Zhang, H.; Tang, J.; Liu, Q. Research progresses on the application of perovskite in adsorption and photocatalytic removal of water pollutants. *J. Hazard. Mater.* **2022**, *442*, 130024. [CrossRef] [PubMed]
37. Leng, Y.; Guo, W.; Su, S.; Yi, C.; Xing, L. Removal of antimony(III) from aqueous solution by graphene as an adsorbent. *Chem. Eng. J.* **2012**, *211–212*, 406–411. [CrossRef]
38. Watkins, R.; Weiss, D.; Dubbin, W.; Peel, K.; Coles, B.; Arnold, T. Investigations into the kinetics and thermodynamics of Sb (III) adsorption on goethite (α-FeOOH). *J. Colloid Interface Sci.* **2006**, *303*, 639–646. [CrossRef]
39. Filella, M.; Belzile, N.; Chen, Y.-W. Antimony in the environment: A review focused on natural waters: II. Relevant solution chemistry. *Earth-Science Rev.* **2002**, *59*, 265–285. [CrossRef]
40. Fu, X.; Song, X.; Zheng, Q.; Liu, C.; Li, K.; Luo, Q.; Chen, J.; Wang, Z.; Luo, J. Frontier Materials for Adsorption of Antimony and Arsenic in Aqueous Environments: A Review. *Int. J. Environ. Res. Public Health* **2022**, *19*, 10824. [CrossRef]
41. Farquhar, M.L.; Charnock, J.M.; Livens, F.R.; Vaughan, D.J. Mechanisms of Arsenic Uptake from Aqueous Solution by Interaction with Goethite, Lepidocrocite, Mackinawite, and Pyrite: An X-ray Absorption Spectroscopy Study. *Environ. Sci. Technol.* **2002**, *36*, 1757–1762. [CrossRef] [PubMed]
42. Saleh, T.A.; Sarı, A.; Tuzen, M. Effective adsorption of antimony(III) from aqueous solutions by polyamide-graphene composite as a novel adsorbent. *Chem. Eng. J.* **2017**, *307*, 230–238. [CrossRef]
43. Chen, H.; Li, T.; Zhang, L.; Wang, R.; Jiang, F.; Chen, J. Pb(II) adsorption on magnetic γ-Fe2O3/titanate nanotubes composite. *J. Environ. Chem. Eng.* **2015**, *3*, 2022–2030. [CrossRef]
44. Xiao, X.; Yang, L.; Zhou, D.; Zhou, J.; Tian, Y.; Song, C.; Liu, C. Magnetic γ-Fe2O3/Fe-doped hydroxyapatite nanostructures as high-efficiency cadmium adsorbents. *Colloids Surfaces A: Physicochem. Eng. Asp.* **2018**, *555*, 548–557. [CrossRef]
45. Zhao, X.; Dou, X.; Mohan, D.; Pittman, C.U.; Ok, Y.S.; Jin, X. Antimonate and antimonite adsorption by a polyvinyl alcohol-stabilized granular adsorbent containing nanoscale zero-valent iron. *Chem. Eng. J.* **2014**, *247*, 250–257. [CrossRef]
46. Zhao, W.; Ren, B.; Hursthouse, A.; Wang, Z. Facile synthesis of nanosheet-assembled γ-Fe2O3 magnetic microspheres and enhanced Sb (III) removal. *Environ. Sci. Pollut. Res.* **2021**, *28*, 19822–19837. [CrossRef]
47. Zhang, H.; Khanal, S.K.; Jia, Y.; Song, S.; Lu, H. Fundamental insights into ciprofloxacin adsorption by sulfate-reducing bacteria sludge: Mechanisms and thermodynamics. *Chem. Eng. J.* **2019**, *378*, 122103. [CrossRef]
48. Tran, H.N. Improper Estimation of Thermodynamic Parameters in Adsorption Studies with Distribution Coefficient KD (qe/Ce) or Freundlich Constant (KF): Considerations from the Derivation of Dimensionless Thermodynamic Equilibrium Constant and Suggestions. *Adsorpt. Sci. Technol.* **2022**, *2022*, 5553212. [CrossRef]
49. Fu, R.; Yang, Y.; Xu, Z.; Zhang, X.; Guo, X.; Bi, D. The removal of chromium (VI) and lead (II) from groundwater using sepiolite-supported nanoscale zero-valent iron (S-NZVI). *Chemosphere* **2015**, *138*, 726–734. [CrossRef]
50. Li, W.; Fu, F.; Ding, Z.; Tang, B. Zero valent iron as an electron transfer agent in a reaction system based on zero valent iron/magnetite nanocomposites for adsorption and oxidation of Sb (III). *J. Taiwan Inst. Chem. Eng.* **2018**, *85*, 155–164. [CrossRef]
51. Qi, Z.; Lan, H.; Joshi, T.P.; Liu, R.; Liu, H.; Qu, J. Enhanced oxidative and adsorptive capability towards antimony by copper-doping into magnetite magnetic particles. *RSC Adv.* **2016**, *6*, 66990–67001. [CrossRef]

Article

Efficient Uptake of Angiotensin-Converting Enzyme II Inhibitor Employing Graphene Oxide-Based Magnetic Nanoadsorbents

Miguel Pereira de Oliveira [1], Carlos Schnorr [2], Theodoro da Rosa Salles [1], Franciele da Silva Bruckmann [3], Luiza Baumann [3], Edson Irineu Muller [3], Wagner Jesus da Silva Garcia [4], Artur Harres de Oliveira [5], Luis F. O. Silva [2] and Cristiano Rodrigo Bohn Rhoden [2,6,*]

1. Laboratório de Materiais Magnéticos Nanoestruturados—LaMMaN, Universidade Franciscana, Santa Maria 97010-030, Brazil
2. Department of Civil and Environmental, Universidad de la Costa, CUC, Calle 58 # 55–66, Barranquilla 080002, Colombia
3. Programa de Pós-graduação em Química, Universidade Federal de Santa Maria—UFSM, Santa Maria 97105-900, Brazil
4. Departamento de Desenho Industrial, Universidade Federal de Santa Maria—UFSM, Santa Maria 97105-900, Brazil
5. Departamento de Física, Universidade Federal de Santa Maria—UFSM, Santa Maria 97105-900, Brazil
6. Programa de Pós-graduação em Nanociências, Universidade Franciscana—UFN, Santa Maria 97010-030, Brazil
* Correspondence: cristianorbr@gmail.com

Citation: de Oliveira, M.P.; Schnorr, C.; da Rosa Salles, T.; da Silva Bruckmann, F.; Baumann, L.; Muller, E.I.; da Silva Garcia, W.J.; de Oliveira, A.H.; Silva, L.F.O.; Rhoden, C.R.B. Efficient Uptake of Angiotensin-Converting Enzyme II Inhibitor Employing Graphene Oxide-Based Magnetic Nanoadsorbents. *Water* **2023**, *15*, 293. https://doi.org/10.3390/w15020293

Academic Editor: Hai Nguyen Tran

Received: 13 December 2022
Revised: 5 January 2023
Accepted: 6 January 2023
Published: 10 January 2023

Copyright: © 2023 by the authors. Licensee MDPI, Basel, Switzerland. This article is an open access article distributed under the terms and conditions of the Creative Commons Attribution (CC BY) license (https://creativecommons.org/licenses/by/4.0/).

Abstract: This paper reports a high efficiency uptake of captopril (CPT), employing magnetic graphene oxide (MGO) as the adsorbent. The graphene oxide (GO) was produced through an oxidation and exfoliation method, and the magnetization technique by the co-precipitation method. The nanomaterials were characterized by FTIR, XRD, SEM, Raman, and VSM analysis. The optimal condition was reached by employing GO·Fe$_3$O$_4$ at pH 3.0 (50 mg of adsorbent and 50 mg L^{-1} of CPT), presenting values of removal percentage and maximum adsorption capacity of 99.43% and 100.41 mg g^{-1}, respectively. The CPT adsorption was dependent on adsorbent dosage, initial concentration of adsorbate, pH, and ionic strength. Sips and Elovich models showed the best adjustment for experimental data, suggesting that adsorption occurs in a heterogeneous surface. Thermodynamic parameters reveal a favorable, exothermic, involving a chemisorption process. The magnetic carbon nanomaterial exhibited a high efficiency after five adsorption/desorption cycles. Finally, the GO·Fe$_3$O$_4$ showed an excellent performance in CPT removal, allowing future application in waste management.

Keywords: adsorption; carbon nanomaterials; magnetite; captopril

1. Introduction

Water contamination by emerging pollutants (EPs) has become an important problem around the world [1]. Amongst these, drugs and personal care products are widely found in water bodies, due to their intrinsic characteristics and incomplete removal by current methods [2]. Additionally, due to their unique properties and xenobiotic-like character, these compounds present a high tendency to bioaccumulation, which leads to the development of diverse damages to human health, animals, and the environment [3]. In addition, high consumption, inadequate disposal, and incomplete metabolism are associated with the high and constant generation of waste. Due to low concentration and traces in water, the conventional systems are not able to remove these compounds, resulting in a secondary source of exposition and contamination [4].

Captopril (CPT) (1-(3-mercapto-2-D-methyl-1-oxopropyl)-L proline is a hydrophilic drug used to inhibit angiotensin-converting enzyme II (ACE II) and is one of the most commonly prescribed medicines for the treatment of high blood pressure and heart failure [5,6]. Although this medicine is effective for control of diseases, around 50% of CPT is excreted by kidneys in unchanged form, contributing to water contamination [7]. Additionally, this pharmaceutical compound may cause diverse side effects and toxicity to organisms [8,9].

In this scenario, diverse methods have been explored for EPs removal from aqueous medium, such as Fenton [10], photocatalyst [11], ozonation [12], and adsorption [13,14]. The adsorption technique is characterized by the capture of contaminant (adsorbate) using a solid surface (adsorbent). This method presents advantages compared with other processes, such as low energy requirements and easy operation, without byproducts and generation of sludge [15]. Regarding the uptake of captopril, recent studies reported its removal using alternative materials from the biomass and lignocellulosic waste [9,16]. Considering the high consumption of captopril and the potential damage to health, and the limited studies that report its removal from the aquatic environment, it is necessary to use and design alternative and novel adsorbents. Additionally, nanotechnology is an excellent tool to enhance the process, using adsorbents at nanoscale, improving the properties and characteristics of conventional adsorbents, as well as the adsorption efficiency [17,18].

Among these, carbon nanomaterials, such as graphene oxide (GO) can be used in the adsorption process, due to the high specific surface area, reactivity, and presence of oxygenated groups [19]. Nonetheless, the unique properties of GO allow combination with other nanoparticles, as magnetic nanoparticles (MNPs), conferring new properties as well as avoiding the filtration and centrifugation steps. Recently, Li et al. [20] synthesized a magnetic nanocomposite with GO and hydroxyapatite for Pb (II) removal from water. The results demonstrate that nanomaterials exhibited a high adsorption capacity and maintained efficiency after several cycles of adsorption/desorption.

This work reports captopril adsorption using magnetic graphene oxide (GO·Fe_3O_4), synthesized through a straightforward method, using only Fe^{2+} as the iron source [21]. The adsorption behavior was evaluated using different experimental variables. The isotherm and kinetic models were used to understand the adsorption equilibrium and kinetic profile. Additionally, the adsorbent regeneration was investigated to determine the lifetime and efficiency after diverse adsorption/desorption processes. The mechanistic hypotheses for captopril adsorption onto graphene oxide-based magnetic adsorbent are also proposed, based on chemical species and magnetite incorporation on the GO surface.

2. Materials and Methods

2.1. Synthesis of Graphene Oxide and Graphene Oxide-Based Magnetic Nanoadsorbent

Graphene oxide was synthesized following Salles et al. [22]. 1 g of graphite in flakes (Sigma-Aldrich®, St. Louis, MO, USA) was added to a flask with 60 mL of H_2SO_4 98% (Synth®) under magnetic stirring for 10 min at room temperature. Sequentially, 6 g of $KMnO_4$ 99% (Synth®) was slowly added over 20 min. The solution was heated at 40 °C and stirred continuously for 5 h. Afterward, 180 mL of distilled H_2O was slowly added and stirring for 12 h at room temperature. Sequentially, the reaction was heated at 40 °C for 2 h. Finally, 300 mL of distilled H_2O and 10 mL of H_2O_2 29% (Synth®) was added into the solution. The yellow solution was washed until pH 5.0 and dried in an oven (DeLeo) at 50 °C for 24 h.

The magnetic graphene oxide was produced by the co-precipitation method [21]. In a 250 mL round-bottom flask containing 120 mL of ultrapure water, 100 mg of GO were added to 1000 mg of $FeCl_2 \geq 99\%$ (Sigma-Aldrich®). Afterwards, ammonium hydroxide (Synth®) was added to reach pH 9.0. Sequentially, the mixture was submitted to ultrasonic radiation (Elma, power 150 W) for 90 min at room temperature. Finally, the product was purified with distilled water and acetone (Synth®), and dried in an oven (DeLeo) at 50 °C for total solvents evaporation.

2.2. Adsorbent Characterization

Graphene oxide and magnetic graphene oxide were characterized by several different techniques. The functional groups were determined through the Fourier Transform spectroscopy (Perkin-Elmer, model Spectro One). The crystallinity was determined by X-ray diffraction (Bruker diffractometer, model D2 Phaser), and Raman Spectroscopy (Renishaw inVia spectrometer system). The morphological characteristics were evaluated using a scanning electron microscopic (Zeiss Sigma 300 VP). The magnetic properties were evaluated with a vibrating sample magnetometer (Stanford Research Systems Model SR830 DSP lock-in amplifier, coupled to a Stanford Research Systems low-noise pre-amplifier model SR560, including a Kepco bipolar operation power supply).

The zero point of charge (pH_{ZPC}) of MGO was evaluated by the following procedure. 10 mg of adsorbent was added to a 100 mL flask containing 50 mL of ultrapure water (Milli-Q®, Darmstadt, Germany) and pH adjusted in the range of (2.0–12.0) using NaOH (0.01 mol L^{-1}) and HCl (0.01 mol L^{-1}). Sequentially, the samples were maintained under stirring (120 rpm) for 24 h at room temperature. Subsequentially, the final pH was measured with a potentiometer (Digimed).

2.3. Adsorption Procedure and Mathematical Modeling

The captopril adsorption was performed in a batch system. The influence of magnetite was initially evaluated employing GO and GO·Fe$_3$O$_4$ to determine the effect of iron nanoparticles in GO surface on the adsorption efficiency. The adsorption equilibrium study was evaluated using 50 mg of GO·Fe$_3$O$_4$ and 50 mg L^{-1} of initial concentration of captopril, pH 3.0, and different temperatures (20 °C, 30 °C, and 40 °C). The adsorption kinetic was performed employing different concentrations of CPT $\geq$ 98% (10–200 mg L^{-1}), 50 mg of adsorbent, at room temperature, and pH 3.0. Thermodynamic parameters were evaluated under the same conditions of the equilibrium study. Besides, this the influence of pH was verified in a pH range between 2.0 and 10.0, using HCl and NaOH solutions (0.1 mol L^{-1}) for adjustment of the pH of the solution. The effect of ionic strength was performed employing a gradual NaCl concentration (0.01–1 mol L^{-1}). Following this, the effect of initial concentration and adsorbent dosage were performed using 10–200 mg L^{-1} of CPT and 12.5–50 mg of GO·Fe$_3$O$_4$, respectively. The adsorbent regeneration was carried out employing NaOH (0.25 mol L^{-1}) as the desorbing agent at room temperature for 1 h and 150 rpm. The residual concentration of CPT was measured using a UV-vis spectrophotometer (Shimadzu-1650PC) at λ = 217 nm. The removal percentage and adsorption capacity at equilibrium are shown in Table 1 [23].

Isotherms models were employed to understand the relationship between the adsorbate and adsorbent molecules. The Langmuir, Freundlich, and Sips models were used to fit the experimental results, and the equations are shown in Table 1 [24].

The kinetic study is useful to explain the mass transfer phenomenon and the effect of contact time on the adsorption rate. The models used in this work were Pseudo-first-order (PFO), Pseudo-second-order (PSO), and Elovich. Thermodynamic parameters were calculated by Van 't Hoff equation (Equation (10)) to determine the adsorption behavior [25,26]. The equilibrium and kinetic data were estimated through non-linear regression, using the Statistical 10 software (StatSoft, Tulsa, OK, USA). The coefficient of determination (R^2), adjusted coefficient of determination (R^2_{adj}), sum of squared errors (SSE), and average relative error (ARE) were used to verify the validity of models, and mean square error (MSE) are shown in Table 1 [27].

The mathematical modeling used to calculate the adjustment parameters are shown in Table 1.

Table 1. Equations of removal percentage, adsorption capacity, adsorption isotherms, kinetics, thermodynamics models, and mathematical modelling.

Removal Percentage and Adsorption Capacity	Mathematical Expression	Parameters
Removal percentage	$R\% = \frac{C_0 - C_e}{C_0} \times 100$ (1)	C_0 is the initial concentration of CPT (mg L^{-1}); C_e is the concentration of CPT at equilibrium (mg L^{-1}), q_e is the adsorption capacity at equilibrium; V is the volume (L); m is the mass of adsorbent material (g)
Adsorption capacity	$q_e = \frac{(C_0 - C_e)\,V}{m}$ (2)	
Isotherm models	Mathematical expression	Parameters
Langmuir	$q_e = \frac{q_{max}\,K_L\,C_e}{1 + K_L\,C_e}$ (3)	q_{max}- maximum amount adsorbed (mg g^{-1}); K_L- is the Langmuir isotherm constant (L g^{-1})
Freundlich	$q_e = K_F \cdot C_e^{1/n}$ (4)	K_F- Freundlich constant ((mg g^{-1}) (L mg^{-1})$^{-1/n}$); n- heterogeneity factor
Sips	$q_e = \frac{q_{Sips}\,K_{Sips}\,C_e^{n_{Sips}}}{1 + K_{Sips}\,C_e^{n_{Sips}}}$ (5)	K_{Sips}- Sips isotherm constant (L mg^{-1}); q_{Sips}- maximum amount adsorbed (mg g^{-1}); n_{Sips}- heterogeneity factor
Kinetic models	Mathematical expression	Parameters
Pseudo-first-order	$q_t = q_1\left(1 - e^{-k_1 \cdot t}\right)$ (6)	k_1- pseudo-first-order kinetic constant (min^{-1})
Pseudo-second-order	$q_t = \dfrac{t}{\left(\frac{1}{k_2 \cdot q_2^2}\right) + \left(\frac{t}{q_2}\right)}$ (7)	k_2- pseudo-second-order kinetic constant (mg g^{-1} min^{-1})
Elovich	$q_e = \frac{1}{\beta}\ln(\alpha\beta t + 1)$ (8)	α- initial adsorption rate (mg g^{-1} min^{-1}); β- desorption constant (g mg^{-1})
Thermodynamic	Mathematical expression	Parameters
Gibbs free energy variation (ΔG^0)	$\Delta G° = -RT\,lnK_e$ (9)	R- gas constant (8.314 J mol^{-1} K^{-1}); T- absolute temperature (K); K_d- thermodynamic equilibrium constant
van't Hoff	$lnK_e = \frac{\Delta S°}{R} + \frac{\Delta H°}{RT}$ (10)	
Relationship of Gibbs free energy (ΔG^0), Enthalpy (ΔH^0) and entropy (ΔS^0) variation	$\Delta G° = \Delta H° - T\Delta S°$ (11)	
Mathematical modeling	Mathematical expression	Parameters
Adjusted coefficient of determination (R^2_{adj})	$R^2_{adj} = 1 - \left[\frac{(1-R^2)(n-1)}{n-k-1}\right]$ (12)	$q_{e,exp}$ and $q_{e,pred}$ are the adsorbed amounts obtained from the experiment and the isotherm and kinetic model, respectively; n is the amount of data; k is the number of parameters in the model
Sum squares error (SSE)	$SSE = \frac{1}{n}\sum_{i=1}^{n}\left(q_{e,exp} - q_{e,pred}\right)^2$ (13)	
Average relative error (ARE)	$ARE = \frac{100}{n}\sum_{i=1}^{n}\left|\frac{q_{e,exp} - q_{e,pred}}{q_{e,exp}}\right|$ (14)	
Mean square error (MSE)	$MSE = \frac{1}{n-p}\sum_{i=1}^{n}\left(q_{e,exp} - q_{e,pred}\right)^2$ (15)	$q_{e,exp}$ is the experimental value; $q_{e,pred}$ is the predicted value; n is the number of experimental values; p is the number of parameters according to the model.

3. Results and Discussion

3.1. Fourier Transform Infrared Spectroscopy (FTIR)

The presence of functional groups was evaluated by FTIR analysis (Figure 1). According to the results, it was possible to observe an intense signal around 3405 cm^{-1}, corresponding to the stretching vibration of OH groups in GO composition [28]. The bands at 1734, 1651, 1342, and 1048 cm^{-1} are related to the C=O stretch vibration, C=C bonds, C-OH vibration, and C-O bonds, respectively [29]. Besides this, for the GO·Fe$_3$O$_4$, it was possible to verify the appearance of band around 617 cm^{-1} characteristics of Fe-O, which suggests the presence of magnetite into GO surface [30,31].

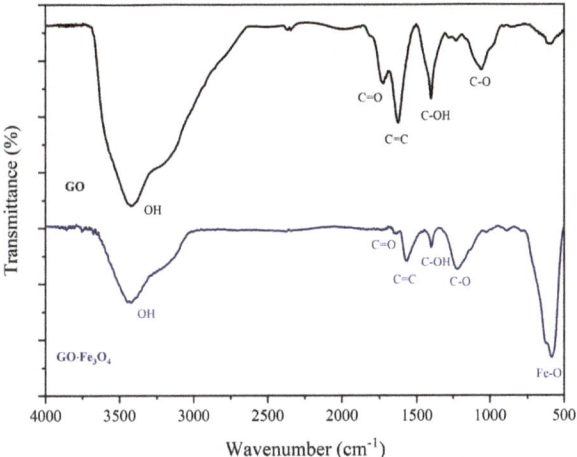

Figure 1. FTIR of GO and GO·Fe$_3$O$_4$.

3.2. X-ray Diffraction (XRD)

The XRD results of graphene oxide and magnetic graphene oxide are shown in Figure 2. The GO diffraction shows the characteristic signal around 2θ ≈ 10.5°, corresponding to the (001) plane, and the absence of a peak at 2θ ≈ 26° indicate the complete oxidation of graphite (precursor material) [32]. The peak at 2θ ≈ 44.2° (100) suggests a small distance between GO layers. For the GO with iron oxide nanoparticles, it was possible to verify the appearance of peaks around 2θ ≈ 30.3°, 35.2°, 43.5°, 57.02°, and 63.3°, corresponding to the indices (220), (311), (400), (511) and (440), typical of magnetite [33]. The position and intensity of all peaks are compatible with Fe$_3$O$_4$ (ICDD card no:88-0315). The absence of GO signal in the GO·Fe$_3$O$_4$ diffractogram is attributed to the surface coated by iron nanoparticles, which is possible to observe in SEM results (Figure 3) [21].

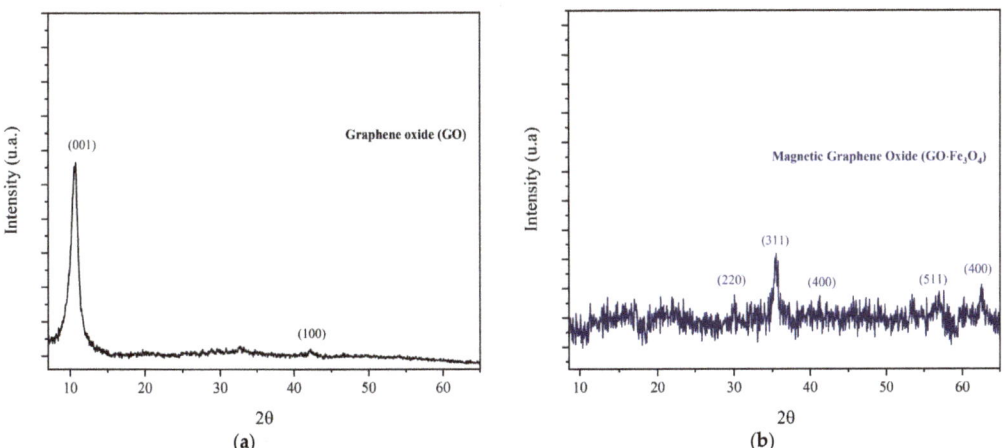

Figure 2. XRD of (**a**) GO and (**b**) GO·Fe$_3$O$_4$.

(a)   (b)

Figure 3. SEM of (**a**) GO and (**b**) GO·Fe_3O_4.

3.3. Scanning Electron Microscopy (SEM)

The morphology of GO and GO·Fe_3O_4 was determined by scanning electron microscopy (SEM) and is shown in Figure 3. For the GO (Figure 3a), it was possible to verify a wrinkles sheet, which is characteristic of a nanomaterial with few layers [34]. In Figure 3b, for the magnetic graphene oxide, the GO surface was covered by magnetite. The high amount of $FeCl_2$ employed in the magnetization process results in a magnetic nanocomposite with high magnetic nanoparticle content [21].

3.4. Raman Spectroscopy

The Raman spectroscopy results are shown in Figure 4. The three bands can be observed for GO andgraphene oxide-based magnetic nanoadsorbent (GO·Fe_3O_4). For both materials, it was possibile to observe the bands at 1366, 1580, and 2698 cm^{-1} that represent the bands D (defects and disorder of graphite structure), G (sp^2 vibration by carbon atoms), and 2D (second order of the D band), which is characteristic of graphene materials [21]. Additionally, the I_D/I_G ratio increased after the magnetization procedure, showing values of 0.96 and 1.22 for GO and GO·Fe_3O_4. The increases are related to the involvement of carbon atoms in partial Fe^{3+} reduction [35].

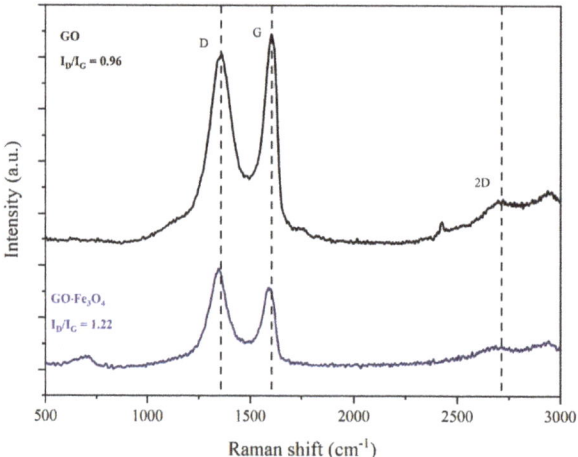

Figure 4. Raman spectroscopy of GO and GO·Fe_3O_4.

Nonetheless, the co-precipitation method is widely used to synthesize magnetic nanoparticles, due to the low energy requirements, low temperature, and high-quality products. In this work, the magnetization process increased the I_D/I_G ratio; however, the low values showed that the MGO was not significantly affected by the procedure. Alongside this, using ammonium hydroxide as a precipitant agent, the graphene structure was preserved, resulting in excellent magnetic graphene oxide surface and low values for I_D/I_G ratio [21]. Zhang et al. [36] synthesized a magnetic nanocomposite, employing a microwave absorbing method with a mixture of Fe^{2+} and Fe^{3+} (1:2 molar ratio). The results of I_D/I_G ratio showed a high disorder in the nanocomposite structure. Nevertheless, Hatel et al. [37] developed GO/Fe_3O_4 nanorods using ultrasonic radiation and NaOH as the precipitating agent. Raman spectroscopy results demonstrated a nanocomposite with a low crystallinity degree when compared to the GO sample.

3.5. Vibrating Sample Magnetometer (VSM)

Figure 5 shows the magnetization profile of magnetic graphene oxide at room temperature. According to the graph, it was possible to verify that $GO\cdot Fe_3O_4$ presented a ferromagnetic behavior; magnetization saturation (M_S), remanent magnetization (M_R), and M_R/M_S ratio values were 45 emu g^{-1}, 3.44 emu g^{-1}, and 0.076, respectively. Similarly, Ghosh et al. [38] developed a magnetic GO functionalized with nitrogen groups, and the materials showed a ferromagnetic behavior.

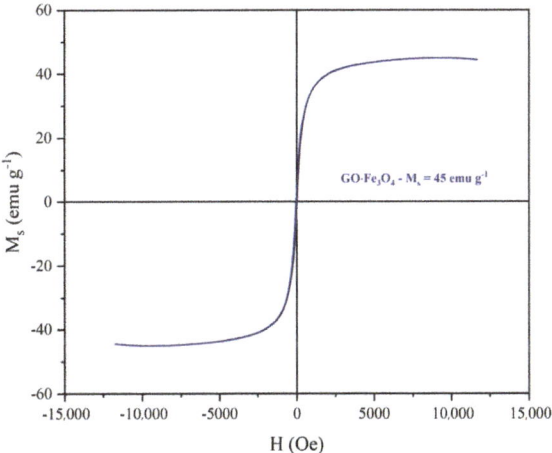

Figure 5. VSM of $GO\cdot Fe_3O_4$.

3.6. Captopril Adsorption

3.6.1. Effect of Magnetite Incorporation onto Graphene Oxide Surface

The synthesis of nanocomposites is an important tool for the design of adsorbent materials with different characteristics and properties. In this study, the effect of the magnetite incorporation on the GO surface on the captopril adsorption behavior was evaluated (Figure 6).

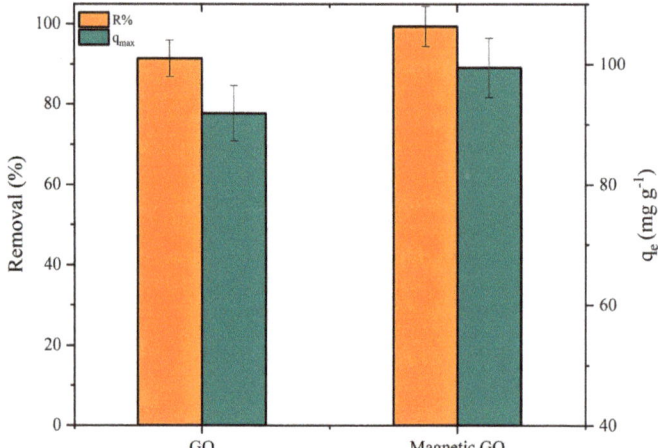

Figure 6. Effect of magnetite incorporation onto the graphene oxide surface. Adsorbent dosage (0.5 g L^{-1}), initial concentration of CPT (50 mg L^{-1}), pH 3.0, and 293.15 K.

According to Figure 6, it was possible to verify that both adsorbents exhibited excellent performance when removing the emerging pollutant. However, the graphene oxide-based magnetic adsorbent presented the highest adsorption capacity and removal percentage values compared to the pristine carbon nanomaterial. The GO and GO·Fe$_3$O$_4$ display removal values of 91.42 and 99.43% and adsorption capacities of 91.84 and 99.41 mg g^{-1}, respectively. The development of composite materials not only improve the physicochemical stability, but also can enhance their efficiency as an adsorbent. Recently, diverse studies have reported the effects of magnetite on the GO surface in terms of the adsorption performance [21,39,40].

3.6.2. Effect of Initial Concentration of CPT and Adsorbent Dosage

The initial adsorbate concentration and the adsorbent dosage play an important role in the adsorption efficiency, due to the higher ability to overcome the mass transfer resistance phenomenon and the more available surface area, respectively. Here, the effect of the initial concentration of CPT and GO·Fe$_3$O$_4$ was investigated in the range of 10–200 mg L^{-1} and 0.125–1.0 g L^{-1}, respectively.

The influence of CPT concentration is shown in Figure 7. Remarkably, the adsorption capacity values improved with the gradual increase in the initial concentration of CPT (e.g., the q$_e$ value raised from 14.68 to 390.31 mg g^{-1} when the CPT concentration was changed from 10 to 200 mg L^{-1}), thus implying that CPT adsorption is favored at higher concentrations. However, the removal percentage displayed a slight decrease with rising concentration. The removal value reduced by around 23% from the lowest to the highest concentration. The improvement in the adsorption performance can be attributed to the higher concentration gradient, impelling the motions of the adsorbate to the boundary layer, favoring the interaction between the liquid phase and the solid phase (surface of the adsorbent) and, therefore, the mass transfer phenomenon [41,42].

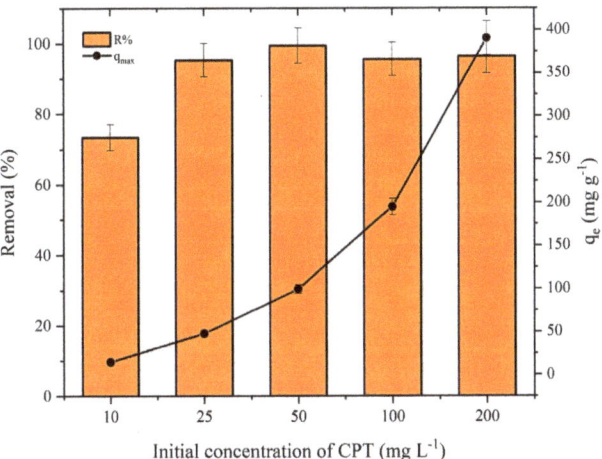

Figure 7. Effect of the initial concentration of CPT (GO·Fe$_3$O$_4$ dosage (0.5 g L^{-1}), C$_0$ = 10–200 mg L^{-1}, pH 3.0, and 293.15 K).

Regarding the effect of adsorbent dosage (Figure 8), it was observed that an initial increase in the GO·Fe$_3$O$_4$ mass lead to an improvement in the removal percentage values. However, at the highest dosages (0.75 and 1.0 g L^{-1}), a slight reduction was verified (the removal value decreased from 99.43 to 88.71% when the adsorbent dosage changed from 0.5 to 1.0 g L^{-1}). On the other hand, the adsorption capacity decreased inversely with the gradual rise in the adsorbent quantity. This is related to the imbalance between available adsorption sites and absorbate molecules, which is caused bt higher adsorbent dosages [43,44].

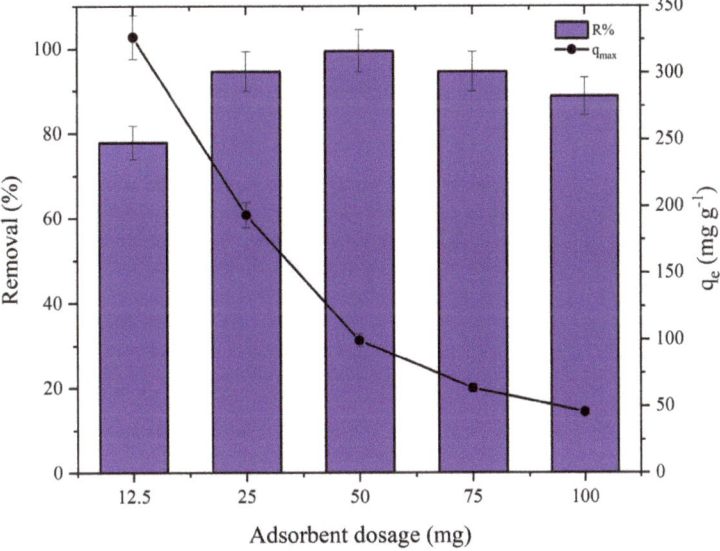

Figure 8. Effect of the adsorbent dosage on CPT adsorption (GO·Fe$_3$O$_4$ dosage (0.125–1.0 g L^{-1}), initial concentration of CPT (50 mg L^{-1}), pH 3.0, and 293.15 K).

3.6.3. Effect of pH and Adsorption Mechanisms

The pH of the solution is an important experimental parameter, due to its influence on chemical speciation and adsorbent surface charge. The effect of pH on CPT adsorption was evaluated in the range of 2–9, keeping other parameters constant. At the same time, the zero potential of charge (pH$_{ZPC}$) was estimated by the typical 11-point experiment.

From the results presented in Figure 9, it was possible to observe that the pH had a significant effect on the adsorption capacity and removal percentage values. The adsorption was reached at pH 3.0 (q_e = 99.41 mg g^{-1} and R% = 99.43). Captopril is a drug used for arterial hypertension that contains an ionizable group (-COOH, pK$_a$ = 3.7), i.e., it has an anionic character in this pH condition [45]. Under acidic conditions, the GO·Fe$_3$O$_4$ presents a positive surface charge (pH$_{ZPC}$ = 7.41), which allows the occurrence of electrostatic interactions between the adsorbent and adsorbate (pH 3.0). In contrast, in the range between 4–7, the adsorbent and captopril have a cationic characteristic, decreasing the affinity of adsorbate by binding sites. At the same time, in acidic media, the H$^+$ ions can compete for the available active sites, justifying the considerable reduction in the adsorption performance. In addition, under alkaline conditions, the adsorbent material exhibits an anionic surface, also affecting the interaction between the systems [21].

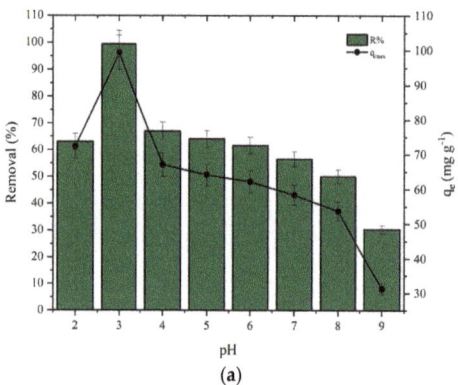
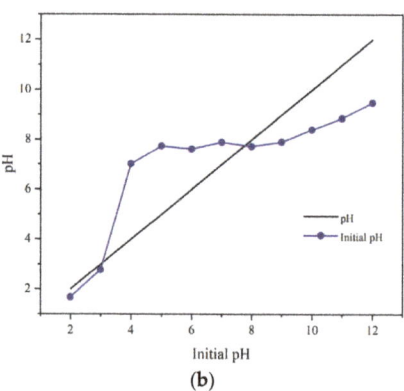

Figure 9. (a) Effect of pH on CPT adsorption (C$_0$ = 50 mg L^{-1}, pH = 2–9, adsorbent dosage = 0.5 g L^{-1}, V = 100 mL, and 293.15 K) and (b) Zero point of charge of GO·Fe$_3$O$_4$.

Adsorption mechanisms are closely related to the chemical structure of the adsorbate, such as functional groups, aromaticity, hydrophilicity, and pH sensitivity (ionizable groups). Additionally, the interaction depends on the charge and specific surface area of the adsorbent material. Thus, some hypotheses regarding the CPT adsorption onto GO·Fe$_3$O$_4$ were proposed (Figure 10). Considering the influence of pH solution on the chemical speciation of adsorbate and the adsorbent surface charge, this suggests that the captopril adsorption phenomenon is mainly governed by electrostatic interactions (optimal condition was reached at pH 3.0) [45,46]. Although attraction forces can act on the CPT removal, the abundance of oxygenated functional groups in nanoadsorbent structure and carboxylic acids on the captopril molecule also allows the occurrence of other interactions, such as hydrogen bonds (Yoshida H-bonding and dipole-dipole hydrogen bonding) [44,47].

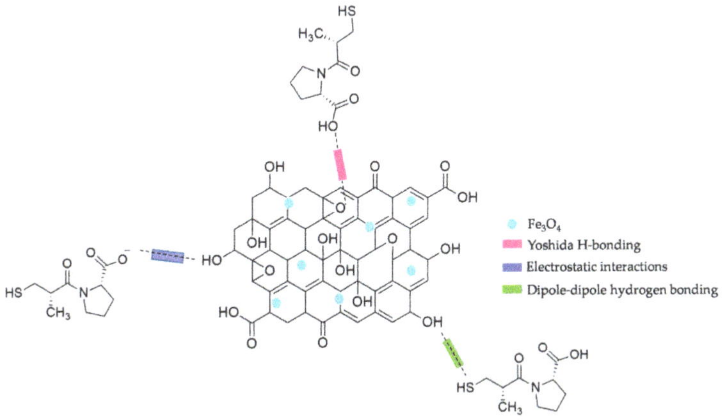

Figure 10. Hypotheses of the captopril adsorption mechanism.

3.6.4. Effect of Ionic Strength

The evaluation of the effect from diverse experimental conditions is an important requirement considering the potential application in wastewater treatment systems. In this study, the influence of ionic strength on CPT adsorption was performed using different concentrations of sodium chloride (0.01–1.0 mol L^{-1}), keeping other parameters constant. As shown in Figure 11, it was possible to observe that an increase in NaCl molarity caused a considerable reduction in the adsorption capacity value. For instance, the system without NaCl presents a q_e of 99.41 mg g^{-1}, while, at a concentration of 1.0 mol L^{-1}, the adsorption capacity was 32.38 mg g^{-1}. The Na$^+$ and Cl$^-$ ions can hamper the binding between the adsorbate and adsorbent due to the competition by adsorption sites. Additionally, the presence of inorganic salts may decrease the specific surface area, as well as reduce the solubility of the adsorbate [44,48,49].

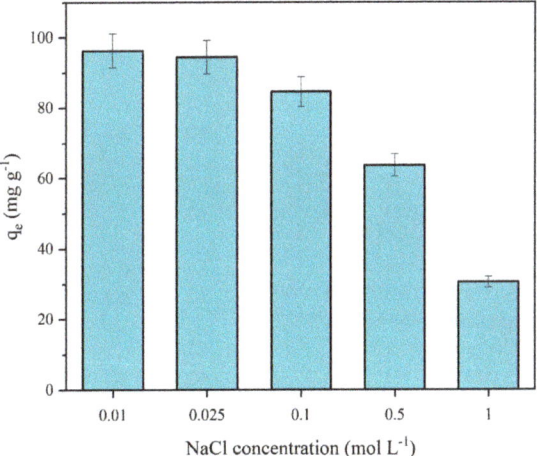

Figure 11. Effect of ionic strenght on CPT adsorption (C_0 = 50 mg L^{-1}, pH = 3.0, adsorbent dosage = 0.5 g L^{-1}, NaCl concentration (0.01–1.0 mol L^{-1}), V = 100 mL, and 293.15 K).

3.6.5. Kinetic Modeling

Kinetic models are valuable mathematical expressions, widely used to understand the adsorption behavior in relation to the experimental time. In this work, three models were employed: pseudo-first-model, pseudo-second-order, and Elovich to construct the kinetic profile and calculate the adjustment parameters for captopril adsorption onto GO·Fe$_3$O$_4$. The kinetic study was conducted at room temperature using different concentrations of adsorbate. Table 2 shows the adjust of experimental data to nonlinear kinetic models.

Table 2. Kinetic parameters for captopril adsorption onto GO·Fe$_3$O$_4$.

Concentration (mg L^{-1})	10	25	50	100	200
Pseudo-first-order model (PFO)					
q_1 (mg g^{-1})	13.44 ± 0.34 [a]	45.69 ± 0.97	96.06 ± 0.91	186.32 ± 1.81	386.58 ± 0.60
k_1 (min^{-1})	0.375 ± 0.11	0.205 ± 0.04	0.606 ± 0.15	0.403 ± 0.05	0.804 ± 0.05
R^2	0.948	0.966	0992	0.991	0.998
R^2_{adj}	0.935	0.957	0.990	0.988	0.997
ARE (%)	5.67	4.28	2.09	1.96	0.25
SSE	0.84	6.17	5.77	6.47	2.51
MSE (mg g^{-1})^2	1.33	9.69	9.06	38.17	3.94
Pseudo-second-order model (PSO)					
q_2 (mg g^{-1})	13.59 ± 0.35	47.99 ± 0.62	97.03 ± 0.87	190.11 ± 1.21	387.53 ± 0.48
k_2 (g mg^{-1} min^{-1})	0.049 ± 0.02	0.007 ± 0.001	0.025 ± 0.01	0.005 ± 0.001	0.023 ± 0.004
R^2	0.966	0.991	0.994	0.997	0.983
R^2_{adj}	0.957	0.988	0.992	0.996	0.978
ARE (%)	5.04	1.97	1.82	1.11	2.59
SSE	0.53	1.51	3.93	7.16	16.35
MSE (mg g^{-1})^2	0.83	3.36	6.17	11.25	9.98
Elovich model					
α (mg g^{-1} min^{-1})	10.48 ± 0.21	15.35 ± 0.59	16.62 ± 1.02	4.57 ± 0.97	3.95 ± 1.02
β (g mg^{-1})	0.997 ± 0.01	0.222 ± 0.05	0.273 ± 0.026	0.118 ± 0.02	0.066 ± 0.002
R^2	0.995	0.994	0.998	0.999	0.999
R^2_{adj}	0.993	0.992	0.997	0.998	0.998
ARE (%)	2.53	2.88	1.02	0.62	0.19
SSE	0.14	2.14	2.41	1.81	1.20
MSE (mg g^{-1})^2	0.23	2.27	4.01	2.84	2.57

[a] (Mean ± standard deviation).

According to the results presented in Table 2, it was possible to assume that the captopril adsorption onto graphene oxide-based magnetic nanoadsorbent was well-described by the Elovich model in all concentrations tested. This because the model exhibited high values for coefficient of determination ($R^2 \geq 0.994$) and low values for error functions (ARE, MSE and SSE). Additionally, the Elovich parameters related to sorption rate (α) and surface coverage (β), respectively, indicated that the adsorption occurs in the heterogeneous surface, i.e., the adsorbent exhibits adsorption sites with different energy levels [24,50].

The kinetic profile of captopril adsorption is shown in Figure 12. As demonstrated by kinetic curves, it was possible to verify that the adsorption presents a characteristic behavior of fast kinetic, with three different stages. In the first minutes, a high amount of adsorbate is removed, followed by a stationary phase (plateau) and, finally, equilibrium is achieved. In addition, the adsorption capacity at equilibrium is proportional to the initial concentration of the contaminant. The increase in adsorption efficiency can be attributed to greater ease when overcoming of the mass transfer phenomenon and effective occupation of binding sites [51].

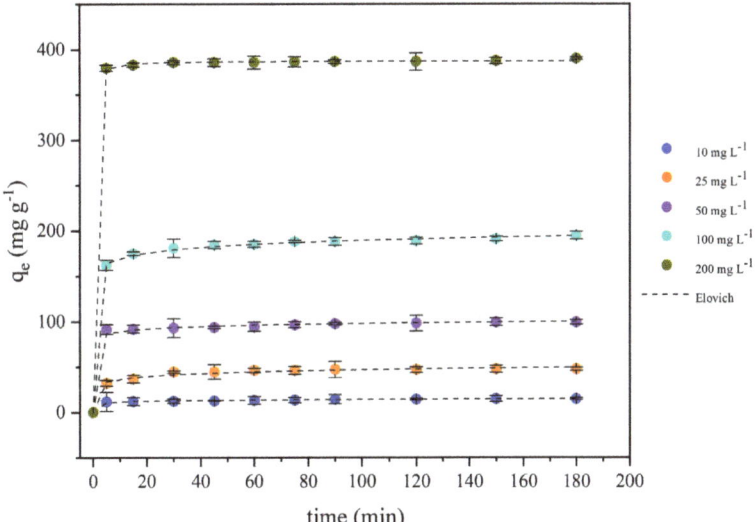

Figure 12. Kinetic curves for CPT adsorption onto GO·Fe$_3$O$_4$.

3.6.6. Adsorption Equilibrium Isotherms and Thermodynamic Study

The isotherm models are useful mathematical expressions that help us to understand the interaction between adsorbate and adsorbent; they are also used to verify the influence of temperature in the adsorption equilibrium. In this present work, Langmuir, Freundlich, and Sips isotherm models were used to predict the best adjustment of equilibrium data at different temperatures. As shown in Table 3, it was possible to determine that the Sips isotherm was the best model to describe the CPT adsorption onto GO·Fe$_3$O$_4$. Concerning values for coefficients of determination (R^2) and error functions (ARE and SSE), the descending order of fit of the models is as follows:

$$\text{Sips} < \text{Freundlich} < \text{Langmuir}$$

The Sips model is a combined equation of the Langmuir and Freundlich isotherms, used to predict the adsorption on the heterogenous surface [52]. Regarding the Sips constant associated to the heterogeneity (ns), it is remarkable that the heterogeneity of the system increased with the temperature (ns = 2.26 at 20 °C and ns = 7.88 at 40 °C). Additionally, the maximum adsorption capacity estimated by this model decreased with the rise in the temperature, indicating that adsorption is an exothermic process (in agreement with the thermodynamic study) [45]. On the other hand, in the Sips constant related to equilibrium (Ks), a considerable decrease was observed with the gradual increase in temperature, i.e., suggesting that the process is disadvantaged at high temperatures [53].

Figure 13 shows the Sips isotherms curves. According to the results, it is possible to observe a characteristic "L-shape" type for 20 and 30 °C, indicating an adsorbent surface with a high presence of binding sites [54]. However, for 40 °C, the isotherm showed an "S-shape" suggesting a solute–solute attraction and/or an adsorption site competition [55].

Table 3. Equilibrium parameters for CPT adsorption onto GO·Fe$_3$O$_4$.

Temperature	20 °C	30 °C	40 °C
	Langmuir		
q_{max} (mg g^{-1})	101.93 ± 1.76 [a]	101.49 ± 3.35	99.28 ± 5.96
K_L (L mg^{-1})	0.771 ± 0.46	0.43 ± 0.055	0.057 ± 0.97
R^2	0.979	0.953	0.979
R^2_{adj}	0.973	0.941	0.973
ARE (%)	3.82	5.69	4.92
SSE	15.71	26.73	8.54
MSE (mg g^{-1})^2	16.38	17.42	19.55
	Freundlich		
K_F ((mg g^{-1}) (L^{-1})$^{-1/n}$	74.51 ± 2.21	5.01 ± 3.32	2.61 ± 1.97
n	10.95 ± 3.29	45.85 ± 2.05	16.54 ± 1.44
R^2	0.986	0.997	0.997
R^2_{adj}	0.982	0.995	0.997
ARE (%)	3.84	1.99	1.01
SSE	10.21	2.64	1.87
MSE (mg g^{-1})^2	9.77	5.67	15.43
	Sips		
q_s (mg g^{-1})	100.41 ± 0.97	90.47 ± 1.19	70.72 ± 2.56
K_s (L mg^{-1})	0.444 ± 0.14	0.255 ± 0.65	0.082 ± 0.001
ns	2.26 ± 1.15	3.91 ± 1.74	7.88 ± 0.68
R^2	0.998	0.999	0.999
R^2_{adj}	0.997	0.998	0.999
ARE (%)	1.24	1.84	0.09
SSE	1.85	0.55	0.01
MSE (mg g^{-1})^2	3.41	2.48	3.01

[a] (Mean ± standard deviation).

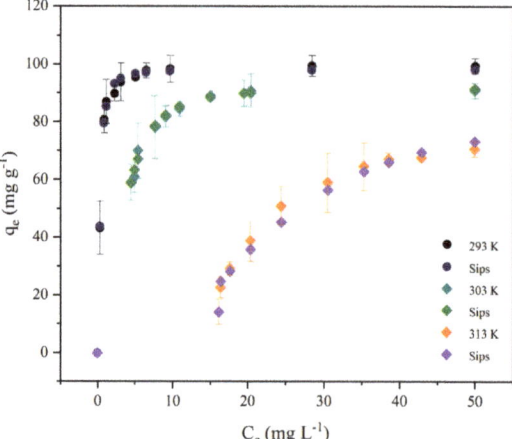

Figure 13. Sips model for CPT adsorption onto GO·Fe$_3$O$_4$.

To understand adsorption behavior under the influence of temperature, this study was performed at different conditions (293.15, 313.15, and 333.15 K). The thermodynamic equilibrium constant (K_e) was calculated using the K_{Sips} (best adjustment for isotherm model), and the captopril molecular weight (217.29 g mol^{-1}) [16,56–59] Thermodynamic parameters for CPT adsorption onto GO·Fe$_3$O$_4$ were calculated by the Van 't Hoff equation and are shown in Table 4. From the values for Gibbs free energy variation (ΔG^0), it was possible to assume that the captopril adsorption was a thermodynamically favorable process [56]. Furthermore, the negative value for enthalpy change (ΔH^0) indicated exothermic adsorption [47]. Additionally, the magnitude (-64.19 kJ mol^{-1}) suggests that the adsorption phenomenon is mainly governed by chemical interactions [60]. Regarding the entropy change (ΔS^0), the negative value is related to a decrease in the randomness of the system.

Table 4. Thermodynamic parameters for CPT adsorption onto GO·Fe$_3$O$_4$.

T(K)	K$_e$	ΔG^0 (kJ mol^{-1})	ΔH^0 (kJ mol^{-1})	ΔS^0 (kJ mol^{-1} K^{-1})
293.15	96476.76	-27.92		
303.15	55408.95	-27.52	-64.19	-0.12
313.15	17817.78	-25.48		

The adsorption performance was compared to previous studies employing different adsorbents and experimental conditions. It is possible to observe, in Table 5, that all the materials employed in CPT removal showed a removal percentage. Karperiski et al. [16] synthesized an activated carbon with ZnCl$_2$ from Caesalpina ferrea (CFAC) and obtained a q$_{max}$ of 535.5 mg g^{-1}. The results showed that the best adsorbent performance was reached by CFCAC.1.5 and the process was spontaneous and favorable for removal of 97.67% of CPT from the aqueous solution. Cunha et al. [9] developed an activated carbon derivate from Butia catarinensis (ABc) with a high surface area (1267 m^2 g^{-1}) and with high efficiency removal (up to 99%). The maximum adsorption capacity was dependent on the temperature, rising with the temperature. In this work, magnetic graphene oxide removed up to 99.0% of the drug, with a high adsorption capacity (100.41 mg g^{-1}) at pH 3.0. Therefore, the magnetite functionalization improved the adsorption process, allowing easy adsorbent regeneration and reusability and avoiding centrifugation/filtration steps.

Table 5. Removal percentage of different adsorbents of CPT.

Adsorbents	pH	Removal (%)	Reference
Activated carbon functionalized with ZnCl$_2$	7.0	97.67	[16]
Activated carbon derivate from Butia catarinensis	7.0	<99.0	[9]
Microporous carbons	Not informed	97.0	[61]
Magnetic graphene oxide	3.0	<99	This work

3.6.7. Regeneration and Reuse

The study of the regeneration of adsorbents is a crucial step in the adsorption processes, aiming at the development and application of materials with high performance and cost-effectiveness. In this work, the regeneration of GO·Fe$_3$O$_4$ was performed using a solution of sodium hydroxide (0.25 mol L^{-1}), stirring at room temperature for 60 min. Thus, five consecutive adsorption-desorption cycles were conducted to estimate the efficiency for several experiments. According to Figure 14, it was possible to observe that the nanoadsorbent exhibits high efficiency, even after diverse cycles of regeneration and reuse. The percentage removal value displayed a slight decrease over time, which can be attributed to incomplete desorption and mass loss during the adsorbent recovery step [26,47].

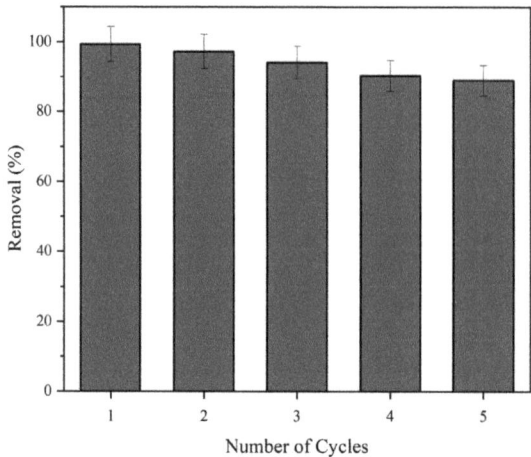

Figure 14. GO·Fe_3O_4 performance after several adsorption cycles.

4. Conclusions

The magnetic graphene oxide was synthesized employing a straightforward method with low energy requirements and easy operation. The FTIR, XRD, Raman, SEM, and VSM techniques demonstrated that the nanocomposite was successfully obtained, exhibiting excellent properties (crystalline structure with low defects) and high value for saturation magnetization (Ms = 45 emu g^{-1}). The graphene oxide-based magnetic nanoadsorbent exhibits higher performance when compared with the pristine carbon nanomaterial. The maximum adsorption capacity and removal percentage (99.43% and 100.41 mg g^{-1}) was reached at pH 3.0, using 50 mg of adsorbent, 50 mg L^{-1} of CPT, and 293.15 K. These results can be attributed to the synergic effect between graphene oxide and iron oxide nanoparticles. The pH dependence and influence of the adsorbent type suggested that the CPT adsorption onto GO·Fe_3O_4 is mainly driven by electrostatic interactions and hydrogen bonds. Sips and Elovich models were well-suited to describe the equilibrium and adsorption kinetics data, assuming a process on a heterogeneous surface. The thermodynamic study inferred that the adsorption was spontaneous, exothermic, and occurred predominantly by chemical mechanisms. Additionally, the GO·Fe_3O_4 demonstrated high efficiency after five adsorption/desorption cycles. From the experimental results, it can be concluded that the magnetic adsorbent is a promising material for wastewater management, especially regarding emerging pollutants like captopril.

Author Contributions: M.P.d.O., C.S.: Methodology, Investigation. T.d.R.S., F.d.S.B., L.B.: Investigation, Writing—Original Draft. E.I.M., W.J.d.S.G.: Visualization, Data Curation. A.H.d.O., L.F.O.S.: Visualization, Writing—Review & Editing. C.R.B.R.: Conceptualization, Writing—Review & Editing, Supervision (cristianorbr@gmail.com). All authors have read and agreed to the published version of the manuscript.

Funding: This research received no external funding.

Data Availability Statement: Not applicable.

Acknowledgments: The authors thank Laboratório de Materiais Magnéticos Nanoestruturados—LaMMaN, Laboratório de Magnetismo e Materiais Magnéticos—LMMM, UFSM, and CAPES, CNPq, FAPERGS (TO 22/2551-0000609-4) for their support.

Conflicts of Interest: The authors declare no conflict of interest.

References

1. Duarte, E.D.V.; Oliveira, M.G.; Spaolonzi, M.P.; Costa, H.P.S.; da Silva, T.L.; da Silva, M.G.C.; Vieira, M.G.A. Adsorption of Pharmaceutical Products from Aqueous Solutions on Functionalized Carbon Nanotubes by Conventional and Green Methods: A Critical Review. *J. Clean. Prod.* **2022**, *372*, 133743. [CrossRef]
2. Ramos, S.; Homem, V.; Alves, A.; Santos, L. A Review of Organic UV-Filters in Wastewater Treatment Plants. *Environ. Int.* **2016**, *86*, 24–44. [CrossRef] [PubMed]
3. Gavrilescu, M.; Demnerová, K.; Aamand, J.; Agathos, S.; Fava, F. Emerging Pollutants in the Environment: Present and Future Challenges in Biomonitoring, Ecological Risks and Bioremediation. *N. Biotechnol.* **2015**, *32*, 147–156. [CrossRef] [PubMed]
4. Wang, L.; Chen, G.; Shu, H.; Cui, X.; Luo, Z.; Chang, C.; Zeng, A.; Zhang, J.; Fu, Q. Facile Covalent Preparation of Carbon Nanotubes / Amine-Functionalized Fe_3O_4 Nanocomposites for Selective Extraction of Estradiol in Pharmaceutical Industry Wastewater. *J. Chromatogr. A* **2021**, *1638*, 461889. [CrossRef]
5. Xi, L.; Zhang, X.; Chen, Y.; Peng, J.; Liu, M.; Huo, D.; Li, G.; He, H. A Fluorescence Turn-on Strategy to Achieve Detection of Captopril Based on Ag Nanoclusters. *Chem. Phys. Lett.* **2022**, *807*, 140085. [CrossRef]
6. Qu, F.; Zhu, G.; Huang, S.; Li, S.; Qiu, S. Effective controlled release of captopril by silylation of mesoporous MCM-41. *ChemPhysChem* **2006**, *7*, 400–406. [CrossRef]
7. Mahmoud, W.M.M.; Kümmerer, K. Captopril and Its Dimer Captopril Disulfide: Photodegradation, Aerobic Biodegradation and Identification of Transformation Products by HPLC-UV and LC-Ion Trap-MS(n). *Chemosphere* **2012**, *88*, 1170–1177. [CrossRef]
8. Da Silva, D.M.; Carneiro da Cunha Areias, M. Voltammetric Detection of Captopril in a Commercial Drug Using a Gold-Copper Metal-organic Framework Nanocomposite Modified Electrode. *Electroanalysis* **2021**, *33*, 1255–1263. [CrossRef]
9. Cunha, M.R.; Lima, E.C.; Lima, D.R.; Da Silva, R.S.; Thue, P.S.; Seliem, M.K.; Sheir, F.; Dos Reis, G.S.; Larsson, S.H. Removal of captopril pharmaceutical from synthetic pharmaceutical-industry wastewaters: Use of activated carbon derived from Butia catarinensis. *J. Environ. Chem. Eng.* **2020**, *8*, 104506. [CrossRef]
10. Alayli, A.; Nadaroglu, H.; Turgut, E. Nanobiocatalyst beds with Fenton process for removal of methylene blue. *Appl. Water Sci.* **2021**, *11*, 32. [CrossRef]
11. Oviedo, L.R.; Muraro, P.C.L.; Pavoski, G.; Espinosa, D.C.R.; Ruiz, Y.P.M.; Galembeck, A.; Rhoden, C.R.B.; da Silva, W.L. Synthesis and Characterization of Nanozeolite from (Agro)Industrial Waste for Application in Heterogeneous Photocatalysis. *Environ. Sci. Pollut. Res. Int.* **2022**, *29*, 3794–3807. [CrossRef] [PubMed]
12. Sani, O.N.; Yazdani, M.; Taghavi, M. Catalytic ozonation of ciprofloxacin using γ-Al_2O_3 nanoparticles in synthetic and real wastewaters. *J. Water Process Eng.* **2019**, *32*, 100894. [CrossRef]
13. Erdem, S.; Öztekin, M.; Açıkel, Y.S. Investigation of tetracycline removal from aqueous solutions using halloysite/chitosan nano-composites and halloysite nanotubes/alginate hydrogel beads. *Environ. Nanotechnol. Monit. Manag.* **2021**, *16*, 100576. [CrossRef]
14. Ciğeroğlu, Z.; Kazan-Kaya, E.S.; El Messaoudi, N.; Fernine, Y.; Américo-Pinheiro, J.H.P.; Jada, A. Remediation of tetracycline from aqueous solution through adsorption on g-C_3N_4-ZnO-$BaTiO_3$ nanocomposite: Optimization, modeling, and theoretical calculation. *J. Mol. Liq.* **2022**, *369*, 120866. [CrossRef]
15. Silveira, C.C.; Botega, C.S.; Rhoden, C.R.B.; Nunes, M.R.S.; Braga, A.L.; Lenardão, E.J. A Facile Synthesis of α-Phenylchalcogeno(S, Se) α,β-Unsaturated Esters from Ethyl α-Bromo-α-Phenylchalcogeno Acetates. *Synth. Commun.* **1998**, *28*, 3371–3380. [CrossRef]
16. Kasperiski, F.M.; Lima, E.C.; Umpierres, C.S.; Dos Reis, G.S.; Thue, P.S.; Lima, D.R.; Dias, S.L.P.; Saucier, C.; Da Costa, J.B. Production of porous activated carbons from Caesalpinia ferrea seed pod wastes: Highly efficient removal of captopril from aqueous solutions. *J. Clean. Prod.* **2018**, *197*, 919–929. [CrossRef]
17. Singh, S.; Kumar, V.; Anil, A.G.; Kapoor, D.; Khasnabis, S.; Shekar, S.; Pavithra, N.; Samuel, J.; Subramanian, S.; Singh, J.; et al. Adsorption and Detoxification of Pharmaceutical Compounds from Wastewater Using Nanomaterials: A Review on Mechanism, Kinetics, Valorization and Circular Economy. *J. Environ. Manag.* **2021**, *300*, 113569. [CrossRef]
18. El Messaoudi, N.; El Mouden, A.; Fernine, Y.; El Khomri, M.; Bouich, A.; Faska, N.; Ciğeroğlu, Z.; Américo-Pinheiro, J.H.P.; Jada, A.; Lacherai, A. Green synthesis of Ag_2O nanoparticles using Punica granatum leaf extract for sulfamethoxazole antibiotic adsorption: Characterization, experimental study, modeling, and DFT calculation. *Environ. Sci. Pollut. Res.* **2022**, *29*, 1–18. [CrossRef]
19. Singh, S.; Anil, A.G.; Khasnabis, S.; Kumar, V.; Nath, B.; Adiga, V.; Kumar Naik, T.S.S.; Subramanian, S.; Kumar, V.; Singh, J.; et al. Sustainable Removal of Cr(VI) Using Graphene Oxide-Zinc Oxide Nanohybrid: Adsorption Kinetics, Isotherms and Thermodynamics. *Environ. Res.* **2022**, *203*, 111891. [CrossRef]
20. Li, R.; Liu, Y.; Lan, G.; Qiu, H.; Xu, B.; Xu, Q.; Sun, N.; Zhang, L. Pb(II) Adsorption Characteristics of Magnetic GO-Hydroxyapatite and the Contribution of GO to Enhance Its Acid Resistance. *J. Environ. Chem. Eng.* **2021**, *9*, 105310. [CrossRef]
21. Rhoden, C.R.B.; Bruckmann, F.d.S.; Salles, T.d.R.; Kaufmann Junior, C.G.; Mortari, S.R. Study from the Influence of Magnetite onto Removal of Hydrochlorothiazide from Aqueous Solutions Applying Magnetic Graphene Oxide. *J. Water Proc. Eng.* **2021**, *43*, 102262. [CrossRef]
22. Salles, T.d.R.; Rodrigues, H.d.B.; Bruckmann, F.d.S.; Alves, L.C.S.; Mortari, S.R.; Rhoden, C.R.B. Graphene Oxide Optimization Synthesis for Application on Laboratory of Universidade Franciscana. *Discip. Sci.* **2020**, *21*, 15–26. [CrossRef]
23. Bruckmann, F.d.S.; Zuchetto, T.; Ledur, C.M.; dos Santos, C.L.; da Silva, W.L.; Binotto Fagan, S.; Zanella da Silva, I.; Bohn Rhoden, C.R. Methylphenidate Adsorption onto Graphene Derivatives: Theory and Experiment. *New J. Chem.* **2022**, *46*, 4283–4291. [CrossRef]

24. Bruckmann, F.S.; Schnorr, C.; Oviedo, L.R.; Knani, S.; Silva, L.F.O.; Silva, W.L.; Dotto, G.L.; Bohn Rhoden, C.R. Adsorption and Photocatalytic Degradation of Pesticides into Nanocomposites: A Review. *Molecules* **2022**, *27*, 6361. [CrossRef]
25. Tran, H.N.; You, S.-J.; Chao, H.-P. Thermodynamic Parameters of Cadmium Adsorption onto Orange Peel Calculated from Various Methods: A Comparison Study. *J. Environ. Chem. Eng.* **2016**, *4*, 2671–2682. [CrossRef]
26. Da Rosa Salles, T.; Da Silva Bruckamann, F.; Viana, A.R.; Krause, L.M.F.; Mortari, S.R.; Rhoden, C.R.B. Magnetic nanocrystalline cellulose: Azithromycin adsorption and in vitro biological activity against melanoma cells. *J. Polym. Environ.* **2022**, *30*, 2695–2713. [CrossRef]
27. Cimirro, F.N.; Lima, C.E.; Cunha, M.R.; Dias, S.L.; Thue, P.S.; Mazzocato, A.C.; Dotto, G.L.; Gelesky, M.A.; Pavan, F.A. Removal of pharmaceutical compounds from aqueous solution by novel activated carbon synthesized from lovegrass (Poaceae). *Environ. Sci. Pollut. Res.* **2020**, *27*, 21442–21454. [CrossRef]
28. Kanta, U.-A.; Thongpool, V.; Sangkhun, W.; Wongyao, N.; Wootthikanokkhan, J. Preparations, Characterizations, and a Comparative Study on Photovoltaic Performance of Two Different Types of Graphene/TiO$_2$ Nanocomposites Photoelectrodes. *J. Nanomater.* **2017**, *2017*, 2758294. [CrossRef]
29. Ossonon, B.D.; Bélanger, D. Synthesis and Characterization of Sulfophenyl-Functionalized Reduced Graphene Oxide Sheets. *RSC Adv.* **2017**, *7*, 27224–27234. [CrossRef]
30. Da Silva Bruckmann, F.; Viana, A.R.; Lopes, L.Q.S.; Santos, R.C.V.; Muller, E.I.; Mortari, S.R.; Rhoden, C.R.B. Synthesis, Characterization, and Biological Activity Evaluation of Magnetite-Functionalized Eugenol. *J. Inorg. Organomet. Polym. Mater.* **2022**, *32*, 1459–1472. [CrossRef]
31. Bruckmann, F.d.S.; Pimentel, A.C.; Viana, A.R.; Salles, T.d.R.; Krause, L.M.F.; Mortari, S.R.; da Silva, I.Z.; Rhoden, C.R.B. Synthesis, Characterization and Cytotoxicity Evaluation of Magnetic Nanosilica in L929 Cell Line. *Discip. Sci.* **2020**, *21*, 1–14. [CrossRef]
32. Ain, Q.T.; Haq, S.H.; Alshammari, A.; Al-Mutlaq, M.A.; Anjum, M.N. The Systemic Effect of PEG-NGO-Induced Oxidative Stress in Vivo in a Rodent Model. *Beilstein J. Nanotechnol.* **2019**, *10*, 901–911. [CrossRef] [PubMed]
33. Liu, J.; Xu, D.; Chen, P.; Yu, Q.; Qiu, H.; Xiong, X. Solvothermal Synthesis of Porous Superparamagnetic RGO@Fe3O4 Nanocomposites for Microwave Absorption. *J. Mater. Sci. Mater. Electron.* **2019**, *30*, 17106–17118. [CrossRef]
34. Kellici, S.; Acord, J.; Ball, J.; Reehal, H.S.; Morgan, D.; Saha, B. A Single Rapid Route for the Synthesis of Reduced Graphene Oxide with Antibacterial Activities. *RSC Adv.* **2014**, *4*, 14858–14861. [CrossRef]
35. Côa, F.; Strauss, M.; Clemente, Z.; Rodrigues Neto, L.L.; Lopes, J.R.; Alencar, R.S.; Souza Filho, A.G.; Alves, O.L.; Castro, V.L.S.S.; Barbieri, E.; et al. Coating Carbon Nanotubes with Humic Acid Using an Eco-Friendly Mechanochemical Method: Application for Cu(II) Ions Removal from Water and Aquatic Ecotoxicity. *Sci. Total Environ.* **2017**, *607–608*, 1479–1486. [CrossRef] [PubMed]
36. Zhang, K.; Zhang, Q.; Gao, X.; Chen, X.; Wang, Y.; Li, W.; Wu, J. Effect of absorbers' composition on the microwave absorbing performance of hollow Fe3O4 nanoparticles decorated CNTs/graphene/C composites. *J. Alloys Compd.* **2018**, *748*, 70–716. [CrossRef]
37. Hatel, R.; Majdoub, S.E.; Bakour, A.; Khenfouch, M.; Baitoul, M. Graphene Oxide/Fe$_3$O$_4$ Nanorods Composite: Structural and Raman Investigation. *J. Phys. Conf. Ser.* **2018**, *1081*, 012006. [CrossRef]
38. Ghosh, B.; Sarma, S.; Pontsho, M.; Ray, S.C. Tuning of Magnetic Behaviour in Nitrogenated Graphene Oxide Functionalized with Iron Oxide. *Diam. Relat. Mater.* **2018**, *89*, 35–42. [CrossRef]
39. Da Silva Bruckmann, F.; Mafra Ledur, C.; Zanella da Silva, I.; Luiz Dotto, G.; Rodrigo Bohn Rhoden, C. A DFT Theoretical and Experimental Study about Tetracycline Adsorption onto Magnetic Graphene Oxide. *J. Mol. Liq.* **2022**, *353*, 118837. [CrossRef]
40. Cheng, Y.; Yang, S.; Tao, E. Magnetic graphene oxide prepared via ammonia coprecipitation method: The effects of preserved functional groups on adsorption property. *Inorg. Chem. Commun.* **2021**, *128*, 108603. [CrossRef]
41. Zeng, K.; Hachem, K.; Kuznetsova, M.; Chupradit, S.; Su, C.H.; Nguyen, H.C.; El-Shafay, A.S. Molecular dynamic simulation and artificial intelligence of lead ions removal from aqueous solution using magnetic-ash-graphene oxide nanocomposite. *J. Mol. Liq.* **2022**, *347*, 118290. [CrossRef]
42. Nuengmatcha, P.; Mahachai, R.; Chanthai, S. Thermodynamic and kinetic study of the intrinsic adsorption capacity of graphene oxide for malachite green removal from aqueous solution. *Orient. J. Chem.* **2014**, *30*, 1463. [CrossRef]
43. Nasiri, A.; Rajabi, S.; Amiri, A.; Fattahizade, M.; Hasani, O.; Lalehzari, A.; Hashemi, M. Adsorption of tetracycline using CuCoFe$_2$O$_4$@ Chitosan as a new and green magnetic nanohybrid adsorbent from aqueous solutions: Isotherm, kinetic and thermodynamic study. *Arab. J. Chem.* **2022**, *15*, 104014. [CrossRef]
44. Da Silva Bruckmann, F.; Schnorr, C.E.; Da Rosa Salles, T.; Nunes, F.B.; Baumann, L.; Müller, E.I.; Silva, L.F.O.; Dotto, G.L.; Bohn Rhoden, C.R. Highly Efficient Adsorption of Tetracycline Using Chitosan-Based Magnetic Adsorbent. *Polymers* **2022**, *14*, 4854. [CrossRef]
45. Pereira, A.V.; Garabeli, A.A.; Schunemann, G.D.; Borck, P.C. Determination of dissociation constant (Ka) of captopril and nimesulide: Analytical chemistry experiments for undergraduate pharmacy. *Quim Nova* **2011**, *34*, 1656–1660. [CrossRef]
46. Zhu, H.; Chen, T.; Liu, J.; Li, D. Adsorption of tetracycline antibiotics from an aqueous solution onto graphene oxide/calcium alginate composite fibers. *RSC Adv.* **2018**, *8*, 2616–2621. [CrossRef]
47. Bruckmann, F.S.; Rossato Viana, A.; Tonel, M.Z.; Fagan, S.B.; Garcia, W.J.D.S.; Oliveira, A.H.D.; Dorneles, L.S.; Mortari, S.R.; Da Silva, W.L.; Da Silva, I.Z.; et al. Influence of magnetite incorporation into chitosan on the adsorption of the methotrexate and in vitro cytotoxicity. *Environ. Sci. Pollut. Res.* **2022**, *29*, 70413–70434. [CrossRef]

48. Ji, L.; Chen, W.; Bi, J.; Zheng, S.; Xu, Z.; Zhu, D.; Alvarez, P.J. Adsorption of tetracycline on single-walled and multi-walled carbon nanotubes as affected by aqueous solution chemistry. *Environ. Toxicol. Chem.* **2010**, *29*, 2713–2719. [CrossRef]
49. Liang, J.; Fang, Y.; Luo, Y.; Zeng, G.; Deng, J.; Tan, X.; Tang, N.; Li, X.; He, X.; Feng, C.; et al. Magnetic nanoferromanganese oxides modified biochar derived from pine sawdust for adsorption of tetracycline hydrochloride. *Environ. Sci. Pollut. Res.* **2019**, *26*, 5892–5903. [CrossRef]
50. Agarry, S.E.; Aworanti, O.A. Kinetics, Isothermal and Thermodynamic Modelling Studies of Hexavalent Chromium Ions Adsorption from Simulated Wastewater onto Parkia biglobosa-Sawdust Derived Acid-Steam Activated Carbon. *Appl. J. Envir. Eng. Sci.* **2017**, *3*, 58–76.
51. De Souza, F.M.; Dos Santos, O.A.A.; Vieira, M.G.A. Adsorption of herbicide 2,4-D from aqueous solution using organo-modified bentonite clay. *Environ. Sci. Pollut. Res.* **2019**, *26*, 18329–18342. [CrossRef] [PubMed]
52. Nunes, F.B.; Da Silva Bruckmann, F.; Da Rosa Salles, T.; Rhoden, C.B.R. Study of phenobarbital removal from the aqueous solutions employing magnetite-functionalized chitosan. *Environ. Sci. Pollut. Res.* **2022**, *29*, 1–14. [CrossRef] [PubMed]
53. Carvajal-Bernal, A.M.; Gomez-Granados, F.; Giraldo, L.; Moreno-Pirajan, J.C. Application of the Sips model to the calculation of maximum adsorption capacity and immersion enthalpy of phenol aqueous solutions on activated carbons. *Eur. J. Chem.* **2017**, *8*, 112–118. [CrossRef]
54. Kalam, S.; Abu-Khamsin, S.A.; Kamal, M.S.; Patil, S. Surfactant Adsorption Isotherms: A Review. *ACS Omega* **2021**, *6*, 32342–32348. [CrossRef]
55. Gago, D.; Chagas, R.; Ferreira, L.M.; Velizarov, S.; Coelhoso, I. A Novel Cellulose-Based Polymer for Efficient Removal of Methylene Blue. *Membranes* **2020**, *10*, 13. [CrossRef]
56. Salvestrini, S.; Ambrosone, L.; Kopinke, F.D. Some mistakes and misinterpretations in the analysis of thermodynamic adsorption data. *J. Mol. Liq.* **2022**, *352*, 118762. [CrossRef]
57. Tran, H.N. Improper Estimation of Thermodynamic Parameters in Adsorption Studies with Distribution Coefficient KD (Qe/Ce) or Freundlich Constant (KF): Considerations from the Derivation of Dimensionless Thermodynamic Equilibrium Constant and Suggestions. *Adsorp. Sci. Technol.* **2022**, *2022*, 5553212. [CrossRef]
58. Lima, E.C.; Hosseini-Bandegharaei, A.; Moreno-Piraján, J.C.; Anastopoulos, I. A Critical Review of the Estimation of the Thermodynamic Parameters on Adsorption Equilibria. Wrong Use of Equilibrium Constant in the Van't Hoof Equation for Calculation of Thermodynamic Parameters of Adsorption. *J. Mol. Liq.* **2019**, *273*, 425–434. [CrossRef]
59. Tran, H.N.; Lima, E.C.; Juang, R.-S.; Bollinger, J.-C.; Chao, H.-P. Thermodynamic Parameters of Liquid–Phase Adsorption Process Calculated from Different Equilibrium Constants Related to Adsorption Isotherms: A Comparison Study. *J. Environ. Chem. Eng.* **2021**, *9*, 106674. [CrossRef]
60. Dotto, G.L.; Moura, J.M.D.; Cadaval, T.R.S.; Pinto, L.A.D.A. Application of chitosan films for the removal of food dyes from aqueous solutions by adsorption. *Chem. Eng. J.* **2013**, *214*, 8–16. [CrossRef]
61. Li, Z.; Wu, D.; Liang, Y.; Xu, F.; Fu, R. Facile Fabrication of Novel Highly Microporous Carbons with Superior Size-Selective Adsorption and Supercapacitance Properties. *Nanoscale* **2013**, *5*, 10824–10828. [CrossRef] [PubMed]

Disclaimer/Publisher's Note: The statements, opinions and data contained in all publications are solely those of the individual author(s) and contributor(s) and not of MDPI and/or the editor(s). MDPI and/or the editor(s) disclaim responsibility for any injury to people or property resulting from any ideas, methods, instructions or products referred to in the content.

Article

Predictive Model Based on K-Nearest Neighbor Coupled with the Gray Wolf Optimizer Algorithm (KNN_GWO) for Estimating the Amount of Phenol Adsorption on Powdered Activated Carbon

Meriem Zamouche [1,*], Mouchira Chermat [1], Zohra Kermiche [1], Hichem Tahraoui [2], Mohamed Kebir [3], Jean-Claude Bollinger [4], Abdeltif Amrane [5] and Lotfi Mouni [6,*]

1. Laboratoire de Recherche sur le Médicament et le Développement Durable (ReMeDD), Faculté de Génie des Procédés, Université de Salah BOUBNIDER Constantine 3, Constantine 25000, Algeria
2. Laboratory of Biomaterials and Transport Phenomena (LBMPT), University of MÉDÉA, Nouveau Pôle Urbain, Médéa 26000, Algeria
3. Research Unit on Analysis and Technological Development in Environment (URADTE-CRAPC), BP 384, Bou-Ismail, Tipaza 42004, Algeria
4. Laboratoire E2Lim, Université de Limoges, 123 Avenue Albert Thomas, 87060 Limoges, France
5. Ecole Nationale Supérieure de Chimie de Rennes, University of Rennes, CNRS, ISCR—UMR6226, 35000 Rennes, France
6. Laboratory of Management and Valorization of Natural Resources and Quality Assurance, SNVST Faculty, Akli Mohand Oulhadj University, Bouira 10000, Algeria
* Correspondence: meriem.zamouche@univ-constantine3.dz (M.Z.); lotfimouni@gmail.com (L.M.)

Abstract: In this work, the adsorption mechanism of phenol on activated carbon from aqueous solutions was investigated. Batch experiments were performed as a function of adsorbent rate, solution temperature, phenol initial concentration, stirring speed, and pH. The optimal operating condition of phenol adsorption were: mass/volume ratio of 0.6 g.L^{-1}, temperature of 20 °C and stirring speed of 300 rpm. The equilibrium data for the adsorption of phenol were analyzed by Langmuir, Freundlich, and Temkin isotherm models. It was found that the Freundlich and Temkin isotherm models fitted well the phenol adsorption on the activated carbon and that the adsorption process is favorable. The Langmuir equilibrium isotherm provides a maximum adsorption of 156.26 mg.g^{-1} at 20 °C. The pseudo-first-order, pseudo-second-order, intraparticle diffusion, and Boyd models were used to fit the kinetic data. The adsorption kinetics data were well described by the pseudo-second-order model. The kinetic was controlled by the external diffusion by macropore and mesopore, as well as by the micropore diffusion. The thermodynamic study revealed the exothermic and spontaneous nature of phenol adsorption on activated carbon with increased randomness at the solid-solution interface. On the other hand, a very large model based on the optimization parameters of phenol adsorption using k-nearest neighbor coupled with the gray wolf optimizer algorithm was launched to predict the amount of phenol adsorption. The KNN_GWO model showed an advantage in giving more precise values related to very high statistical coefficients (R = 0.9999, R^2 = 0.9998 and R$^2_{adj}$ = 0.9998) and very low statistical errors (RMSE = 0, 0070, MSE = 0.2347 and MAE = 0.2763). These advantages show the efficiency and performance of the model used.

Keywords: adsorption; phenol; activated carbon; modeling; k-nearest neighbor; gray wolf optimizer

1. Introduction

Water is the major constituent of our planet and is of great importance for living beings (humans, animals, and plants) for the balance of the aquatic ecosystems. Water is covering about 70% of the earth's surface, which is why we call the earth the Blue Planet. However, 97.5% of this water is salty; it represents the oceans, the seas, and some groundwater, while only 2.5% of all the water on earth is fresh water.

About 2.1% of the world's freshwater is in the form of polar ice and permanent snow in the mountains, so it is not available for consumption or agricultural uses. The very small remaining percentage of freshwater is estimated at 0.4%, of which one-third is groundwater hidden in the earth's crust and often difficult to access [1]. This low percentage of this vital source has reduced and continues to be threatened year after year, as the consequences of climate change impact the planet, on the one hand, and on the other hand, by pollution caused by industrial and agricultural activities. In addition, urban wastes include many toxic substances that contribute strongly to the degradation of freshwater quality and limited access to water resources.

Water pollution is a major concern and is tied to the top worry of states and organizations for the protection of the aquatic environment. Among the emerging chemicals of concern and greatest risks to the aquatic environment are phenol and phenolic compounds. These tend to enter and persist extensively in the environmental ecosystem, accumulate and exert toxic effects on humans and animals [2].

The presence of phenol in the aquatic environment is mainly due to discharges from the plastics, dyes, pulp, and paper industries [3]. Phenols are also released from the industries of solid fuel gasification, crude oil processing, phytosanitary chemicals, pharmaceutical plants, perfume-manufacturing industries, disinfectants, and insecticides [4]. Due to their wide use, phenols or phenolic compounds are released with the effluents of these industries. This often leads to the contamination of groundwater and surface water. Phenol and phenolic compounds are not easily biodegradable, due to their resistance and complex structure; they constitute a major menace to public health and the aquatic environment [5].

Phenol is toxic to aquatic organisms, plants, and humans at concentrations of 5 $\mu g \cdot L^{-1}$ [6,7]. Intoxication and ingestion of phenol into the human body can occur through the skin, inhalation, or ingestion of phenol-contaminated products (water, foods that have been treated with phenol, etc.). Inhalation of phenol vapors can cause pulmonary edema, while prolonged contact with the skin can cause pulmonary edema and burns. Chronic exposure to phenol causes oral irritation, visual disturbances, diarrhea, deep necrosis, cardiac dysrhythmias, respiratory distress, metabolic acidosis, renal failure, and methemoglobinemia; in the most complicated cases, it causes cardiovascular shock, coma, and can even result in death [6–9]. Therefore, the World Health Organization (WHO) recommends that the maximum allowable concentration of phenol in drinking water should be less than 1 $\mu g \cdot L^{-1}$ [6,8].

Thus, the elimination of this generated toxic pollutant from the industries requires the implementation of reliable and efficient treatment processes. The treatment of water contaminated by phenol is an environmental challenge, and it has been investigated by many authors using several processes; the biological treatment is efficient only for low concentrations of phenols, which are lower than the discharging limits required and not toxic for the microbial flora [4].

However, the choice of such a process is based on its efficiency to remove different large concentrations of pollutants from wastewater with an economical treatment cost. The processes often used are the membrane technique, electrocoagulation, ion exchange, precipitation, flotation, and oxidation processes [10–13]. However, these processes have some limits and weaknesses, such generation of highly toxic byproducts or the high cost. On the other hand, adsorption is an alternative, and very effective, classic technique for the elimination of different kinds of pollution [14–18].

Adsorption has been conducted in this work and seems to be the most appropriate for the removal of phenol from contaminated water. Nevertheless, adsorption remains complex [19], because it is a non-linear process, both time-consuming and expensive compared to digital modeling techniques, especially if dealing with large areas. Recently, many authors have studied the parameters affecting water quality using an artificial intelligence approach [20].

Several works deal with great success in the simulation and prediction approaches of environmental parameters and can be used for management decisions [21], such as:

Artificial neural networks [22–24], support vector machine [24,25], multiple linear regressions [23], and adaptive neuro-fuzzy inference System [24].

To evaluate the suggested approach, the machine learning process is considered as one of the advanced methods capable of efficiently solving complex optimization problems [19]. Notably, k-nearest neighbor (KNN) uses many sets of nearest neighbors for processing [19]. The KNN searches for the nearest neighbor match using the pre-processed data and this data is used for the prediction process [19]. Moreover, the use of optimization algorithms remains an effective process methodology for advanced evaluation progression with time consumption [19]. Among these algorithms is the gray wolf optimization algorithm (GWO), a bio-inspired algorithm coupled with a very popular prediction-based classifier, named as KNN algorithm, to categorize data with data made up of datasets [19]. The effectiveness of the prediction process can be actualized with the experience of the data to validate the degree of effectiveness of the properties measured through the optimization process in the product [19].

In our work, powdered activated carbon (PAC) has been considered for elimination of phenol from aqueous solution. Activated carbon adsorption (PAC) is a highly efficient and flexible process for water treatment, and has been successfully used to reduce environmental pollution. The adsorption mechanism of phenol on activated carbon is affected by several factors, including not only the characteristics of the PAC but also the physical and chemical characteristics of the phenol [26].

The main goal of this study is to investigate the adsorption capacity of PAC in the field of wastewater treatment for removing phenol as a toxic molecule using artificial intelligence that will allow the prediction and optimization of parameters that have a significant effect on the phenol removal mechanism.

Isotherm blindness study follow-up: In the end, k-nearest neighbor coupled with the gray wolf optimizer algorithm (KNN_GWO) was used to create an extended model to overcome traditional methods and to predict the amount of phenol adsorbed on PAC. The GWO algorithm has been used to solve a wide variety of optimization problems such as improving machine-learning performance by optimizing model hyper-parameters. It has been followed by an application executable on windows to facilitate optimization by GWO and prediction by KNN_GWO. To date and to our knowledge, the prediction of phenol adsorption quantity using KNN_GWO has never been investigated before.

2. Materials and Methods

2.1. Reagents

All chemicals used were of analytical grade and were used as received (from Sigma-Aldrich, St. Louis, Missouri, USA) without further purification: Phenol (99.9%) was used as an adsorbate. Sodium hydroxide (NaOH) (98%) was supplied by Labosi. Hydrochloric acid (HCl) (37%) and sodium thiosulfate ($Na_2S_2O_3$) were purchased from Panreac, Barcelona, Spain. Sodium bicarbonate ($NaHCO_3$) (99.5%), iodine (I_2) (99%), sodium carbonate (Na_2CO_3) (99.5%) and methylene blue (99%) were supplied by Biochem Chemopharma, Cosne-Cours-sur-Loire, France, and powdered activated carbon (PAC) was supplied by Biochem Chemopharma.

2.2. Solution Preparation

The stock solution at a concentration of 1 g/L^{-1}, was prepared by dissolving 1 g of phenol powder accurately weighted in one liter of distilled water. The solution was agitated until complete dissolution of phenol. The experimental solutions were obtained by diluting the stock solutions to the desired initial concentrations.

2.3. Adsorption Kinetics Experiments

The adsorption kinetics were performed in discontinuous mode at a constant temperature of 20 °C ± 1 °C. The temperature was kept constant by a water bath (Kottermann Labortechnik, Uetze, Germany). Therefore, a mass volume ratio of 0.6 ((g) adsorbent. (L^{-1})

phenol solution) at a concentration of 25 mg.L^{-1} was stirred by a stirrer and heating plate (IKAMAG, Snijders, Tilburg, The Netherlands) at 300 rpm. During the adsorption process, a volume of 4 mL was sampled and filtered with a Millipore filter (0.45 µm). The recovered filtrate was immediately analyzed by spectrophotometer (Shimadzu mini 1601, Kyoto, Japan) at a wavelength of 272 nm. The amount of phenol adsorbed (mg.g^{-1}) at time t, and the pollutant removal efficiency (%) were calculated by the following equations:

$$q_t = \frac{(C_0 - C_t)}{m} V \tag{1}$$

$$E(\%) = \frac{(C_0 - C_t)}{C_0} 100\% \tag{2}$$

where: q_t (mg.g^{-1}) is the amount adsorbed at time t, V (L) is the volume of the solution, m (g) is the mass of the adsorbent, C_0 and C_t are the initial and residual concentration of the adsorbate in solution (mg.L^{-1}), and E (%) is percent removal of pollutant.

To examine the effect of the operating parameters on the amount of phenol removed, the experiments were carried out at different adsorbent rates from 0.4 to 1.6 g.L^{-1}; the temperature of the solution was varied from 20 to 50 °C and the stirring speed from 200 to 600 tr.min^{-1}. The effect of the initial concentration of phenol was studied by varying the phenol concentration from 10 to 200 mg.L^{-1}, while the pH effect is conducted in the range of pH from 2 to 12. The pH value was adjusted by hydrochloric acid (0.1 N) or sodium hydroxide (0.1 N). The pseudo-first-order and pseudo-second-order kinetic models have been widely used to describe the adsorption rate [27–29], and they are given by following the equations:

$$\frac{dq}{dt} = k_1 (q_e - q_t) \tag{3}$$

$$\frac{dq_t}{dt} = k_2 (q_e - q_t)^2 \tag{4}$$

where q_t (mg.g^{-1}) is the adsorption capacity at time t (min), k_1 is the rate constant of pseudo-first-order adsorption (1.min^{-1}), k_2 (g.mg^{-1}.min^{-1}) is the pseudo-second-order rate constant, and q_e (mg.g^{-1}) is the equilibrium adsorption capacity.

In order to understand the adsorption mechanism of phenol on the PAC and to identify the limiting steps in the adsorption mechanism process, the data kinetics of phenol adsorption on activated carbon were modeled by Weber–Morris intra-particle diffusion [27–29] and Boyd's [27–29] kinetic models, which are expressed as:

$$q_t = K_D \sqrt{t} + c_d \tag{5}$$

$$B_t = -0.4977 - \ln(1 - F) \tag{6}$$

K_D is the intraparticle diffusion parameter; c_d is the thickness of the boundary layer, and F is the extent of exchange at time t, which is expressed in terms:

$$F = q/q_e \tag{7}$$

2.4. Adsorption Isotherms

The equilibrium data of adsorption isotherm of phenol on activated carbon at different temperature were modeled by Langmuir [30,31], Freundlich [31], and Temkin [31] isotherms, whose mathematical expressions are expressed as:

$$q_e = \frac{q_m \cdot K_L \cdot C_e}{1 + K_L \cdot C_e} \tag{8}$$

$$q_e = K_F \cdot C_e^{\frac{1}{n}} \tag{9}$$

$$\theta = \frac{q_e}{q_m} = \left(\frac{RT}{b_T}\right).\ln(A_T.C_e) \qquad (10)$$

where C_e is the phenol concentration at equilibrium (mg.L^{-1}), qe is the adsorption capacity at equilibrium (mg.g^{-1}), q_m is the maximum adsorption capacity (mg.g^{-1}), K_L is the Langmuir constant (mg.L^{-1}), K_F is the Freundlich adsorbent capacity, n is the heterogeneity factor, R is the universal gas constant (8.314 J.mol^{-1}K^{-1}), b_T and A_T are characteristic constants of the Temkin isotherm, and θ is the recovery rate.

2.5. K-Nearest Neighbor

K-nearest neighbor (KNN) is a well-known method in machine learning, where it has recently been applied for the classification and parametric estimation analysis of difficult-to-evaluate unknown probabilities [19,20]. The KNN is inferred to classify the model, which has an idea of the k nearest neighbor rule [19,20], by proposing a new classification technique based on the distance between distributions [19,20]. The concept of KNN is to classify individual data that contains the nearest neighbor majority [19,20]. On the other hand, the KNN consists of classifiers and regressions for the prediction process [19,20], since the classification involves identifying and classifying groups based on the characteristics of the data [19,20]. In addition, for regression, one needs to use existing data to predict future data [19,20].

In this work, the KNN model was used to create an extended model to overcome traditional methods for predicting the amount of phenol adsorbed onto the best adsorbent (PCA). Another advantage is that the KNN model can easily handle large amounts of noisy data from dynamic and nonlinear systems. For this purpose, the results of the adsorbed quantity of phenol obtained at each instant by the optimization of the adsorption parameters were collected in a single database; the database collected with the dimension [376:7] includes seven parameters considered as inputs and one output parameter. The independent parameters were the contact time of each sample ($\times 1$), the phenol concentration ($\times 2$), the mass of adsorbent ($\times 3$), the stirring speed ($\times 4$), the temperature ($\times 5$), and the pH of the solution ($\times 6$). On the other hand, output (y) was the quantity of adsorbed phenol obtained at each instant.

To obtain the optimal result from KNN, a strategy based on design and optimization was followed. From the start, the database was normalized once in the interval [-1, $+1$] using MATLAB "mapminmax" function. Then, a small database of size [50:7] was hidden for later use to test the performance of the best obtained model. Therefore, the rest of the size database [326:7] was split in half: 80% of the dataset for training and 20% of validation samples. Subsequently, 11 distance metrics were optimized (Euclidean, Chebychev, Minkowski, Mahalanobis, Cosine, Correlation, Spearman, Hamming, Jaccard, Cityblock, Seuclidean) with their distance weighting functions (equal, inverse and squared inverse). Knowing this, each distance metric was also optimized for its specific parameters (number of neighbors and exponent in the case of "Minkowski" cubic distance metrics).

For this, the gray wolf optimizer (GWO) algorithm was coupled with KNN (KNN_GWO) to optimize the specific parameters of each distance metric, since this optimization of the algorithm has been used to solve a wide variety of optimization problems [19,21]. Among these solved problems are different machine learning optimization problems, fundamental values, and penalty parameters, which were sent by the GWO to train machine learning using training data [19,21]. This last point was confirmed by some studies that proved that machine learning performance was achieved when hyperparameters were optimized using the GWO algorithm [19,21].

At the end, to evaluate the performance and to select the best model, statistical criteria were used, including: the correlation coefficient (R), coefficient of determination (R^2), adjusted coefficient (R_{adj}^2), root mean square error (RMSE), mean square error (MSE) and mean absolute error (MAE). These criteria are calculated considering the following equations [22–25].

$$R = \frac{\sum_{i=1}^{N}\left(y_{exp} - \overline{y}_{exp}\right)\left(y_{pred} - \overline{y}_{pred}\right)}{\sqrt{\sum_{i=1}^{N}\left(y_{exp} - \overline{y}_{exp}\right)^2 \sum_{i=1}^{N}\left(y_{pred} - \overline{y}_{pred}\right)^2}} \quad (11)$$

$$R_{adj}^2 = 1 - \frac{(1-R^2)(N-1)}{N-K-1} \quad (12)$$

$$RMSE = \sqrt{\left(\frac{1}{N}\right)\left(\sum_{i=1}^{N}\left[\left(y_{exp} - \overline{y}_{exp}\right)\right]^2\right)} \quad (13)$$

$$MSE = \sqrt{\left(\frac{1}{N}\right)\left(\sum_{i=1}^{N}\left[\left(y_{exp} - y_{pred}\right)\right]^2\right)} \quad (14)$$

$$MAE = \sqrt{\left(\frac{1}{N}\right)\left(\sum_{i=1}^{N}\left|y_{exp} - y_{pred}\right|\right)} \quad (15)$$

where N is the number of data samples; K is the number of variables (inputs); y_{exp} and y_{pred} are the experimental and the predicted values, respectively; and $\overline{y}_{exp}$ and $\overline{y}_{pred}$ are the average values of the experimental and the predicted values, respectively [22–25].

3. Results and Discussions

3.1. Phenol

Phenol has a chemical formula C_6H_5OH with a molecular weight of 94.11 g.mol^{-1}. It is a weak acid with pKa = 9.99, a mass density 1.07 g.cm^{-3} and an aqueous solubility of 83.11 mg.L^{-1} at 293.15 K [18].

3.2. Characterization of PAC

The powdered active carbon (PAC) used in this study was the same one that was prepared and used previously for the adsorption of ketoprofen [32]. Its characterization results are summarized in the Table 1.

Table 1. Results of PAC characterization.

Parameters	Value
pH$_{PZC}$	7.5
$S_{(BET)}$ (m^2.g^{-1})	463.4
Ø(pore size) (Å)	27.36
$\tilde{v}$ (total pore volume) (cm^3.g^{-1})	0.13
Iodine number (mg.g^{-1})	752
Methylene blue number (mg.g^{-1})	244
Strong acid carboxylic group (meq.g^{-1})	0.026
Hydroxyl and phenol groups (meq.g^{-1})	0.725

The iodine number estimates the capacity of activated carbon to adsorb molecules of microporous size (less than or equal to that of the iodine molecule, i.e., 0 to 20 Å) [33]. The methylene blue (MB) index is used to estimate the amount of both the mesopores and macropores, which is indicative of a high adsorption ability of larger molecules. The iodine and the methylene blue numbers found were equal to 752 and 244 mg.g^{-1}, respectively. It is quite clear that the iodine index is higher than the methylene blue one, which indicates that the surface of the powdered activated carbon is mostly composed of micropores rather than of macro- or mesopores. Thus, this PAC has a strong adsorptive ability for microporous particles. This finding is coherent, since the phenol molecule is microporous (diameter of phenol = 0.56 nm), therefore it can easily be adsorbed by microporous adsorbents.

3.3. Effect of Operating Parameters

3.3.1. Effect of the Adsorbent Mass to Volume

To examine the influence of the adsorbent mass on the phenol adsorption, the ratio of activated carbon mass to solution volume was investigated in the range 0.4 to 1.6 g.L^{-1}, while maintaining the solution temperature at 20 °C and the phenol concentration at 25 mg.L^{-1}. According to Figure 1, the equilibrium adsorption capacity per gram of adsorbent decreased from 57.57 to 15.14 mg.g^{-1}, when the adsorbent mass to volume ratio increased from 0.4 g to 1.6 g.L^{-1}, while the percentage removal increased from 95.83 to 99.94%. The increase in phenol removal was due to the increase in the adsorption surface area with the disposal of a greater number of adsorption sites [32]. It was also noted that the adsorption was very fast at the beginning, which may be related to the large concentration gradient of phenol and powdered activated carbon; then the equilibrium was achieved due to the saturation of adsorbent sites. For all the studied adsorbent mass to solution volume ratios, a rapid removal of phenol (99.94%) occurred in a few minutes ($t_e \leq 5$ min). Phenol removal increased up to the ratio of 0.6 g.L^{-1}, and then did not significantly change. Thus, the optimal mass/volume ratio chosen to continue this work was 0.6 g.L^{-1}.

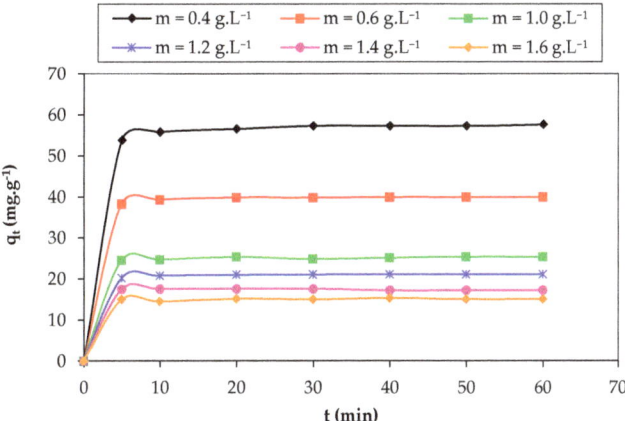

Figure 1. Effect of the mass volume ratio of PAC on the removal of phenol. (C_0 = 25 mg.L^{-1}, w = 300 rpm, T = 20 °C).

3.3.2. Effect of the Initial Concentration of Phenol

To study the effect of the initial concentration of phenol by adsorption on PAC, the initial concentrations of phenol was varied from 10 to 200 mg.L^{-1}. The obtained results, shown in Figure 2, indicate that the amount of adsorbate fixed on the PAC increased from 16.31 to 162.54 mg.g^{-1} with the increase in the initial concentration of phenol from 10 to 200 mg.L^{-1}. The initial concentration provides a significant driving force required to overcome all mass transfer resistances of all the phenol molecules between the aqueous and the solid phases [34]. It is also shown that the removal yield decreased from 99.94 to 48.92% with increasing the initial concentration of phenol. The increase in the initial phenol concentration enhances the interaction between phenol and the active sites on the surface of the activated carbon, increasing the removal of phenol. This can be attributed to the saturation of the adsorption sites for increasing phenol concentrations [34,35].

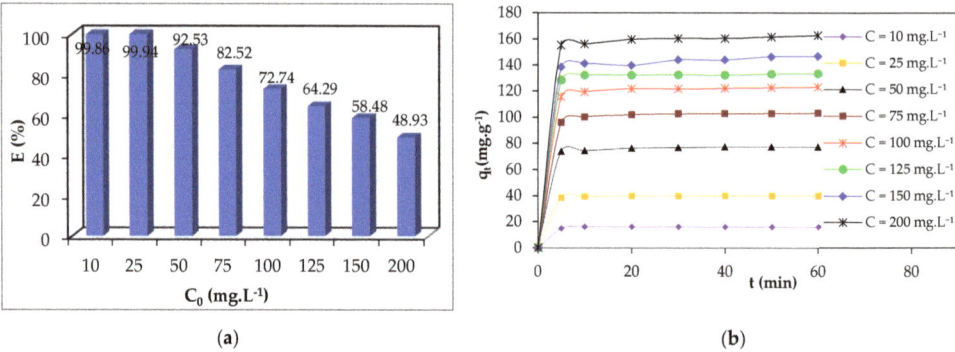

Figure 2. Percent removal of phenol by PAC as a function of the initial concentration (**a**), amount of phenol adsorbed as a function of time at different phenol initial concentration (**b**) (r = 0.6 g.L^{-1}, w = 300 rpm, T = 20 °C).

3.3.3. Effect of Temperature

Temperature is an indicator of the adsorption nature, whether it is an exothermic or an endothermic process [35]. Figure 3 shows that the equilibrium adsorption capacity of PAC decreased from 40.02 to 38.96 mg.g^{-1} with an increase of the temperature from 20 to 50 °C, on the one hand, and the removal percentage decreased from 99.94 to 94.80%, on the other hand. According to the study carried out by Abdelwahab and Amin [35], the decrease in the removal efficiency for increasing temperatures is due to the weakening of the adsorption forces between the active sites on the surface of the adsorbent and the adsorbate species and probably the damage of the active sites on the adsorbent [35]. The weakening of the adsorption forces results in a decrease in the adsorption efficiency. Consequently, the reduction in the rate of adsorbed phenol with increasing temperature reveals that the adsorption is an exothermic process [34]. The maximum removal was observed at the optimum temperature of 20 °C; further increase in temperature is inappropriate as it leads to the decrease in the percentage removal.

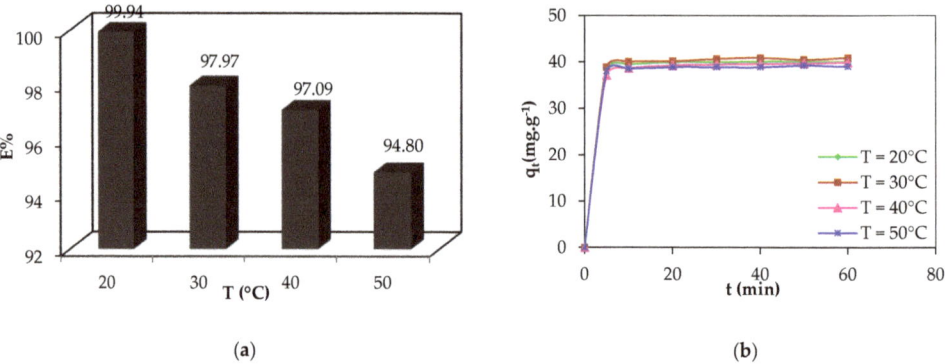

Figure 3. Percentage of elimination of phenol (**a**) and amount adsorbed of phenol (**b**) by PAC as a function of temperature. (r = 0.6 g.L^{-1}, w = 300 rpm, C$_0$ = 25 mg.L^{-1}).

3.3.4. Effect of the Stirring Speed

The effect of the stirring speed was evaluated from 200 to 600 rpm. The results obtained are illustrated in Figure 4, showing that the stirring speed did not affect the phenol adsorption, most likely due to the very fast adsorption that was achieved in the first contact minutes. Thus, the speed of 300 rpm was chosen as an optimal; this speed is sufficient to favor the contact between the PAC particles and the phenol molecules without slowing down the adsorption forces [36].

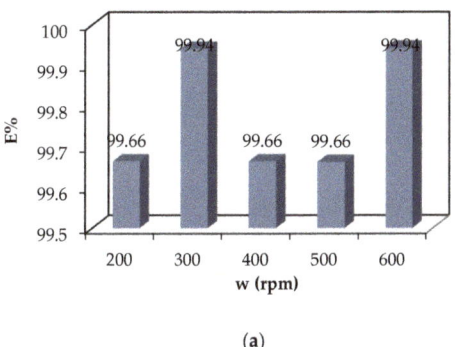 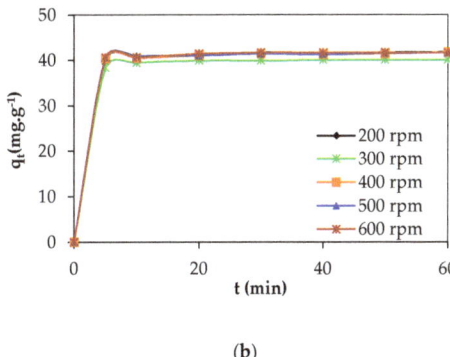

(a) (b)

Figure 4. Evolution of the percent removal (**a**) and amount adsorbed of phenol (**b**) at equilibrium as a function of the stirring speed (r = 0.6 g·L^{-1}, T = 20 °C, C_0 = 25 mg·L^{-1}).

3.3.5. Effect of the pH

In order to study the influence of this parameter on the phenol adsorption by PAC, pH values in the range of 2 to 12 were considered.

Figure 5 shows that the amount of phenol adsorbed by the PAC remained stable within the range pH of 2–8, with an almost total elimination. Beyond pH 8, phenol percent removal decreased until reaching a value of 57.34% at pH 12.

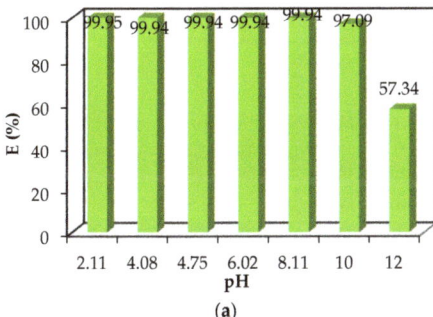 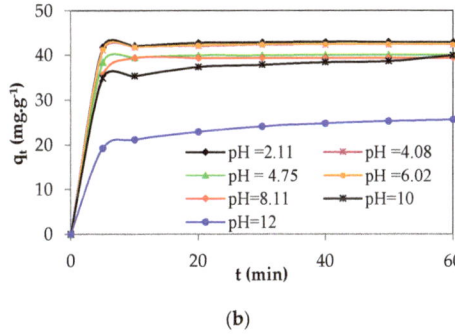

(a) (b)

Figure 5. Evolution of the percentage removal (**a**) and amount adsorbed of phenol (**b**) at equilibrium as a function of the pH of the solution (r = 0.6 g·L^{-1}, T = 20 °C, C_0 = 25 mg·L^{-1}, w = 300 rpm).

In the pH range between 2 and 8, an almost invariable total elimination has been observed. In this pH range, the surface of PAC is positively charged, since pH < pH_{PZC} = 7.5 (Table 1); while phenol, is neutral when pH < pKa = 9.99. Thus, the adsorption of phenol on the PAC surface is favored by hydrogen bonding, according to Luz-Asunción et al. [37] (Figure 6a).

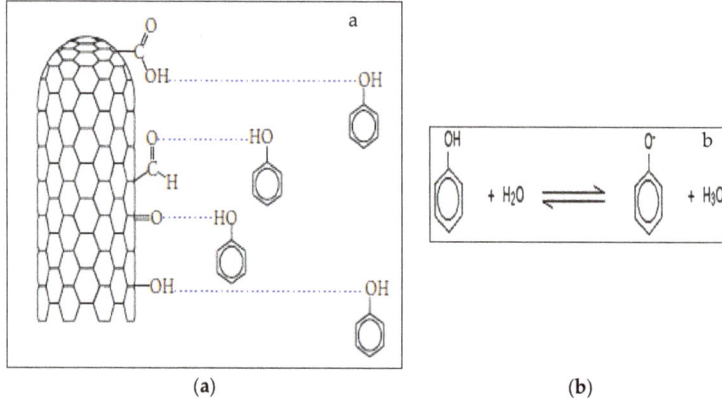

Figure 6. (a) Interaction mechanism between AC surface functions and phenol. (b) Phenol dissociation at pH > pKa.

It has been reported [37] that the presence of carboxylic and hydroxyl groups on the surface of the adsorbent promotes the removal of phenol by hydrogen bonding. The Boehm titration performed on PAC showed that strong carboxylic acid, and hydroxyl and phenol groups are favored to be adsorbed on the surface of the PAC, increasing the probability that the phenol adsorption occurs by hydrogen bonding.

As mentioned previously [37], phenol is a weak acid with an acid dissociation constant (pKa) equal to 9.99; at pH > pKa it dissociates into phenolate ions (Figure 6b) [38]. Therefore, at pH > pKa, the concentration of negatively charged phenolate ions increases in solution and the surface of the PAC is negatively charged (pH > pH_{PZC} = 7.5), hence, electrostatic repulsions occur between the negative surface charge of the PAC and the phenolate anions in solution, which caused the decrease in adsorption efficiency at pH values above 8.

3.4. Kinetics Study

The study of adsorption kinetics provides a description for the rate of adsorbate uptake and gives information of the mechanism controlling the rate of adsorption [39]. The kinetics parameters of Lagergren, Blanchard, and Webber and Morris for different concentrations of adsorbate at a 20 °C are gathered in Tables 2 and 3.

Table 2. Kinetics parameters of Lagergren and Blanchard for different initial concentrations of phenol.

C_0 (mg.L^{-1})	q_{eexp} (mg.g^{-1})	Lagergren			Blanchard		
		k_1 (1.min^{-1})	q_{etheo} (mg.g^{-1})	R^2	k_2 (g.mg^{-1}.min^{-1})	q_{etheo} (mg.g^{-1})	R^2
10	16.31	0.290	0.004	0.477	0.339	16.36	1
25	40.02	0.246	13.335	0.896	0.135	40.16	1
50	77.34	0.379	141.58	0.854	0.036	78.125	1
75	103.1	0.095	18.24	0.774	0.025	104.2	1
100	123.1	0.196	101.25	0.632	0.021	123.5	1
125	133.2	0.081	13.56	0.589	0.038	133.3	1
150	146.95	0.191	159.62	0.624	0.01	147.0	0.999
200	162.7	0.069	26.08	0.648	0.014	163.9	0.999

Table 3. Kinetics parameters of Weber and Morris for different initial concentrations of phenol.

C_0 (mg.L^{-1})	10	25	50	75	100	125	150	200
K_{d1} (mg.mg^{-1} min$^{-1/2}$)	5.457	13.338	25.283	33.723	40.285	44.635	47.874	53.08
C_{d1} (mg.g^{-1})	0.666	1.938	3.917	4.614	5.757	6.439	7.124	8.30
R_1^2	0.951	0.931	0.923	0.939	0.933	0.932	0.928	0.921
K_{d2} (mg.mg^{-1} min$^{-1/2}$)	-1×10^{-14}	0.044	0.506	0.351	0.450	0.349	2.051	0.893
C_{d2} (mg.g^{-1})	16.315	39.704	74.135	100.45	119.36	130.34	131.56	155.4
R_2^2	-7×10^{-15}	0.773	0.998	0.913	0.902	0.714	0.902	0.886

According to the results assembled in Table 2, the coefficients of correlation obtained by the pseudo-first-order model were very low, showing a bad correlation; the maximum capacities given by the pseudo-first-order model were not in agreement with those obtained experimentally, accounting for the poor fit obtained. On the other hand, the obtained R^2 values for the Blanchard model were very good ($\approx$1); the equilibrium-adsorbed amounts calculated theoretically by this pseudo-second-order model were almost identical to those found experimentally. Consequently, the adsorption process of phenol by PAC was found to be more suitable with the pseudo-second-order model.

The application of the Weber and Morris model to the kinetics data gave more or less acceptable coefficients correlations R^2 (>0.902) (Table 3), with some exceptions for the two concentrations of 10 and 20 mg.L^{-1}.

The linear plot did not pass through the origin ($c_d \neq 0$), which indicates that the intra-particle diffusion was not the only mechanism controlling the kinetics adsorption of phenol [40]. The obtained lines were multi linear consisting of two segments; the first linear area represented the external diffusion by macropore and mesopore, while the second segment of the fitting slope represented the micropore diffusion by the intraparticle diffusion [18,41].

The application of Boyd model to the experimental data shows that lines did not cross the origin, which indicates that the extra-particle diffusion was not the limiting stage of the adsorption process of phenol by the PAC.

3.5. Isotherms Study

Adsorption isotherms of phenol on powdered activated carbon were performed at temperatures ranging from 20° to 40 °C. Figure 7 shows the variation of the equilibrium adsorbent amount (q_e) as a function of the equilibrium adsorbate concentration (C_e). The isotherm plots of phenol adsorption on PAC show that the quantity of phenol adsorbed at equilibrium increased with the increase in the phenol equilibrium concentration; a slowing down was then observed, which corresponds to the equilibrium saturation of the adsorption sites.

Moreover, the obtained isotherms show that the adsorbed amount at equilibrium decreased with the increase of temperature from 20 to 40 °C. This decrease in the adsorbed amount suggests a weak adsorption interaction between the PAC surface and phenol, which favors the physisorption nature [42].

To evaluate the adsorption mechanisms, the experimental equilibrium data were fitted by different adsorption isotherm models, namely Langmuir, Freundlich, and Temkin, at different temperatures (Figure 8). The fitting parameters are reported in Table 4.

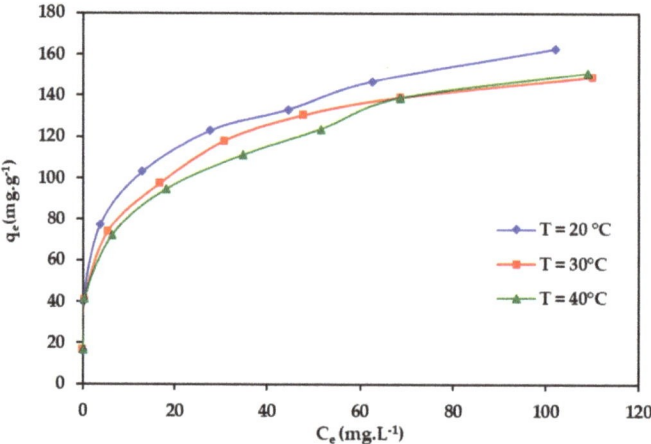

Figure 7. Phenol adsorption isotherms by PAC at different temperatures of 20, 30, and 40 °C.

Table 4. Fitting parameters by nonlinear regression expression of Langmuir, Freundlich, and Temkin isotherms at different temperatures.

Type	Langmuir				Freundlich			Temkin		
	q_m (mg·g^{-1})	K_L (L·mg^{-1})	R_L	R^2	K_F [(mg·g^{-1})/((mg/L)$^{1/n}$)]	n	R^2	b_T (J·mol^{-1})	A_T (L·g^{-1})	R^2
T = 20 °C	156.26	0.196	0.170	0.976	61.31	4.79	0.992	42.60	361.0	0.962
T = 30 °C	146.99	0.174	0.187	0.981	49.88	4.16	0.998	39.80	74.96	0.963
T = 40 °C	150.8	0.118	0.001	0.975	47.64	4.08	0.998	43.00	88.00	0.958

From Table 4, the R^2 obtained by Langmuir was more or less satisfactory ($R^2 \leq 0.975$). The values of maximum adsorption capacity values determined by Langmuir were in the range from 146.9 to 156.26 mg·g^{-1}. These values were comparable to those reported by Fierro [43], where the obtained maximum adsorption capacities varied from 74 to 137.36 mg·g^{-1} for different commercial activated carbon used to adsorb phenol from aqueous solution [43].

The "favorability" of the Langmuir isotherm for an initial concentration C_0 is verified by the Hall parameter (R_L), which is given by [44,45]:

$$R_L = \frac{1}{1 + K_L * C_0} \quad (16)$$

where:

C_0: initial concentration of phenol (mg·L^{-1})

K_L: the constant related to the energy of adsorption (Langmuir constant)

R_L: value indicates the adsorption nature to be either unfavorable if $R_L > 1$, linear if $R_L = 1$, favorable if $0 < R_L < 1$ and irreversible if $R_L = 0$.

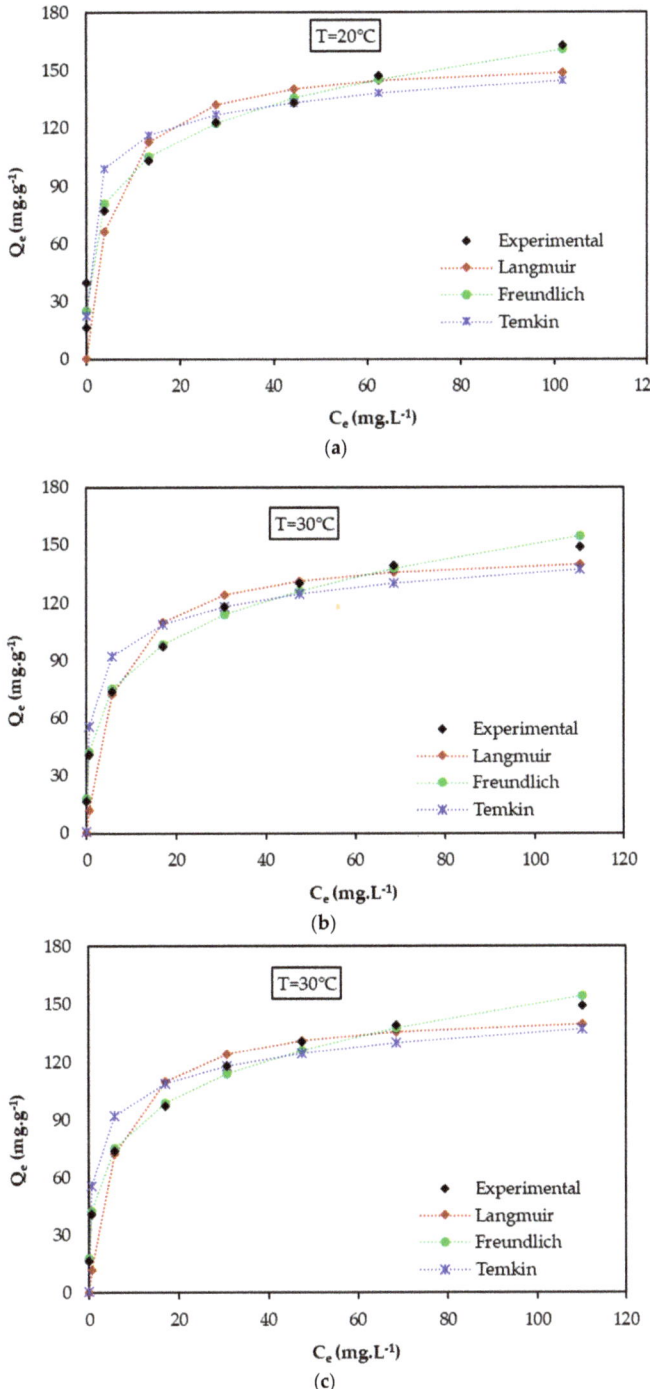

Figure 8. Nonlinear fitting of the equilibrium data by Langmuir, Freundlich, and Temkin isotherms at different temperatures (**a**) T = 20 °C; (**b**) T = 30 °C, (**c**) T = 40 °C.

From the results reported in Table 4, the calculated R_L parameters were between 0 and 1, indicating that the Langmuir isotherm is favorable [44,45].

The value of the Freundlich parameter n was greater than 1 (n > 4) for all temperature studies, indicating that the adsorption of phenol by PAC is favorable and physical [45,46]. It is in agreement with the findings of adsorption isotherms for different temperatures, since the amount adsorbed at equilibrium was inversely proportional to the temperature.

The correlation coefficients of Freundlich are overall very satisfactory, and they are higher than those obtained by Langmuir, suggesting that the adsorption of phenol by PAC is well described by the Freundlich model.

The Temkin isotherm can model the equilibrium data of phenol adsorption on PAC, due to the satisfactory values obtained for the correlation coefficients. The Temkin constant (b_T) related to the adsorption energy was positive at all studied temperatures, which shows that phenol adsorption by powdered activated carbon is exothermic [32].

The results of the modeling equilibrium data reveal that the Langmuir, Freundlich, and Temkin isotherms well fitted the experimental data. It is also concluded that the adsorption is physical, favorable, and may be an exothermic type.

3.6. Thermodynamics Study

The thermodynamic study of phenol adsorption over PAC was conducted by calculating the Gibbs free energy change ($\Delta G°$), enthalpy change ($\Delta H°$), and entropy change ($\Delta S°$) as function of temperature. The Gibbs free energy of adsorption ($\Delta G°$) can be expressed as [47,48]:

$$\Delta G° = -RT \ln K_L° \quad (17)$$

where R is the universal gas constant (8.314 J.mol^{-1}K^{-1}), T the temperature (K), and $K_L°$ is the (dimensionless) "thermodynamic" Langmuir constant for the adsorption process.

The value of K_L (L.mg^{-1}), calculated from Langmuir model (Equation (8)) and given in Table 4 (Section 3.5), should be converted to molar concentration unity considering the standard state $C° = 1$ mol.L^{-1} [47,48]:

$$K_L° = K_L \left(L.mg^{-1} \right) \times 1000 \left(mg.g^{-1} \right) \times M_{Phenol} \left(g.mol^{-1} \right) \times C° \left(mol.L^{-1} \right) \quad (18)$$

where M_{Phenol} is the phenol molar mass = 94.11 g.mol^{-1} and the factor 1000 to convert g to mg.

The enthalpy ($\Delta H°$) and entropy ($\Delta S°$) parameters were estimated from the classical relationships [47–49]:

$$\Delta G° = \Delta H° - T\Delta S° \quad (19)$$

$$\ln K_L° = \frac{\Delta S°}{R} - \frac{\Delta H°}{RT} \quad (20)$$

The plot was drawn between ln $K_L°$ versus 1/T and the determined slope and intercept was used to calculated $\Delta H°$ (kJ.mol^{-1}) and $\Delta S°$ (J.mol^{-1}), respectively (see Table 5).

Table 5. Thermodynamic parameters of phenol adsorption.

T (°K)	K_L (L.mg^{-1})	$K_L°$ (×10^4) (Dimensionless)	Ln $K_L°$	$\Delta G°$ (kJ.mol^{-1})	$\Delta H°$ (KJ.mol^{-1})	$\Delta S°$ (J.mol^{-1})
293	0.196	1.84	9.823	−24.04		
303	0.174	1.64	9.704	−24.20	−19.23	16.433
313	0.118	1.11	9.315	−24.37		

The value of $\Delta H°$ was negative, indicating that the process of phenol adsorption by PAC is exothermic and the adsorption behavior may be of physical nature [50]. The value of $\Delta H° < 40$ kJ.mol^{-1}, suggesting a physically driven process [49]. The positive and small value of $\Delta S°$ indicated that the adsorption proceeds with an increase in the number of species at solid–solution interfaces, showing an increased disorder and randomness at the solid–solution interface during the adsorption of phenol on the activated carbon [51,52]. The variation of $\Delta G°$ was negative for all the temperatures studied, which confirmed the feasibility and spontaneous nature of the phenol adsorption by PAC [51,53].

3.7. KNN Model

The goal in this work, as previously reported, was to develop the KNN model. The 11 distance metrics (Euclidean, Chebychev, Minkowski, Mahalanobis, Cosine, Correlation, Spearman, Hamming, Jaccard, Cityblock, Seuclidean) were optimized with their distance weighting functions (equal, inverse and squared inverse). From this, each metric distance was optimized for its specific parameters by GWO (number of neighbors and exponent in the case of cubic "Minkowski" distance metrics).

Table 6 shows the results of optimizing the 11 distance metrics and their weighting functions, distance. It also shows the statistical coefficients and errors (R, R^2, R^2_{adj}, RMSE, MSE, MAE) for training data, validation data, and for all data (training+validation) according to specific parameters (num neighbors and exponent). In addition, the GWO settings were used for optimization.

Table 6. Performances of the different KNN models tested.

GWO					Max_iteration = 100 SearchAgents_no = 50					
Distance	Distance Weight	Num Neighbors	Exponent	loss	$R/R^2/R^2_{adj}$			RMSE/MSE/MAE		
					Train	VAL	ALL	Train	VAL	ALL
Euclidean	Equal	8	/	0.5356	0.7938 0.6302 0.6199	0.7183 0.5159 0.4565	0.8002 0.6403 0.6324	0.3553 0.1262 0.1606	0.1448 0.0210 0.0659	0.3244 0.3317 0.3605
	Inverse	8	/	0.0271	1.0000 1.0000 1.0000	0.6632 0.4398 0.3711	0.9896 0.9792 0.9788	0.0010 0.0000 0.0001	0.1650 0.0272 0.0808	0.0737 0.2768 0.3237
	Squared Inverse	15	/	0.0271	1.0000 1.0000 1.0000	0.7221 0.5214 0.4627	0.9919 0.9838 0.9835	0.0010 0.0000 0.0001	0.1463 0.0214 0.0725	0.0653 0.2455 0.2905
bychev	Equal	12	/	0.6102	0.7405 0.5483 0.5358	0.6528 0.4262 0.3557	0.7482 0.5597 0.5500	0.3988 0.1591 0.1964	0.1556 0.0242 0.0718	0.3636 0.2202 0.2836
	Inverse	7	/	0.0237	1.0000 1.0000 1.0000	0.7036 0.4951 0.4330	0.9900 0.9802 0.9797	0.0010 0.0000 0.0001	0.1629 0.0266 0.0842	0.0728 0.2153 0.2767
	Squared Inverse	11	/	0.0305	1.0000 1.0000 1.0000	0.8005 0.6408 0.5967	0.9932 0.9864 0.9861	0.0010 0.0000 0.0001	0.1332 0.0177 0.0662	0.0595 0.2977 0.3528
Minkowski	Equal	11	4	0.5627	0.7172 0.5143 0.5009	0.8218 0.6753 0.6355	0.7322 0.5361 0.5259	0.4156 0.1727 0.1920	0.1149 0.0132 0.0426	0.3754 0.2907 0.3370
	Inverse	14	2	0.0271	1.0000 1.0000 1.0000	0.7220 0.5214 0.4626	0.9921 0.9843 0.9840	0.0010 0.0000 0.0001	0.1426 0.0203 0.0627	0.0637 0.2129 0.2830
	Squared Inverse	29	2	0.0237	1.0000 1.0000 1.0000	0.7511 0.5642 0.5107	0.9927 0.9854 0.9851	0.0010 0.0000 0.0001	0.1387 0.0192 0.0647	0.0619 0.2603 0.3026

Table 6. Cont.

GWO					Max_iteration = 100 SearchAgents_no = 50						
Distance	Distance Weight	Num Neighbors	Exponent	loss	$R/R^2/R^2_{adj}$			RMSE/MSE/MAE			
					Train	VAL	ALL		Train	VAL	ALL
Mahalanobis	Equal	3	/	0.3932	0.8902 0.7925 0.7868	0.7330 0.5372 0.4804	0.8920 0.7956 0.7911		0.2531 0.0640 0.0732	0.1360 0.0185 0.0507	0.2344 0.2661 0.3091
	Inverse	4	/	0.0271	1.0000 1.0000 1.0000	0.6717 0.4512 0.3838	0.9909 0.9818 0.9814		0.0010 0.0000 0.0001	0.1537 0.0236 0.0697	0.0686 0.2444 0.3133
	Squared Inverse	25	/	0.0203	1.0000 1.0000 1.0000	0.6895 0.4755 0.4110	0.9904 0.9810 0.9805		0.0010 0.0000 0.0001	0.1584 0.0251 0.0724	0.0707 0.1528 0.2224
Cosine	Equal	6	/	0.5220	0.7670 0.5883 0.5770	0.6574 0.4321 0.3624	0.7714 0.5950 0.5861		0.3746 0.1403 0.1598	0.1533 0.0235 0.0698	0.3421 0.3488 0.3663
	Inverse	7	/	0.0305	1.0000 1.0000 1.0000	0.7365 0.5424 0.4862	0.9928 0.9856 0.9853		0.0010 0.0000 0.0001	0.1363 0.0186 0.0595	0.0609 0.2628 0.3039
	Squared Inverse	152	/	0.0339	1.0000 1.0000 1.0000	0.7427 0.5516 0.4965	0.9919 0.9838 0.9834		0.0010 0.0000 0.0001	0.1460 0.0213 0.0702	0.0652 0.1785 0.2333
Correlation	Equal	5	/	0.4881	0.7687 0.5909 0.5796	0.6676 0.4457 0.3776	0.7756 0.6015 0.5928		0.3757 0.1412 0.1632	0.1499 0.0225 0.0725	0.3428 0.3025 0.3452
	Inverse	67	/	0.0339	1.0000 1.0000 1.0000	0.7197 0.5180 0.4588	0.9916 0.9832 0.9829		0.0010 0.0000 0.0001	0.1481 0.0219 0.0723	0.0662 0.2443 0.3136
	Squared Inverse	101	/	0.0373	1.0000 1.0000 1.0000	0.7284 0.5306 0.4729	0.9921 0.9843 0.9840		0.0010 0.0000 0.0001	0.1432 0.0205 0.0635	0.0640 0.2359 0.2746
Spearman	Equal	3	/	0.5559	0.6513 0.4242 0.4082	0.6004 0.3605 0.2819	0.6662 0.4438 0.4315		0.4508 0.2032 0.2098	0.1797 0.0323 0.1062	0.4113 0.2929 0.3173
	Inverse	6	/	0.4542	0.6569 0.4315 0.4158	0.6158 0.3792 0.3030	0.6723 0.4520 0.4399		0.4359 0.4359 0.1791	0.1740 0.0303 0.0930	0.3977 0.3259 0.3532
	Squared Inverse	4	/	0.4780	0.6534 0.4270 0.4111	0.6373 0.4062 0.3333	0.6684 0.4468 0.4346		0.4392 0.1929 0.1848	0.1725 0.0298 0.0960	0.4004 0.3133 0.3516
Hamming	Equal	3	/	0.4881	0.5071 0.2572 0.2366	0.9238 0.8534 0.8354	0.5241 0.2747 0.2587		0.5326 0.2837 0.2700	0.0737 0.0054 0.0232	0.4777 0.3612 0.3989
	Inverse	5	/	0.0339	1.0000 1.0000 1.0000	0.9678 0.9365 0.9288	0.9991 0.9983 0.9982		0.0010 0.0000 0.0001	0.0471 0.0022 0.0169	0.0210 0.2165 0.2718
	Squared Inverse	4	/	0.0441	1.0000 1.0000 1.0000	0.9894 0.9790 0.9764	0.9997 0.9994 0.9994		0.0010 0.0000 0.0001	0.0275 0.0008 0.0114	0.0123 0.2746 0.3129

Table 6. Cont.

GWO					Max_iteration = 100 SearchAgents_no = 50					
Distance	Distance Weight	Num Neighbors	Exponent	loss	$R/R^2/R^2_{adj}$			RMSE/MSE/MAE		
					Train	VAL	ALL	Train	VAL	ALL
Jaccard	Equal	3	/	0.5051	0.5119 0.2620 0.2416	0.9522 0.9066 0.8952	0.5282 0.2790 0.2631	0.5395 0.2910 0.2823	0.0583 0.0034 0.0233	0.4834 0.3731 0.4090
	Inverse	3	/	0.0475	1.0000 1.0000 1.0000	0.9703 0.9414 0.9342	0.9992 0.9984 0.9984	0.0010 0.0000 0.0001	0.0452 0.0020 0.0179	0.0202 0.2700 0.3193
	Squared Inverse	**4**	**/**	**0.0201**	**1.0000 1.0000 1.0000**	**0.9975 0.9942 0.9935**	**0.9999 0.9998 0.9998**	**0.0010 0.0000 0.0001**	**0.0156 0.0002 0.0078**	**0.0070 0.2347 0.2763**
Cityblock	Equal	4	/	0.4441	0.8309 0.6905 0.6819	0.7616 0.5801 0.5285	0.8364 0.6995 0.6929	0.3200 0.1024 0.1231	0.1288 0.0166 0.0529	0.2921 0.3097 0.3459
	Inverse	5	/	0.0271	1.0000 1.0000 1.0000	0.7373 0.5436 0.4875	0.9924 0.9848 0.9844	0.0010 0.0000 0.0001	0.1407 0.0198 0.0637	0.0628 0.2579 0.2991
	Squared Inverse	18	/	0.0237	1.0000 1.0000 1.0000	0.7286 0.5308 0.4732	0.9922 0.9844 0.9840	0.0010 0.0000 0.0001	0.1424 0.0203 0.0636	0.0636 0.2245 0.2733
Seuclidean	Equal	5	/	0.4542	0.7971 0.6353 0.6252	0.7683 0.5902 0.5399	0.8042 0.6467 0.6389	0.3537 0.1251 0.1434	0.1270 0.0161 0.0518	0.3215 0.3367 0.3586
	Inverse	13	/	0.0271	1.0000 1.0000 1.0000	0.7397 0.5471 0.4915	0.9918 0.9837 0.9834	0.0010 0.0000 0.0001	0.1460 0.0213 0.0688	0.0652 0.2176 0.2901
	Squared Inverse	23	/	0.0237	1.0000 1.0000 1.0000	0.7707 0.5940 0.5442	0.9926 0.9852 0.9848	0.0010 0.0000 0.0001	0.1400 0.0196 0.0653	0.0625 0.2424 0.2879

The results of Table 6 show that most of the models have coefficients (R, R^2, R^2_{adj}) equal to 1 with very low statistical errors (almost 0) in the learning phase (except distance weight "equal" of each distance, and the "Spearman" distance with the three distance weights, "Equal, Inverse and Inverse Squared", which were very small).

This is why the choice of the best model was based on the coefficients and the statistical errors of the validation phase.

According to the results reported in Table 6, the "Jaccard" distance with distance weight "Squared Inverse" led to the best results for the coefficients and the statistical errors compared to the other models in the validation phase, R, R^2, R^2_{adj}, RMSE, MSE, and MAE were 0.9971, 0.9942, 0.9935, 0.0156, 0.0002, and 0.0078, respectively.

This result was confirmed by the statistical coefficients and errors obtained for all the data (the training and validation data): R, R^2, R^2_{adj}, RMSE, MSE, and MAE were 0.9999, 0.9998, 0.9998, 0.0070, 0.2347, and 0.2763, respectively. In addition, the statistical coefficients and errors of the learning phase were: R = 1.0000, R^2 = 1.0000, R^2_{adj} = 1.0000, RMSE = 0.0010, MSE = 0.0000, and MAE = 0.0001.

From these results, the model optimized by the distance "Jaccard" with distance weight "Squared Inverse" was considered as the best one, and it is presented graphically in Figure 9.

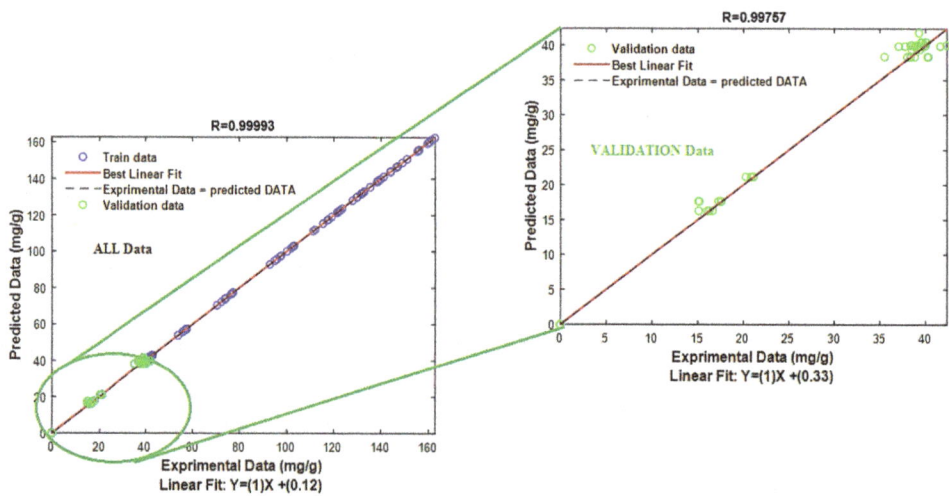

Figure 9. Comparison between experimental and predicted values.

3.7.1. Model Performance Test

In order to test the performance of the KNN_ GWO model, which was obtained by the distance "Jaccard" with distance weight "Squared Inverse", an interpolation was carried out. For this, a previously cached database containing 50 experimental data points was used, and this database was not used in training the model. The results are presented in Table 7 in terms of coefficients and statistical errors.

Table 7. Model test performance.

$R/R^2/R^2_{adj}$	RMSE/MSE/MAE
ALL	ALL
0.9984	0.0152
0.9968	2.3172×10^{-4}
0.9963	0.0051

Table 7 shows very high coefficients, showing the performance and efficiency of our model.

This result was represented graphically (Figure 10) in terms of the experimental values and the predicted values.

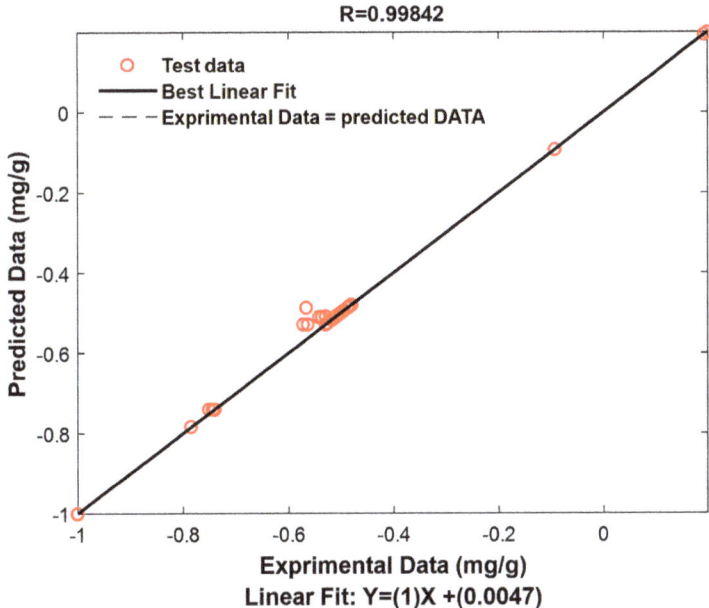

Figure 10. Comparison between experimental and predicted values to assess performance.

3.7.2. Residues Study

The method of residuals was used in this work in order to study the relationship of the experimental values with the predicted values, on the one hand, and the efficiency and performance of the selected model, on the other hand [2,8]. For this purpose, the experimental values and the predicted values were plotted in parallel as a function of samples (Figure 11A) for all data (including data training, data validation, and data test performance) [8–10].

On the other hand, the error was calculated by the difference between the experimental values and the predicted values for all data (including data training, data validation, and data test performance). This error was plotted graphically (Figure 11B–D) by three methods (residue, frequency, and instances) [19–21,54,55].

Figure 11A shows the high convergence between the experimental values and the predicted values, confirming once again the effectiveness of the obtained model.

Moreover, Figure 11B shows small errors obtained in the three phases. Indeed, in the training phase, the error did not exceed 2 mg.g^{-1}; in the validation phase, the error values were in the intervals [−7–3], and the error values of the test phase were in the intervals [−7–0]. It can be therefore concluded that the errors obtained from our model are very small. This was again confirmed at the examination of Figure 11C and 11D, which shows that a very high frequency was obtained with zero error for the former (Figure 11C) and that most errors were 0, regardless of training, validation, and testing data for the latter (Figure 11D).

From these results and interpretations, the efficiency and performance of our model based on the distance "Jaccard" with distance weight "Squared Inverse" is demonstrated.

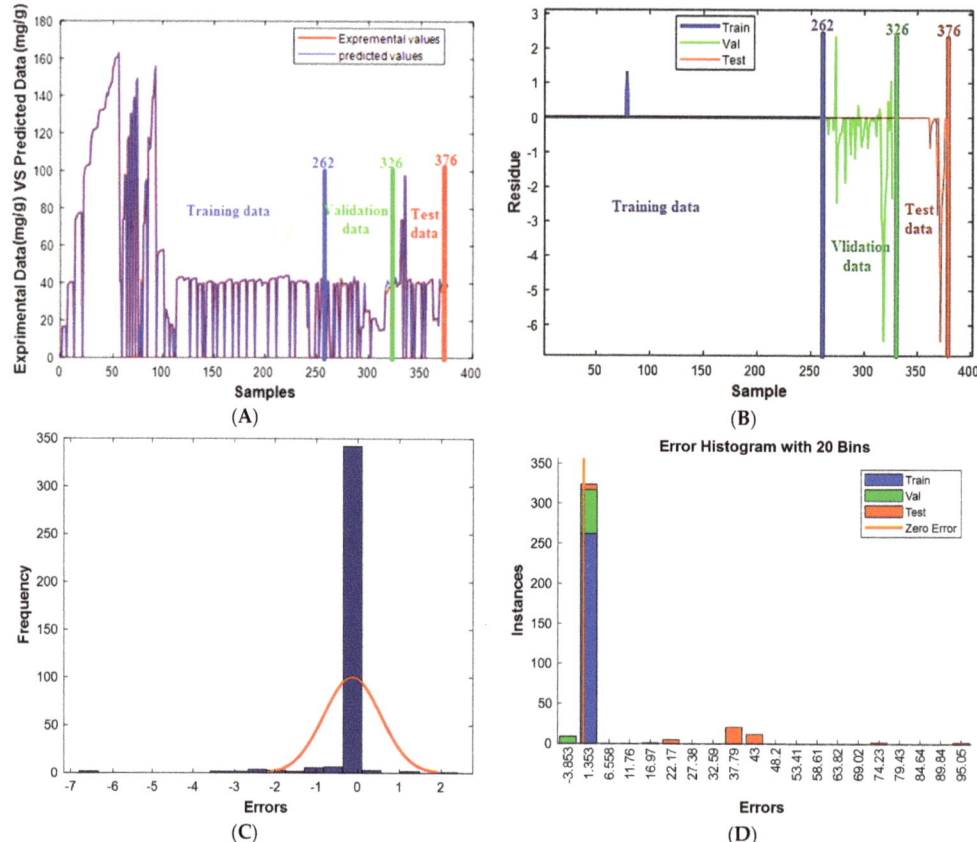

Figure 11. Residuals relating to the models established by the different techniques according to the estimated values: (**A**) Relationship between experimental data and the predicted data of samples, (**B**) residues relating to the models established, (**C**) frequency distribution of errors, and (**D**) instances distribution of errors.

3.7.3. Optimization of the Optimal Analysis and Validation

An optimization method was run using GWO and PSO to find the optimum of the independent parameters to obtain the maximum adsorption quantity of phenol. The optimization results are displayed in Table 8.

Table 8. Results of adsorption quantity of phenol under the optimal conditions.

GWO: $\times 1 = 60$ min, $\times 2 = 200$ mg.L^{-1}, $\times 3 = 0.1504$ g.L^{-1}, $\times 4 = 306{,}683$ rpm, $\times 5 = 20\ °C$, $\times 6 = 4.7526$	
Predicted quantity of adsorbed phenol (mg.g^{-1})	163.4230
PSO: $\times 1 = 60$ min, $\times 2 = 200$ mg.L^{-1}, $\times 3 = 0.1506$ g.L^{-1}, $\times 4 = 308.793$ rpm, $\times 5 = 20\ °C$, $\times 6 = 4.7447$	
Predicted quantity of adsorbed phenol (mg.g^{-1})	163.4230
Experimental quantity of adsorbed phenol (mg.g^{-1})	162.7543
Error (mg.g^{-1})	0.6687

Table 8 shows that the optima obtained by the two algorithms, GWO and particle swarm (PSO), were almost equal (Figure 11A,B). By comparing the predicted values of adsorbed phenol obtained by the optimal conditions of GWO and PSO with the experimental value, it is seen that the results obtained are the same (Figure 11C), with very small error (0.6687 mg.g^{-1}), confirming again the efficiency and performance of the KNN_GWO model.

4. Conclusions

In this study, phenol, which is considered a very hazardous substance, was efficiently removed by powdered activated carbon. The effect of operating parameters on the efficiency removal of phenol by powdered activated carbon revels that the maximum removal (99.94%) of phenol was reached at pH 8, adsorbent rate of 0.6 g.L^{-1}, and temperature of 20 °C when using an initial concentration of phenol of 20 mg.L^{-1}.

Batch adsorption experiments allowed the determination of adsorption isotherm parameters. The results of the equilibrium data modeling show that the Langmuir, Freundlich, and Temkin isotherms fit the experimental data very well. The maximum adsorption capacity values found by Langmuir were 146.9, 150.8, and 156.26 mg.g^{-1} for temperature values 20, 30, and 40°C respectively.

The adsorption kinetic was well described by the pseudo-second-order model, while the intra-particle diffusion was not the only mechanism controlling the kinetics adsorption of phenol; kinetic data was controlled by the external diffusion (by macropore and mesopore) and the micropore diffusion. The thermodynamic study revealed that the adsorption of phenol by activated carbon is spontaneous, exothermic, and physical, and proceeds with an increase randomness at the solid–solution interface.

The KNN_GWO model showed an advantage of giving more precise values related to very high statistical coefficients (close to 1) and very low statistical errors (close to 0). Moreover, the efficiency of the model was confirmed by several techniques, including test interpolation, residual studies, and by validation of the optimal input conditions. Hence, the results obtained showed a very high efficiency and performance.

Author Contributions: Conceptualization, M.Z., M.C., Z.K. and L.M.; methodology, Z.K., M.C. and M.Z.; validation, J.-C.B., A.A., H.T. and L.M.; formal analysis, M.Z. and M.K.; investigation, M.Z.; resources, L.M. and M.K.; data curation, Z.K., M.C. and L.M.; writing—review and editing, L.M., A.A. and J.-C.B.; visualization, M.Z., L.M. and J.-C.B.; supervision, L.M., A.A. and J.-C.B.; project administration, L.M. and M.Z. All authors have read and agreed to the published version of the manuscript.

Funding: This research received no external funding.

Data Availability Statement: Not applicable.

Acknowledgments: The authors wish to thank all who assisted in conducting this work.

Conflicts of Interest: The authors declare no conflict of interest.

References

1. Mishra, R.; Dubey, S. Fresh Water Availability and It's Global Challenge. *Int. J. Eng. Sci. Invent. Res. Dev.* **2015**, *2*, 351–407.
2. Bruce, R.M.; Santodonato, J.; Neal, M.W. Summary Review of the Health Effects Associated with Phenol. *Toxicol. Ind. Health* **1987**, *3*, 535–568. [CrossRef] [PubMed]
3. Chimuka, L.; Nefale, F.; Masevhe, A. Determination of phenols in water samples using a supported liquid membrane extraction probe and liquid chromatography with photodiode array detection. *S. Afr. J. Chem.* **2007**, *60*, 102–108.
4. Puszkarewicz, A.; Kaleta, J.T.; Papciak, D. Adsorption of Phenol from Water on Natural Minerals. *J. Ecol. Eng.* **2018**, *19*, 132–138. [CrossRef]
5. Dehmani, Y.; Khalki, O.E.; Mezougane, H.; Abouarnadasse, S. Comparative study on adsorption of cationic dyes and phenol by natural clays. *Chem. Data Collect.* **2021**, *33*, 100674. [CrossRef]
6. Dehmani, Y.; Dridi, D.; Lamhasni, T.; Abouarnadasse, S.; Chtourou, R.; Lima, E.C. Review of phenol adsorption on transition metal oxides and other adsorbents. *J. Water Process. Eng.* **2022**, *49*, 102965. [CrossRef]
7. Lammini, A.; Dehbi, A.; Omari, H.; Elazhari, K.; Mehanned, S.; Bengamra, Y.; Dehmani, Y.; Rachid, O.; Alrashdi, A.A.; Gotore, O.; et al. Experimental and theoretical evaluation of synthetized cobalt oxide for phenol adsorption: Adsorption isotherms, kinetics, and thermodynamic studies. *Arab. J. Chem.* **2022**, *15*, 104368. [CrossRef]

8. Dehmani, Y.; Lainé, J.; Daouli, A.; Sellaoui, L.; Bonilla-Petriciolet, A.; Lamhasni, T.; Abouarnadasse, S.; Badawi, M. Unravelling the adsorption mechanism of phenol on zinc oxide at various coverages via statistical physics, artificial neural network modeling and ab initio molecular dynamics. *Chem. Eng. J.* **2023**, *452*, 139171. [CrossRef]
9. Ahangar, R.M.; Farmanzadeh, D. Theoretical study for exploring the adsorption behavior of aniline and phenol on pristine and Cu-doped phosphorene surface. *Appl. Surf. Sci.* **2023**, *614*, 156194. [CrossRef]
10. Dougna, A.A.; Gombert, B.; Kodom, T.; Djaneye-Boundjou, G.; Boukari, S.O.B.; Leitner, N.K.V.; Bawa, L.M. Photocatalytic removal of phenol using titanium dioxide deposited on different substrates: Effect of inorganic oxidants. *J. Photochem. Photobiol. A Chem.* **2015**, *305*, 67–77. [CrossRef]
11. Yasmina, M.; Mourad, K.; Mohammed, S.H.; Khaoula, C. Treatment Heterogeneous Photocatalysis; Factors Influencing the Photocatalytic Degradation by TiO2. *Energy Procedia* **2014**, *50*, 559–566. [CrossRef]
12. Suzuki, H.; Araki, S.; Yamamoto, H. Evaluation of advanced oxidation processes (AOP) using O3, UV, and TiO2 for the degradation of phenol in water. *J. Water Process Eng.* **2015**, *7*, 54–60. [CrossRef]
13. García-Ballesteros, S.; Mora, M.; Vicente, R.; Vercher, R.F.; Sabater, C.; Castillo, M.A.; Amat, A.M.; Arques, A. A new methodology to assess the performance of AOPs in complex samples: Application to the degradation of phenolic compounds by O3 and O3/UV-A–Vis. *Chemosphere* **2019**, *222*, 114–123. [CrossRef]
14. Chedri Mammar, A.; Mouni, L.; Bollinger, J.-C.; Belkhiri, L.; Bouzaza, A.; Assadi, A.A.; Belkacemi, H. Modeling and optimization of process parameters in elucidating the adsorption mechanism of Gallic acid on activated carbon prepared from date stones. *Sep. Sci. Technol.* **2020**, *55*, 3113–3125. [CrossRef]
15. Ververi, M.; Goula, A.M. Pomegranate peel and orange juice by-product as new biosorbents of phenolic compounds from olive mill wastewaters. *Chem. Eng. Process. Process Intensif.* **2019**, *138*, 86–96. [CrossRef]
16. Sahu, J.N.; Karri, R.R.; Jayakumar, N.S. Improvement in phenol adsorption capacity on eco-friendly biosorbent derived from waste Palm-oil shells using optimized parametric modelling of isotherms and kinetics by differential evolution. *Ind. Crops Prod.* **2021**, *164*, 113333. [CrossRef]
17. Franco, D.S.P.; Georgin, J.; Netto, M.S.; Allasia, D.; Oliveira, M.L.S.; Foletto, E.L.; Dotto, G.L. Highly effective adsorption of synthetic phenol effluent by a novel activated carbon prepared from fruit wastes of the Ceiba speciosa forest species. *J. Environ. Chem. Eng.* **2021**, *9*, 105927. [CrossRef]
18. Park, K.-H.; Balathanigaimani, M.S.; Shim, W.-G.; Lee, J.-W.; Moon, H. Adsorption characteristics of phenol on novel corn grain-based activated carbons. *Microporous Mesoporous Mater.* **2010**, *127*, 1–8. [CrossRef]
19. Adithiyaa, T.; Chandramohan, D.; Sathish, T. Optimal prediction of process parameters by GWO-KNN in stirring-squeeze casting of AA2219 reinforced metal matrix composites. *Mater. Today Proc.* **2020**, *21*, 1000–1007. [CrossRef]
20. Parul Sinha, P.S. Comparative Study of Chronic Kidney Disease Prediction using KNN and SVM. *Int. J. Eng. Res. Technol.* **2015**, *4*, 608–612. [CrossRef]
21. Mirjalili, S.; Mirjalili, S.M.; Lewis, A. Grey Wolf Optimizer. *Adv. Eng. Softw.* **2014**, *69*, 46–61. [CrossRef]
22. Tahraoui, H.; Amrane, A.; Belhadj, A.-E.; Zhang, J. Modeling the organic matter of water using the decision tree coupled with bootstrap aggregated and least-squares boosting. *Environ. Technol. Innov.* **2022**, *27*, 102419. [CrossRef]
23. Tahraoui, H.; Belhadj, A.-E.; Amrane, A.; Houssein, E.H. Predicting the concentration of sulfate using machine learning methods. *Earth Sci. Inform.* **2022**, *15*, 1023–1044. [CrossRef]
24. Tahraoui, H.; Belhadj, A.-E.; Hamitouche, A.-e.; Bouhedda, M.; Amrane, A. Predicting the concentration of sulfate (SO42-) in drinking water using artificial neural networks: A case study: Médéa-Algeria. *Desalin. Water Treat.* **2021**, *217*, 181–194. [CrossRef]
25. Tahraoui, H.; Belhadj, A.; Moula, N.; Bouranene, S.; Amrane, A. Optimisation and Prediction of the Coagulant Dose for the Elimination of Organic Micropollutants Based on Turbidity. *Kem. Ind.* **2021**, *70*, 675–691. [CrossRef]
26. Xie, B.; Qin, J.; Wang, S.; Li, X.; Sun, H.; Chen, W. Adsorption of Phenol on Commercial Activated Carbons: Modelling and Interpretation. *Int. J. Environ. Res. Public Health* **2020**, *17*, 789. [CrossRef]
27. Yao, J.; Wen, J.; Li, H.; Yang, Y. Surface functional groups determine adsorption of pharmaceuticals and personal care products on polypropylene microplastics. *J. Hazard. Mater.* **2022**, *423*, 127131. [CrossRef]
28. Lima, E.C.; Sher, F.; Guleria, A.; Saeb, M.R.; Anastopoulos, I.; Tran, H.N.; Hosseini-Bandegharaei, A. Is one performing the treatment data of adsorption kinetics correctly? *J. Environ. Chem. Eng.* **2021**, *9*, 104813. [CrossRef]
29. Simonin, J.-P. On the comparison of pseudo-first order and pseudo-second order rate laws in the modeling of adsorption kinetics. *Chem. Eng. J.* **2016**, *300*, 254–263. [CrossRef]
30. Guo, X.; Wang, J. Comparison of linearization methods for modeling the Langmuir adsorption isotherm. *J. Mol. Liq.* **2019**, *296*, 111850. [CrossRef]
31. Tran, H.N.; You, S.-J.; Hosseini-Bandegharaei, A.; Chao, H.-P. Mistakes and inconsistencies regarding adsorption of contaminants from aqueous solutions: A critical review. *Water Res.* **2017**, *120*, 88–116. [CrossRef]
32. Zamouche, M.; Mouni, L.; Ayachi, A.; Merniz, I. Use of commercial activated carbon for the purification of synthetic water polluted by a pharmaceutical product. *Desalin. Water Treat.* **2019**, *172*, 86–95. [CrossRef]
33. Atunwa, B.T.; Dada, A.O.; Inyinbor, A.A.; Pal, U. Synthesis, physiochemical and spectroscopic characterization of palm kernel shell activated carbon doped AgNPs (PKSAC@AgNPs) for adsorption of chloroquine pharmaceutical waste. *Mater. Today Proc.* **2022**, *65*, 3538–3546. [CrossRef]

34. Girish, C.R.; Ramachandra Murty, V. Adsorption of Phenol from Aqueous Solution Using Lantana camara, Forest Waste: Kinetics, Isotherm, and Thermodynamic Studies. *Int. Sch. Res. Not.* **2014**, *2014*, 201626. [CrossRef]
35. Abdelwahab, O.; Amin, N.K. Adsorption of phenol from aqueous solutions by Luffa cylindrica fibers: Kinetics, isotherm and thermodynamic studies. *Egypt. J. Aquat. Res.* **2013**, *39*, 215–223. [CrossRef]
36. El Gaidoumi, A.; Benabdallah, A.; Lahrichi, A.; Kherbeche, A. Adsorption of phenol in aqueous medium by a raw and treated moroccan pyrophyllite. *J. Mater. Environ. Sci.* **2015**, *6*, 2247–2259.
37. De la Luz-Asunción, M.; Sánchez-Mendieta, V.; Martínez-Hernández, A.L.; Castaño, V.M.; Velasco-Santos, C. Adsorption of Phenol from Aqueous Solutions by Carbon Nanomaterials of One and Two Dimensions: Kinetic and Equilibrium Studies. *J. Nanomater.* **2015**, *2015*, 405036. [CrossRef]
38. Gundogdu, A.; Duran, C.; Senturk, H.B.; Soylak, M.; Ozdes, D.; Serencam, H.; Imamoglu, M. Adsorption of Phenol from Aqueous Solution on a Low-Cost Activated Carbon Produced from Tea Industry Waste: Equilibrium, Kinetic, and Thermodynamic Study. *J. Chem. Eng. Data* **2012**, *57*, 2733–2743. [CrossRef]
39. Hudaib, B. Treatment of real industrial wastewater with high sulfate concentrations using modified Jordanian kaolin sorbent: Batch and modelling studies. *Heliyon* **2021**, *7*, e08351. [CrossRef]
40. Weber, W.J.; Morris, J.C. Kinetics of Adsorption on Carbon from Solution. *J. Sanit. Eng. Div.* **1963**, *89*, 31–59. [CrossRef]
41. Kiki, C.; Qiu, Y.; Wang, Q.; Ifon, B.E.; Qin, D.; Chabi, K.; Yu, C.-P.; Zhu, Y.-G.; Sun, Q. Induced aging, structural change, and adsorption behavior modifications of microplastics by microalgae. *Environ. Int.* **2022**, *166*, 107382. [CrossRef]
42. Jnr, M.H.; Spiff, A.I. Effects of temperature on the sorption of Pb2+ and Cd2+ from aqueous solution by Caladium bicolor (Wild Cocoyam) biomass. *Electron. J. Biotechnol.* **2005**, *8*, 43–50.
43. Fierro, V.; Torné-Fernández, V.; Montané, D.; Celzard, A. Adsorption of phenol onto activated carbons having different textural and surface properties. *Microporous Mesoporous Mater.* **2008**, *111*, 276–284. [CrossRef]
44. Dada, A.O.; Olalekan, A.P.; Olatunya, A.M.; Dada, O.J.I.J.C. Langmuir, Freundlich, Temkin and Dubinin–Radushkevich Isotherms Studies of Equilibrium Sorption of Zn 2+ Unto Phosphoric Acid Modified Rice Husk. *J. Appl. Chem.* **2012**, *3*, 38–45.
45. Kumar, P.; Das, S. Kinetics and adsorption isotherm model of 2-thiouracil adsorbed onto the surface of reduced graphene oxide-copper oxide nanocomposite material. *J. Mol. Struct.* **2022**, *1268*, 133723. [CrossRef]
46. Dawood, S.; Sen, T.K. Removal of anionic dye Congo red from aqueous solution by raw pine and acid-treated pine cone powder as adsorbent: Equilibrium, thermodynamic, kinetics, mechanism and process design. *Water Res.* **2012**, *46*, 1933–1946. [CrossRef]
47. Tran, H.N. Improper estimation of thermodynamic parameters in adsorption studies with distribution coefficient KD (qe/Ce) or Freundlich constant (KF): Conclusions from the derivation of dimensionless thermodynamic equilibrium constant and suggestions. *Adsorpt. Sci. Technol.* **2022**, *2022*, 5553212. [CrossRef]
48. Mouni, L.; Belkhiri, L.; Bollinger, J.-C.; Bouzaza, A.; Assadi, A.; Tirri, A.; Dahmoune, F.; Madani, K.; Remini, H. Removal of Methylene Blue from aqueous solutions by adsorption on Kaolin: Kinetic and equilibrium studies. *Appl. Clay Sci.* **2018**, *153*, 38–45. [CrossRef]
49. Imessaoudene, A.; Cheikh, S.; Bollinger, J.-C.; Belkhiri, L.; Tiri, A.; Bouzaza, A.; El Jery, A.; Assadi, A.; Amrane, A.; Mouni, L. Zeolite Waste Characterization and Use as Low-Cost, Ecofriendly, and Sustainable Material for Malachite Green and Methylene Blue Dyes Removal: Box-Behnken Design, Kinetics, and Thermodynamics. *Appl. Sci.* **2022**, *12*, 7587. [CrossRef]
50. Mohammed, N.A.S.; Abu-Zurayk, R.A.; Hamadneh, I.; Al-Dujaili, A.H. Phenol adsorption on biochar prepared from the pine fruit shells: Equilibrium, kinetic and thermodynamics studies. *J. Environ. Manag.* **2018**, *226*, 377–385. [CrossRef]
51. Lima, H.H.C.; Maniezzo, R.S.; Llop, M.E.G.; Kupfer, V.L.; Arroyo, P.A.; Guilherme, M.R.; Rubira, A.F.; Girotto, E.M.; Rinaldi, A.W. Synthesis and characterization of pecan nutshell-based adsorbent with high specific area and high methylene blue adsorption capacity. *J. Mol. Liq.* **2019**, *276*, 570–576. [CrossRef]
52. Kumbhar, P.; Narale, D.; Bhosale, R.; Jambhale, C.; Kim, J.-H.; Kolekar, S. Synthesis of tea waste/Fe3O4 magnetic composite (TWMC) for efficient adsorption of crystal violet dye: Isotherm, kinetic and thermodynamic studies. *J. Environ. Chem. Eng.* **2022**, *10*, 107893. [CrossRef]
53. Kasbaji, M.; Mennani, M.; Grimi, N.; Barba, F.J.; Oubenali, M.; Simirgiotis, M.J.; Mbarki, M.; Moubarik, A. Implementation and physico-chemical characterization of new alkali-modified bio-sorbents for cadmium removal from industrial discharges: Adsorption isotherms and kinetic approaches. *Process Biochem.* **2022**, *120*, 213–226. [CrossRef]
54. Tahraoui, H.; Belhadj, A.; Hamitouche, A.-E. Prediction of the Bicarbonate Amount in Drinking Water in the Region of Médéa Using Artificial Neural Network Modelling. *Kem. Ind.* **2020**, *69*, 595–602. [CrossRef]
55. Bousselma, A.; Abdessemed, D.; Tahraoui, H.; Amrane, A. Artificial Intelligence and Mathematical Modelling of the Drying Kinetics of Pre-treated Whole Apricots. *Kem. Ind.* **2021**, *70*, 651–667. [CrossRef]

Disclaimer/Publisher's Note: The statements, opinions and data contained in all publications are solely those of the individual author(s) and contributor(s) and not of MDPI and/or the editor(s). MDPI and/or the editor(s) disclaim responsibility for any injury to people or property resulting from any ideas, methods, instructions or products referred to in the content.

Article

Nitrogen-Doped Core-Shell Mesoporous Carbonaceous Nanospheres for Effective Removal of Fluorine in Capacitive Deionization

Yubo Zhao [1], Kexun Li [2], Bangsong Sheng [3], Feiyong Chen [1,4,*] and Yang Song [1,4,*]

1. Resources and Environment Innovation Institute, Shandong Jianzhu University, Jinan 250101, China
2. College of Environmental Science and Engineering, Nankai University, Tianjin 300071, China
3. The Second Construction Limited Company of China Construction Eighth Engineering Division, Jinan 250011, China
4. Huzhou Nanxun District Jianda Ecological Environment Innovation Center, Shandong Jianzhu University, Jinan 250101, China
* Correspondence: ctokyo@hotmail.com (F.C.); songyang20@sdjzu.edu.cn (Y.S.)

Abstract: Fluorine pollution of wastewater is a global environmental problem. Capacitive deionization has unique advantages in the defluorination of fluorine-containing wastewater; however, the low electrosorption capacity significantly restricts its further development. To overcome this limitation, nitrogen-doped core-shell mesoporous carbonaceous nanospheres (NMCS) were developed in this study based on structural optimization and polarity enhancement engineering. The maximal electrosorption capacity of NMCS for fluorine reached 13.34 mg g^{-1}, which was 24% higher than that of the undoped counterpart. NMCS also indicated excellent repeatability evidenced by little decrease of electrosorption capacity after 10 adsorption-regeneration cycles. According to material and electrochemical measurements, the doping of nitrogen into NMCS resulted in the improvement of physicochemical properties such as conductivity and wettability, the amelioration of pore structure and the transformation of morphology from yolk-shell to core-shell structure. It not only facilitated ion transportation but also improved the available adsorption sites, and thus led to enhancement of the defluorination performance of NMCS. The above results demonstrated that NMCS would be an excellent electrode material for high-capacity defluorination in CDI systems.

Keywords: fluorine removal; capacitive deionization; mesoporous carbon; nitrogen doping

Citation: Zhao, Y.; Li, K.; Sheng, B.; Chen, F.; Song, Y. Nitrogen-Doped Core-Shell Mesoporous Carbonaceous Nanospheres for Effective Removal of Fluorine in Capacitive Deionization. *Water* **2023**, *15*, 608. https://doi.org/10.3390/w15030608

Academic Editor: Hai Nguyen Tran

Received: 10 January 2023
Revised: 23 January 2023
Accepted: 30 January 2023
Published: 3 February 2023

Copyright: © 2023 by the authors. Licensee MDPI, Basel, Switzerland. This article is an open access article distributed under the terms and conditions of the Creative Commons Attribution (CC BY) license (https:// creativecommons.org/licenses/by/ 4.0/).

1. Introduction

Fluorine (F$^-$) is an indispensable trace element for the human body to maintain normal physiological activities. However, excessive intake of fluorine can be harmful to teeth, bones, central nervous system and reproductive system [1]. In recent years, with the rapid development of industries, increasing amounts of fluorine-containing wastewater have been produced in industrial processes such as metal smelting, steel and cement production, aluminum electrolysis, ceramics, pharmaceuticals and semiconductor manufacture [2]. The fluorine content of fluorine-containing wastewater is generally more than 100 mg L^{-1}, and even reaches several thousand mg L^{-1} in some specific industries [3]. Due to the high migration ability of fluorine pollution, the direct discharge of fluorine-containing wastewater into the environment/water bodies will easily result in pollution of groundwater and drinking water sources, leading to fluorosis in drinkers. Therefore, the fluorine-containing wastewater must be defluoridated before it is permitted to be discharged into the environment/water bodies.

At present, the methods of adsorption, precipitation, reverse osmosis and ion exchange are commonly used for the purification of fluorine [4–7]. These methods have disadvantages such as high cost and secondary pollution. In recent years, capacitive deionization (CDI) has emerged as a promising electrochemical defluorination technology. The principle of CDI fluorine removal is based on double-electric-layer adsorption [8,9]. In CDI, as the fluid flows through the electrodes under the action of electric field, ions in the fluid can migrate to and be absorbed to the oppositely charged electrode to form a double-electric-layer. After adsorption saturation, the electrodes can be regenerated by being short-circuited, since the ions adsorbed to the electrodes can be re-released into the fluid in this state. Certain distinct advantages, such as simple operation, low cost, low energy consumption and environmentally friendly effects, make CDI competitive as an alternative defluorination technology [10,11]. CDI originated in the mid-1960s, and for many years the research on CDI has been mainly focused on desalination of seawater and brackish water, with little attention being paid to the defluorination applications of CDI. Tang et al. studied the defluorination performance of CDI using activated carbon in batch mode and verified the feasibility of CDI for defluorination application in brackish groundwaters [12]. Epshtein et al. fabricated flow electrodes using SiO_2 and constructed flow-electrode CDI. They successfully removed fluorine from acidic wastewater [13]. In the study by Gaikwad et al., defluorination and removal of chromium were simultaneously achieved using activated carbon in CDI [14]. Nevertheless, research on the removal of fluorine using CDI is still in its infancy, evidenced by the limited amount of available literature. The relatively low electrosorption performance of CDI cannot meet the standards for commercial application. For example, Park et al. reported that the defluorination capacity of the reduced graphene oxide/hydroxyapatite composite was only 0.19 mg g^{-1} at 1.2 V [15]. Tang et al. observed the defluorination capacity of activated carbon ranging from 0.4 to 0.8 mg g^{-1} at 1.2 V at flow rates ranging from 30 to 100 mL min^{-1} [16]. Gaikwad et al. prepared activated carbon derived from tea waste biomass, which showed a defluorination capacity of 0.74 mg g^{-1} with an initial fluorine of 10 mg L^{-1} and a defluorination capacity of 2.49 mg g^{-1} with an initial fluorine of 10 mg L^{-1}, respectively [17]. Thus, more research is needed for the further improvement of electrosorption performances.

As the core component of CDI, the electrode plays a crucial role in affecting the defluorination performance of CDI. Porous carbons are generally used to fabricate the CDI electrode due to their low fabrication cost, favorable porous structure, excellent electrical conductivity and remarkable electrochemical stability [18–20]. Among porous carbons, the mesoporous carbons have been considered to be superior candidates because of their larger aperture, which can facilitate ion transport and, hence, enhance the availability of surface area [21,22]. Therefore, the mesoporous carbons are generally considered to possess more accessible active adsorption sites for fluorine as compared with the microporous carbons. However, the pure carbonaceous materials reported in the literature exhibited low defluorination performance [15,16]. Recently, researchers have attempted to modify pure carbonaceous carbons in order to improve the defluorination performance. The modification methods have mainly focused on metal/metal oxide/metal hydroxide doping, such as $Ti(OH)_4$ [23], La(III) [24,25], TiO_2 [26], and so on. However, as yet there has been no breakthrough improvement in the defluorination performance. It has been reported that pure carbonaceous materials generally display inferior wettability and electrical conductivity, which could significantly hinder the adsorption of ions in the carbon framework [27,28]. Heteroatomic doping, such as nitrogen doping, has been proven to be an effective way to enhance the wettability and electrical conductivity of carbonaceous material in CDI desalination applications [29,30]. However, little attention has been paid to the modification of porous carbons by heteroatomic doping to improve the defluorination performance of CDI. The amelioration of wettability is mainly associated with the introduction of nitrogenous polar functional groups into the carbon framework, which displays benign affinity to aqueous solvents through nitrogen doping. The improvement of electrical conductivity through nitrogen doping is mainly due to the electron-rich peculiarity of nitrogen, which can bring more electrons to the delocalized π-system of the carbon framework. Based on the above facts, nitrogen-doped mesoporous carbons are extremely attractive.

In this study, the nitrogen-doped mesoporous carbonaceous nanospheres (NMCS) with core-shell structure were prepared via a gradient sol-gel method for efficient defluorination in CDI systems. As a comparison, mesoporous pure carbonaceous nanospheres without heteroatom doping (MCS) were also prepared. The defluorination performance was studied in a series of batch mode-recirculating mode experiments, and the important contribution of nitrogen doping to the defluorination performance of carbon materials was analyzed in depth using multiple material characterizations and electrochemical analyses.

2. Materials and Methods

2.1. Materials Synthesis and Electrode Fabrication

The precursors of MCS and NMCS were prepared via a gradient sol-gel method [31]. In a typical procedure, 1.5 g of resorcinol (Guangfu Fine Chemical Research Institute, Tianjin, China) and 1.5 g of cetyltrimethylammonium bromide (CTAB, Guangfu Fine Chemical Research Institute, Tianjin, China) were dissolved in a mixed solvent of ammonia solution (25 wt.%, Fengchuan, Tianjin, China) and ethanol (Riolon, Tianjin, China) with a volume ratio of 5:3. After magnetically stirring for 30 min at room temperature, the mixture was added with 7.5 mL of tetraethyl orthosilicate (TEOS, Jiangtian, Tianjin, China) and 2.1 mL of formaldehyde (37 wt.%, Ailan, Shanghai, China). The mixture continued to be magnetically stirred for 24 h at room temperature and was then transferred to Teflon-lined autoclaves to carry hydrothermal reaction for 24 h at 100 °C. To prepare NMCS, the above precursors were ground with urea (Jiangtian, Tianjin, China) for 2 h using a ball mill for intensive mixing. Subsequently, the resulting products were carbonized at 700 °C 180 min in a nitrogen atmosphere. As a comparison, MCS was prepared by directly carbonizing the hydrothermal precursors without adding urea. Next, the carbonizing products were soaked in 10 wt.% hydrofluoric acid (HF, 40 wt.%, Fengchuan, Tianjin, China) for 24 h to remove impurities. Finally, the obtained products were washed with deionized water and ethanol and dried in vacuum overnight.

The as-acquired MCS and NMCS powders were mixed with carbon black and PTFE with a mass ratio of 8:1:1. The mixture was dispersed in ethanol solution and ultrasonically stirred to form a homogenized slurry. The electrode was fabricated by pressing the slurry onto a titanium mesh current collector using a noodle machine. Finally, the electrodes were dried in vacuum in order to remove the residual ethanol.

2.2. Materials Characterization

The morphology of as-acquired MCS and NMCS were characterized by scanning electron microscope (SEM, Hitachi S-3500N, Hitachi, Tokyo, Japan) equipped with an energy-dispersive X-ray spectrometer (EDS) and high-resolution transmission electron microscope (TEM, JEM-ARM200F, JEOL, Tokyo, Japan). The structure of MCS and NMCS was analyzed by X-ray diffraction (XRD, D/max-2500, Rigaku, Tokyo, Japan). The porous structure of MCS and NMCS was analyzed by the Brunauer–Emmett–Teller (BET) method using a density functional theory (DFT) model. The defect structure and graphitization degree of MCS and NMCS were analyzed by Raman spectroscopy (SR-500I-A, Time Tagger, Wuhan, China). The elements and compositions of MCS and NMCS were analyzed by X-ray photoelectron spectroscopy (XPS, ESCALAB 250 XI, Thermo Scientific Escalab, Waltham, MA, USA). The wettability of MCS and NMCS was measured by the drop shape analysis system (Kruss DSA 100S, KRUSS, Shanghai, China).

2.3. Electrochemical Measurements

The electrochemical measurements in this study included cyclic voltammetry (CV), galvanostatic charging–discharging (GCD) and electrochemical impedance spectroscopy (EIS). All measurements were performed in 1 mol L^{-1} NaF solution using a potentiostat (chi760e, CH Instruments, Shanghai, China). A three-electrode system was employed with a carbonaceous electrode as the working electrode, a platinum gauze electrode as the counter electrode and a saturated calomel electrode as the reference electrode.

The measurements of EIS were with a frequency range of 0.1 to 100,000 Hz. For CV measurements, the potential window was ±0.5 V, and the scan rate varied from 10 to

100 mV·s^{-1}. Based on the CV curves, the specific capacitance can be ascertained according to the following equation:

$$C = \frac{S}{2\Delta Vvm} \quad (1)$$

where C is the specific capacitance (F g^{-1}), S is the area of the CV curve, ΔV is the voltage window (V), v is the scan rate (V s^{-1}) and m is the mass of active material on the working electrode (g).

In GCD measurements, a series of current densities including 0.2, 0.5 and 1 A g^{-1} were employed with a potential window of ±0.5 V. Based on the GCD curves, the charge/discharge capacitance can be determined according to the following equation:

$$C = \frac{I\Delta t}{m\Delta V} \quad (2)$$

where I/m is the current density (A g^{-1}), Δt is the charge/discharge time (s) and ΔV is the voltage window (V).

2.4. Defluorination Experiment

The defluorination experiment was conducted in a self-made CDI cell in a batch mode-recirculating mode. As shown in Figure S1, the CDI cell was composed of a rectangular plastic chamber with a pair of parallel electrodes attached to plastic plates inside. Each electrode had an area of around 7 cm^2 and a thickness of around 30 μm. In the defluorination experiment, the adsorption period and the electrode regeneration period alternated. The adsorption period proceeded with 100 mL of NaF solution circulating in the CDI cell through a peristaltic pump with a voltage being applied between the two electrodes. The circulating rate of the circulating solution was constant at 10 mL·min^{-1}, while the fluorine concentration of the circulating solution varied from 250 to 1000 mg L^{-1}. The conductivity of the circulating solution was continuously monitored through a conductivity meter (IN-ESA, DDS-11A, INESA, Shanghai, China), and the concentration of the circulating solution was ascertained based on the calibration curve between the conductivity and concentration, as shown in Figure S2. The electrode regeneration period started when the conductivity of the circulating solution remained unchanged, which indicated the adsorption equilibrium of fluorine onto the electrodes. In this period, the voltage between the two electrodes was removed and the electrodes were in a short circuit.

The electrosorption capacity at a typical time can be calculated according to the following equation:

$$\Gamma_t = \frac{(C_0 - C_t)V}{m} \quad (3)$$

where C_0 is the initial fluorine concentration of the circulating solution (mg L^{-1}), C_t is the fluorine concentration of the circulating solution at adsorption time of t (mg L^{-1}), V is the volume of the circulating solution (L) and m is the mass of active material (g).

The electrosorption rate at a typical time can be calculated according to the following equation:

$$v_t = \frac{\Gamma_t}{t} \quad (4)$$

where Γ_t is the electrosorption capacity at adsorption time of t (mg g^{-1}) and t is the adsorption time (min).

3. Results and Discussion

3.1. Material Characterizations

Figure 1 shows the morphological features of MCS and NMCS. As depicted in the SEM image, MCS exhibited a regular spherical morphology with rough surfaces and with a diameter ranging from 900 to 1400 nm, as shown in Figure 1a. According to Figure 1b, NMCS retained the regular spherical morphology after nitrogen doping while displaying much smoother surfaces and smaller size dimensions. The TEM image shown in Figure 1c indicates the unique yolk-shell structure of MCS. Interestingly, as shown in Figure 1d, the doping of nitrogen resulted in the structural transformation of NMCS from yolk-shell to core-shell spheres. The thickness of the shells of both NMCS and MCS were in the range of 100–150 nm. High-resolution TEM images were further employed to depict the microstructure of the core and shell of NMCS. It is evident from Figure 1e–g that the core had a much looser structure, while more homogeneous and smaller-sized pores were distributed in the core. EDS mapping images of NMCS, as shown in Figure S3, demonstrated that nitrogen was introduced into both the core and shell of NMCS and was evenly distributed across the core and shell.

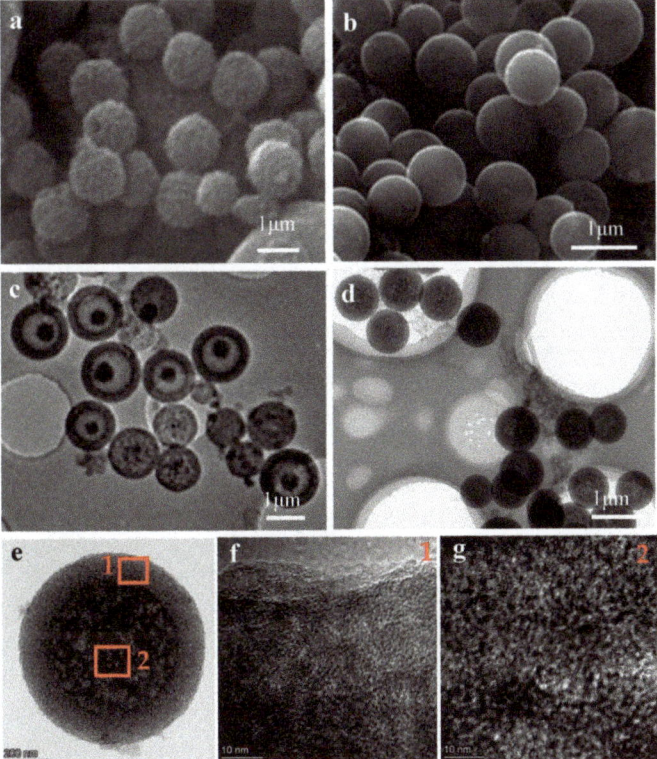

Figure 1. (**a**) SEM image of MCS; (**b**) SEM image of NMCS; (**c**) TEM image of MCS; (**d**) TEM image of NMCS; (**e–g**) high-resolution TEM images of NMCS.

Figure 2 shows the structural features of MCS and NMCS. Raman spectroscopy was employed to analyze the features of NMCS and MCS. In Figure 2a, two prominent peaks appeared at around 1350 cm^{-1} (D band) and at 1595 cm^{-1} (G band). The D band was reported to result from the Csp3 hybridization and could be assigned to defects and disordered structures [32]. The G band was reported to be derived from the Csp2 hybridization and could correspond to the crystalline graphitic structure [33]. The relative intensity ratio of the D band and the G band (ID/IG) is extensively used as an indicator to reflect the disorder degree of carbonaceous materials. The high ID/IG values of NMCS (0.999) and MCS (0.982) demonstrated their disorder features in which abundant defects were distributed. Nevertheless, NMCS displayed a higher ID/IG value as compared with MCS, indicating that more structural disorders and surface defects were formed in the framework of NMCS caused by the doping of nitrogen [34]. The generation of surface defects in carbonaceous materials can promote ion diffusion, provide more accessible site for ion adsorption and, thus, is beneficial for ion removal.

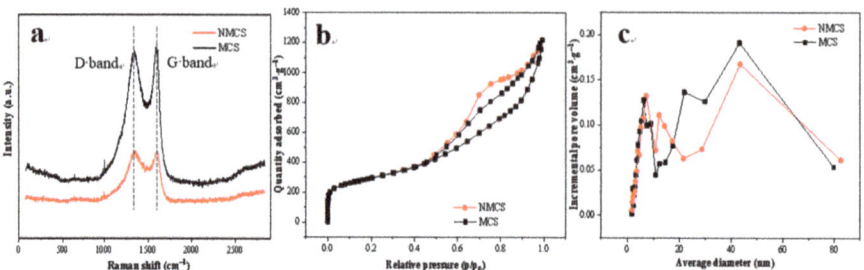

Figure 2. (a) Raman spectra of MCS and NMCS; (b) N_2 adsorption/desorption isotherms of MCS and NMCS; (c) pore size distribution curves of MCS and NMCS.

The pore structures of MCS and NMCS were analyzed based on N_2 adsorption/desorption isotherms. As shown in Figure 2b, both isotherms of NMCS and MCS show obvious hysteresis loops, which can be assigned to type-IV adsorption/desorption isotherm pattern. The type-IV adsorption/desorption isotherm pattern has been reported to be a typical characteristic of the mesoporous structure [35]. The pore size distribution curves of MCS and NMCS are shown in Figure 2c. The porous structure of MCS and MMCS is mainly composed of mesopores, as evidenced by the fact that the pore size is mainly distributed in the range of 2~50 nm. The average pore diameter of MCS and NMCS was 43.2 and 43.7 nm, respectively. As listed in Table 1, the specific surface area of MCS derived from mesopores was 771 $m^2\ g^{-1}$, which was around 72.9% of the total specific surface area (1058 $m^2\ g^{-1}$). After nitrogen doping, the total specific surface area of NMCS (1049 $m^2\ g^{-1}$) was almost unchanged. However, the specific surface area derived from mesopores increased to 819 $m^2\ g^{-1}$, which was 78.1% of the total specific surface area. Similar to specific surface area results, according to Table 1 the pore volume of both MCS and NMCS is primarily derived from mesopores. The above results further illustrate the domination of mesoporous structure in the MCS and NMCS frameworks. The mesopores would promote ion diffusion in the carbon frameworks and provide more accessible sites for ion adsorption, since their large apertures can reduce the resistance for ion transfer to the interior structure [36]. Therefore, the higher specific surface area derived from mesopores endowed NMCS with more available adsorption sites as compared with MCS in spite of similar total adsorption sites, which would contribute to the improvement of CDI performance.

Table 1. Pore structural parameters of MCS and NMCS.

Sample	Specific Surface Area ($m^2\ g^{-1}$)			Pore Volume ($cm^3\ g^{-1}$)		
	S_{BET}	S_{mic}	S_{meso}	V_t	V_{mic}	V_{meso}
MCS	1058	287	771	1.89	0.12	1.77
NMCS	1049	230	819	1.78	0.10	1.68

Figure 3 shows the compositions of MCS and NMCS. The FTIR spectrum of MCS shown in Figure 3a indicates two distinct absorption bands at around 1590 and 1250 cm^{-1}, which can be ascribed to the stretching vibration of C=O and C-O-C, respectively [37]. In the spectrum of NMCS, an additional absorption band at around 3430 cm^{-1}, corresponding to the stretching vibration of N-H, is observed according to Figure 3a, confirming the successful doping of nitrogen into NMCS [38]. The introduction of nitrogen-containing functional groups would significantly change the wettability of carbonaceous material. As shown in Figure S4, the contact angle of NMCS decreased from 35.77° to 17.86° after nitrogen doping, which is associated with the benign affinity of nitrogen-containing functional groups to aqueous solution. The amelioration of wettability of NMCS would enhance the utilization of the specific surface area, thus, more sites can be available for ion adsorption [39]. The content of nitrogen doped was quantitatively analyzed by XPS measurement. As shown in Figure 3b, the survey spectrum of NMCS indicts a distinct N1s peak, but it almost disappears in the survey spectrum of MCS. According to Table 2, the nitrogen content of NMCS was 4.03%, which was significantly higher than that of MCS (0.31%). The above results further suggest the successful doping of nitrogen from urea. As shown in Figure 3c,d, the high-resolution N1s spectra of NMCS and MCS can be fitted into three peaks at around 398, 400 and 401 eV, corresponding to three configurations of nitrogen, i.e., pyridinic-N, pyrrolic-N and graphitic-N. It has been widely reported in the literature that pyridinic-N and pyrrolic-N are beneficial to boost capacitance through generating pseudocapacitance based on faradic reactions, and that graphitic-N contributes to improve the conductivity of carbonaceous materials [40,41]. According to Table 2, all the contents of pyridinic-N, pyrrolic-N and graphitic-N in NMCS were significantly increased as compared with MCS. Therefore, the electricity conductivity and capacitance behavior of NMCS would be enhanced significantly, as verified by the electrochemical measurements discussed in Section 3.2.

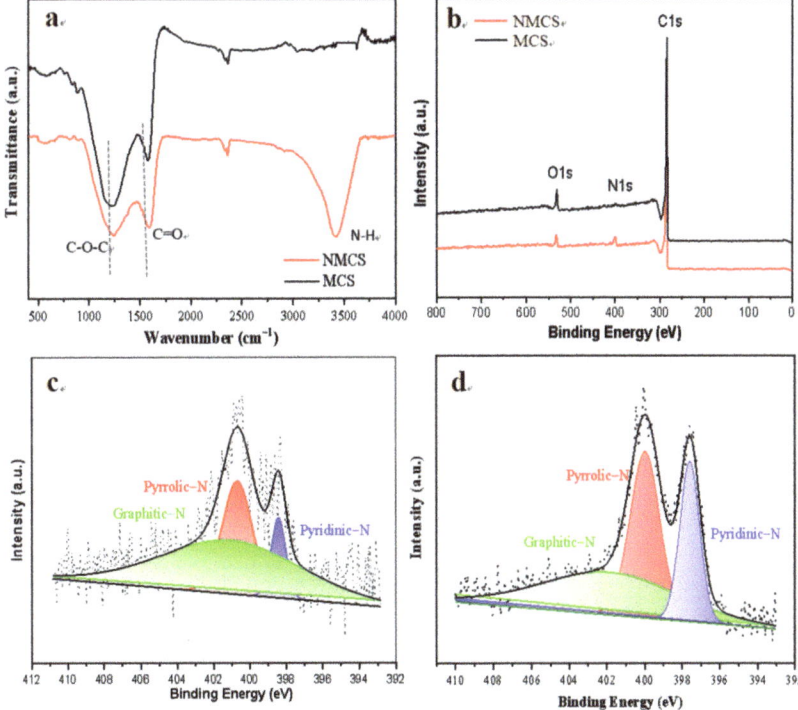

Figure 3. (**a**) FTIR spectra of MCS and NMCS; (**b**) XPS survey spectra of MCS and NMCS; (**c**) high-resolution N1s spectra of MCS; (**d**) high-resolution N1s spectra of NMCS.

Table 2. Nitrogen content of MCS and NMCS.

Sample	N (%)	Fraction of Nitrogen (%)			Content of Nitrogen (%)		
		Pyridinic-N	Pyrrolic-N	Graphitic-N	Pyridinic-N	Pyrrolic-N	Graphitic-N
MCS	0.31	1058	287	771	1.89	0.12	1.77
NMCS	4.03	1049	230	819	1.78	0.10	1.68

Based on the results outlined and discussed above, both MCS and NMCS have superior mesoporous structures which can provide abundant available sites for ion adsorption. The doping of nitrogen into NMCS resulted in morphologic transformation, structural amelioration and improved physicochemical properties such as conductivity and wettability. It not only facilitated ion transportation, but also improved the available sites for ion adsorption and, thus, would boost the defluorination performances in CDI systems.

3.2. Electrochemical Analysis

Figure 4 shows the electrochemical performances of MCS and NMCS. As shown in Figure 4a,b, the CV curves of both MCS and NMCS were close to rectangle at low scan rates, while they gradually changed into a fusiform shape with the increasing scanning rate due to the polarization effect of the electrodes [28]. The absence of redox peaks and excellent symmetry of CV curves demonstrates the ideal electrical-double-layer capacitance behaviors of MCS and NMCS [42]. According to Equation (1), the specific capacitance of MCS and NMCS at different scan rates was calculated, and the results are shown in Figure 4c. Apparently, the specific capacitance of both MCS and NMCS decreased with the increase of scan rate, which is attributed to the improved utilization of interior pores at lower scan rates. Nevertheless, NMCS displayed substantially higher specific capacitance as compared with MCS at all scan rates employed. The capacitance behaviors of MCS and NMCS were further analyzed by GCD curves. As shown in Figure 4d,e, both MCS and NMCS exhibited linear curves, implying the dominant contribution of electrical-double-layer adsorption to capacitance behaviors [43]. The high symmetry of GCD curves indicates the excellent repeatability of MCS and NMCS. The charge/discharge capacitance of MCS and NMCS under different current densities was calculated according to Equation (2), and the results are shown in Figure 4f. Evidently, the charge/discharge capacitance of NMCS was higher than that of MCS under all current densities measured, in accordance with the CV results. The superior capacitance behavior of NMCS is possibly associated with the facilitation of ion transportation and the improvement of available adsorption sites resulting from the doping of nitrogen.

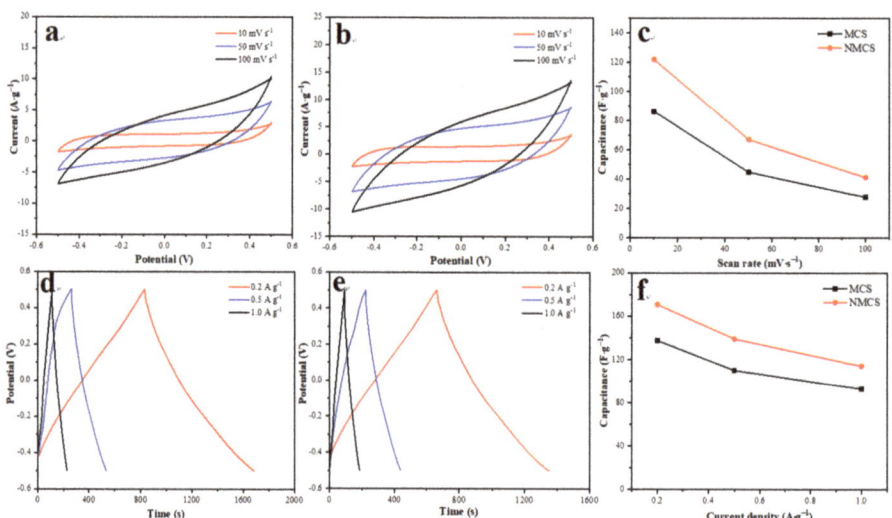

Figure 4. (**a**) CV curves of MCS with different scan rates; (**b**) CV curves of NMCS with different scan rates; (**c**) specific capacitance results of MCS and NMCS; (**d**) GCD curves of MCS with different current densities; (**e**) GCD curves of NMCS with different current densities; (f) charge/discharge capacitance results of MCS and NMCS.

Figure 5 shows the EIS curves of MCS and NMCS with a frequency range of 0.1–100,000 Hz. The semicircle of MCS and NMCS in the high-frequency region is associated with the charge-transfer process within the microstructure. The inclined line of MCS and NMCS in the low-frequency region is related to the ion-diffusion process inside the framework. The charge-transfer resistance (R_{ct}) of MCS and NMCS was fitted to be 2.449 and 1.616 Ω, respectively, based on the equivalent circuit model shown in Figure 5. The low R_{ct} of NMCS implies that the doping of nitrogen reduces the barrier for ion to diffusion and improves the electrical conductivity to facilitate ion transport [44]. Based on the above results, NMCS exhibited superior capacitance and resistance behaviors as compared with their undoped counterparts, which is favorable for effective defluorination in CDI experiments.

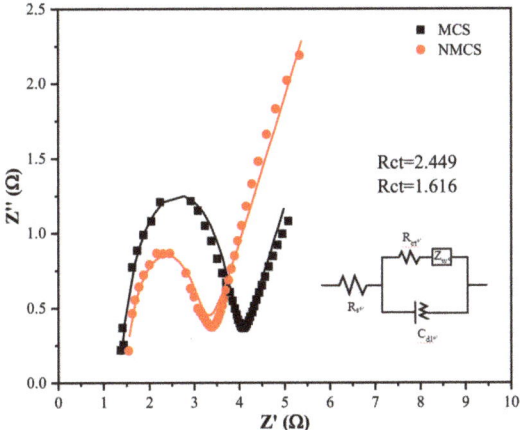

Figure 5. EIS curves of MCS and NMCS.

3.3. CDI Performance Analysis

The defluorination performances of MCS and NMCS were studied with the initial fluorine concentration of 1000 mg L^{-1} at an applied voltage of 1.2 V. Figure 6a shows the change of conductivity with the electrosorption time in a typical cycle. The conductivity of MCS and NMCS initially decreased sharply and gradually reached a plateau with the increase of electrosorption time. In this study, MCS and NMCS reached adsorption saturation at around 30 min, as evidenced by the almost unchanged conductivity at this time. Nevertheless, NMCS indicated rapider and greater decline in conductivity than that of MCS, implying the boosted defluorination performance of NMCS after doping of nitrogen. Figure 6b shows the Kim plots of MCS and NMCS with the initial fluorine concentration of 1000 mg L^{-1} at an applied voltage of 1.2 V. The plot of NMCS is located on the upper right as compared with MCS, suggesting the superior electrosorption capacity and electrosorption rate of NMCS in CDI defluorination application. According to Figure 6b, NMCS exhibited an electrosorption capacity at saturation of 10.77 mg g^{-1} and maximal electrosorption rate of 2.20 mg g^{-1} s^{-1}. The electrosorption capacity of NMCS at saturation was 13.34 mg g^{-1}, and the maximal electrosorption rate of NMCS was 3.13 mg g^{-1} s^{-1}, which showed significant improvement as compared with NMCS. Additionally, the electrosorption capacity of MCS and NMCS under different operating conditions was examined. As shown in Figure 6c,d, NMCS always displayed higher electrosorption capacity as compared with MCS with various initial fluorine concentrations and voltages. These results elucidated the importance of nitrogen doping to improve the defluorination performances of NMCS. Additionally, the electrosorption capacity and adsorption saturation times of NMCS were compared with porous carbons reported in the literature. As shown in Table S1, NMCS showed higher electrosorption capacity and shorter adsorption saturation time as compared with commercial activated carbons [14,16], biomass-derived activated carbons [17,45] and activated carbons modified with metal oxides [26]. The above results imply the significant advantages of NMCS in defluorination applications.

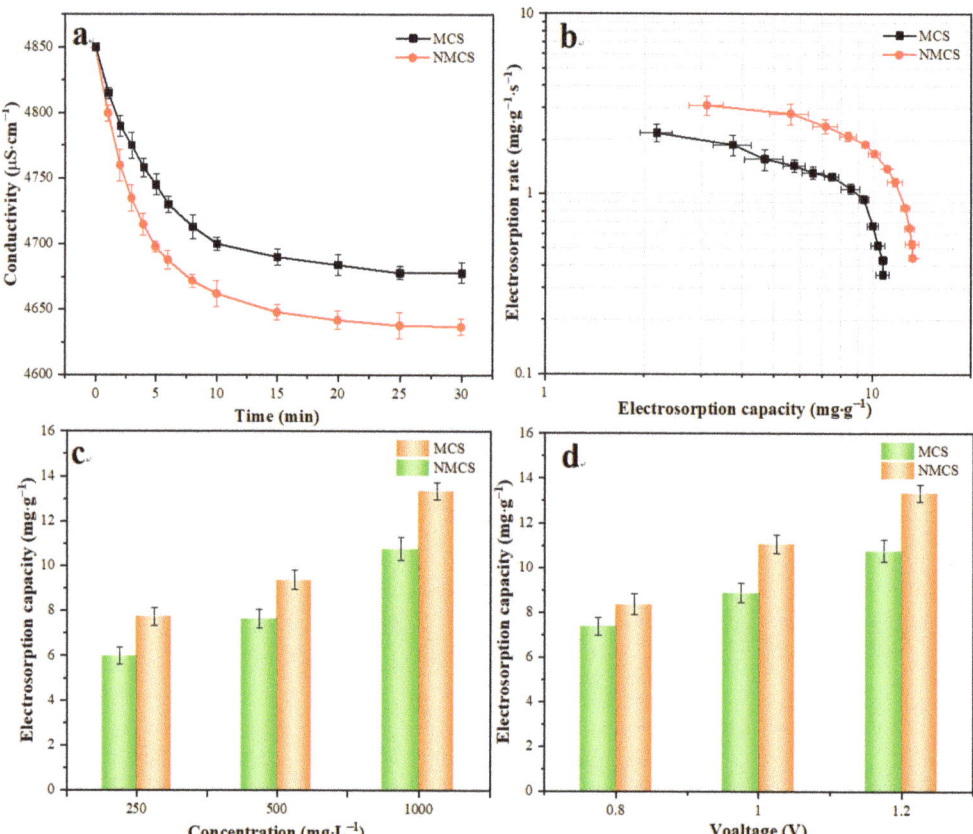

Figure 6. (**a**) Conductivity change of circulating solution of MCS and NMCS with the initial fluorine concentration of 1000 mg L^{-1} at 1.2 V; (**b**) the corresponding Kim plots of MCS and NMCS with the initial fluorine concentration of 1000 mg L^{-1} at 1.2 V; (**c**) electrosorption capacity of MCS and NMCS with different initial fluorine concentrations; (**d**) electrosorption capacity of MCS and NMCS with different voltages.

The adsorption–regeneration experiments were repeated for 10 cycles under the same experimental conditions to investigate the repeatability of the NMCS electrode. An amount of 1000 mg L^{-1} of the initial fluorine concentration and 1.2 V of constant voltage were employed. As shown in Figure 7, more than 94% of electrosorption capacity in the initial cycle was retained after 10 cycles, indicating the outstanding cycling stability of the NMCS electrode. In the cycling experiments, the effluent pH was constant at around 7.0. This implies that faradaic side reactions such as water splitting did not occur during the adsorption–regeneration cycling. After 10 adsorption–regeneration cycles, the NMCS electrode was taken out and characterized by TEM. As shown in Figure S6, the regular spherical morphology with distinct core-shell structure was completely retained, further revealing the remarkable stability of NMCS during adsorption–regeneration cycling. The superior CDI performance outlined and discussed above demonstrates that NMCS would be a promising electrode material for effective defluorination.

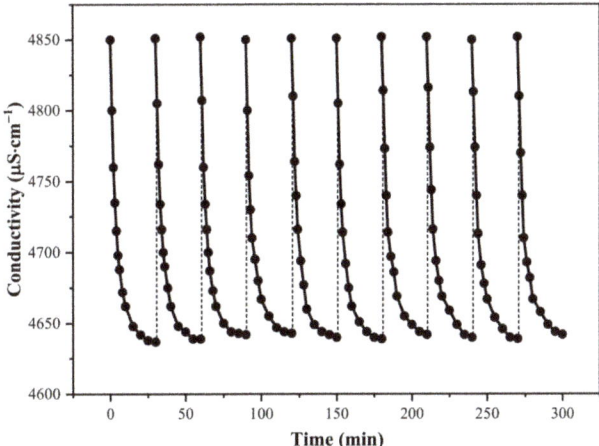

Figure 7. Cycling stability of NMCS.

4. Conclusions

In this study, we prepared NMCS via a gradient sol-gel method to effectively remove fluorine in CDI systems. NMCS achieved a high electrosorption capacity of 13.34 mg g^{-1} with the initial fluorine concentration of 1000 mg L^{-1} at 1.2 V and also excellent repeatability with little decrease of electrosorption capacity after 10 cycles. The superior defluorination performance of NMCS as compared with its undoped counterpart was associated with improved physicochemical properties, such as conductivity and wettability, the amelioration of pore structure and the transformation of morphology due to nitrogen doping. Nitrogen doping into NMCS was found to not only facilitate ion transportation, but to also improve the available adsorption sites. Therefore, NMCS would be an excellent electrode material for effective defluorination in CDI systems.

Supplementary Materials: The following supporting information can be downloaded at: https://www.mdpi.com/article/10.3390/w15030608/s1, Figure S1: Schematic diagram of the defluorination system (a) and the CDI unit (b); Figure S2: Standard curve of conductivity and fluorine concentration; Figure S3: EDS elemental mapping images equipped with TEM of NMCS; Figure S4: The water contact angle of MCS (a) and NMCS (b); Figure S5: Conductivity change of circulating solution of MCS (a) and NMCS (b) with different initial fluorine concentrations; conductivity change of circulating solution of MCS (c) and NMCS (d) with different voltages; Figure S6: TEM image of NMCS after 10 adsorption–regeneration cycles; Table S1: Comparison of defluorination capacity and time between NMCS and reported materials in the literature.

Author Contributions: Y.Z.: methodology, original draft preparation, funding acquisition. K.L.: conceptualization, reviewing and editing. B.S.: reviewing and editing. F.C.: supervision, review and editing, funding acquisition. Y.S.: methodology, data curation, reviewing and editing. All authors have read and agreed to the published version of the manuscript.

Funding: This research was funded by Shandong Provincial Natural Science Foundation (No. ZR2022QE088), Shandong Postdoctoral Science Foundation (No. SDCX-ZG-202202028), Shandong Top Talent Special Foundation, Doctoral Research Fund Project in Shandong Jianzhu University (No. X22005Z), National Key Research and Development Program of China (No. 2022YFE0105800), Nanxun Collaborative Innovation Center Key Research Project (No. JZ2022ZH01).

Data Availability Statement: All data that support the findings of this study are available from the corresponding author upon reasonable request.

Conflicts of Interest: The authors declare no conflict of interest.

References

1. Smith, G.E. Fluoride, the environment, and human health. *Perspect. Biol. Med.* **1986**, *29*, 560–572. [CrossRef] [PubMed]
2. Wang, M.; Li, X.; He, W.; Li, J.; Zhu, Y.; Liao, Y.; Yang, J.; Yang, X. Distribution, health risk assessment, and anthropogenic sources of fluoride in farmland soils in phosphate industrial area, southwest China. *Environ. Pollut.* **2019**, *249*, 423–433. [CrossRef] [PubMed]
3. Wan, K.; Huang, L.; Yan, J.; Ma, B.; Huang, X.; Luo, Z.; Zhang, H.; Xiao, T. Removal of fluoride from industrial wastewater by using different adsorbents: A review. *Sci. Total Environ.* **2021**, *773*, 145535. [CrossRef] [PubMed]
4. Velazquez-Jimenez, L.H.; Vences-Alvarez, E.; Flores-Arciniega, J.L.; Flores-Zuñiga, H.; Rangel-Mendez, J.R. Water defluoridation with special emphasis on adsorbents-containing metal oxides and/or hydroxides: A review. *Sep. Purif. Technol.* **2015**, *150*, 292–307. [CrossRef]
5. Wu, Q.; Liang, D.; Lu, S.; Wang, H.; Xiang, Y.; Aurbach, D.; Avraham, E.; Cohen, I. Advances and perspectives in integrated membrane capacitive deionization for water desalination. *Desalination* **2022**, *542*, 116043. [CrossRef]
6. Nunes-Pereira, J.; Lima, R.; Choudhary, G.; Sharma, P.R.; Ferdov, S.; Botelho, G.; Sharma, R.K.; Lanceros-Méndez, S. Highly efficient removal of fluoride from aqueous media through polymer composite membranes. *Sep. Purif. Technol.* **2018**, *205*, 1–10. [CrossRef]
7. Pan, B.; Xu, J.; Wu, B.; Li, Z.; Liu, X. Enhanced removal of fluoride by polystyrene anion exchanger supported hydrous zirconium oxide nanoparticles. *Environ. Sci. Technol.* **2013**, *47*, 9347–9354. [CrossRef]
8. Nordstrand, J.; Dutta, J. Theory of bipolar connections in capacitive deionization and principles of structural design. *Electrochim. Acta* **2022**, *430*, 141066. [CrossRef]
9. Wang, R.; Sun, K.; Zhang, Y.; Qian, C.; Bao, W. Dimensional optimization enables high-performance capacitive deionization. *J. Mater. Chem. A* **2022**, *10*, 6414–6441. [CrossRef]
10. Hu, C.-C.; Hsieh, C.-F.; Chen, Y.-J.; Liu, C.-F. How to achieve the optimal performance of capacitive deionization and inverted-capacitive deionization. *Desalination* **2018**, *442*, 89–98. [CrossRef]
11. Suss, M.E.; Porada, S.; Sun, X.; Biesheuvel, P.M.; Yoon, J.; Presser, V. Water desalination via capacitive deionization: What is it and what can we expect from it? *Energy Environ. Sci.* **2015**, *8*, 2296–2319. [CrossRef]
12. Tang, W.; Kovalsky, P.; He, D.; Waite, T.D. Fluoride and nitrate removal from brackish groundwaters by batch-mode capacitive deionization. *Water Res.* **2015**, *84*, 342–349. [CrossRef]
13. Epshtein, A.; Nir, O.; Monat, L.; Gendel, Y. Treatment of acidic wastewater via fluoride ions removal by SiO_2 particles followed by phosphate ions recovery using flow-electrode capacitive deionization. *Chem. Eng. J.* **2020**, *400*, 125892. [CrossRef]
14. Gaikwad, M.S.; Balomajumder, C. Simultaneous electrosorptive removal of chromium(VI) and fluoride ions by capacitive deionization (CDI): Multicomponent isotherm modeling and kinetic study. *Sep. Purif. Technol.* **2017**, *186*, 272–281. [CrossRef]
15. Park, G.; Hong, S.P.; Lee, C.; Lee, J.; Yoon, J. Selective fluoride removal in capacitive deionization by reduced graphene oxide/hydroxyapatite composite electrode. *J. Colloid Interface Sci.* **2012**, *581*, 396–402. [CrossRef]
16. Tang, W.; Kovalsky, P.; Cao, B.; Waite, T.D. Investigation of fluoride removal from low-salinity groundwater by single-pass constant-voltage capacitive deionization. *Water Res.* **2016**, *99*, 112–121. [CrossRef]
17. Gaikwad, M.S.; Balomajumder, C. Tea waste biomass activated carbon electrode for simultaneous removal of Cr(VI) and fluoride by capacitive deionization. *Chemosphere* **2017**, *184*, 1141–1149. [CrossRef]
18. Xu, X.; Tan, H.; Wang, Z.; Wang, C.; Pan, L.; Kaneti, Y.V.; Yang, T.; Yamauchi, Y. Extraordinary capacitive deionization performance of highly-ordered mesoporous carbon nanopolyhedra for brackish water desalination. *Environ. Sci. Nano* **2019**, *6*, 981. [CrossRef]
19. Wang, H.; Wei, D.; Gang, H.; He, Y.; Zhang, L. Hierarchical porous carbon from synergistic "pore-on-pore" strategy for efficient capacitive deionization. *ACS Sustain. Chem. Eng.* **2020**, *8*, 1129–1136. [CrossRef]
20. Baroud, T.N.; Giannelis, E.P. High salt capacity and high removal rate capacitive deionization enabled by hierarchical porous carbons. *Carbon* **2018**, *139*, 614–625. [CrossRef]
21. Baroud, T.N.; Giannelis, E.P. Role of mesopore structure of hierarchical porous carbons on the electrosorption performance of capacitive deionization electrodes. *ACS Sustain. Chem. Eng.* **2019**, *7*, 7580–7596. [CrossRef]
22. Gao, T.; Li, H.; Zhou, F.; Gao, M.; Liang, S.; Luo, M. Mesoporous carbon derived from ZIF-8 for high efficient electrosorption. *Desalination* **2019**, *451*, 133–138. [CrossRef]
23. Li, Y.; Zhang, C.; Jiang, Y.; Wang, T.-J. Electrically enhanced adsorption and green regeneration for fluoride removal using $Ti(OH)_4$-loaded activated carbon electrodes. *Chemosphere* **2018**, *200*, 554–560. [CrossRef] [PubMed]
24. Martinez-Vargas, D.R.; Rangel-Mendez, J.R.; Chazaro-Ruiz, L.F. Fluoride electrosorption by hybrid La(III)-activated carbon electrodes under the influence of the La(III) content and the polarization profile. *J. Environ. Chem. Eng.* **2022**, *10*, 106926. [CrossRef]
25. Martinez-Vargas, D.R.; Larios-Dur'an, E.R.; Chazaro-Ruiz, L.F.; Rangel-Mendez, J.R. Correlation between physicochemical and electrochemical properties of an activated carbon doped with lanthanum for fluoride electrosorption. *Sep. Purif. Technol.* **2021**, *268*, 118702. [CrossRef]
26. Wu, P.; Xia, L.; Dai, M.; Lin, L.; Song, S. Electrosorption of fluoride on TiO_2-loaded activated carbon in water. *Colloids Surf. A Physicochem. Eng. Asp.* **2016**, *502*, 66–73. [CrossRef]
27. He, D.; Niu, J.; Dou, M.; Ji, J.; Huang, Y.; Wang, F. Nitrogen and oxygen co-doped carbon networks with a mesopore-dominant hierarchical porosity for high energy and power density supercapacitors. *Electrochim. Acta* **2017**, *238*, 310–318. [CrossRef]

28. Zhao, Y.; Zhang, Y.; Tian, P.; Wang, L.; Li, K.; Lv, C.; Liang, B. Nitrogen-rich mesoporous carbons derived from zeolitic imidazolate framework-8 for efficient capacitive deionization. *Electrochim. Acta* **2019**, *321*, 134665. [CrossRef]
29. Zong, M.; Zhang, Y.; Li, K.; Lv, C.; Tian, P.; Zhao, Y.; Liang, B. Zeolitic imidazolate framework-8 derived two-dimensional N-doped amorphous mesoporous carbon nanosheets for efficient capacitive deionization. *Electrochim. Acta* **2020**, *329*, 135089. [CrossRef]
30. Yang, L.; Hussain, I.; Qi, J.; Chao, L.; Li, J.; Shen, J.; Sun, X.; Han, W.; Wang, L. N-doped hierarchical porous carbon derived from hypercrosslinked diblock copolymer for capacitive deionization. *Sep. Purif. Technol.* **2016**, *165*, 190–198.
31. Yan, T.; Liu, J.; Lei, H.; Shi, L.; Zhang, D. Capacitive deionization of saline water using sandwich-like nitrogen-doped graphene composites via a self-assembling strategy. *Environ. Sci. Nano* **2018**, *5*, 2722–2730. [CrossRef]
32. Zhang, W.; Xu, J.; Hou, D.; Yin, J.; Liu, D.; He, Y. Hierarchical porous carbon prepared from biomass through a facile method for supercapacitor applications. *J. Colloid Interface Sci.* **2018**, *530*, 338–344. [CrossRef]
33. Wang, K.; Xu, M.; Gu, Y.; Gu, Z.; Fan, Q.H. Symmetric supercapacitors using urea-modified lignin derived N-doped porous carbon as electrode materials in liquid and solid electrolytes. *J. Power Sources* **2016**, *332*, 180–186. [CrossRef]
34. Gao, T.; Du, Y.; Li, H. Preparation of nitrogen-doped graphitic porous carbon towards capacitive deionization with high adsorption capacity and rate capability. *Sep. Purif. Technol.* **2015**, *165*, 190–198. [CrossRef]
35. Li, G.-X.; Hou, P.-X.; Zhao, S.-Y.; Liu, C.; Cheng, H.-M. A flexible cotton-derived carbon sponge for high-performance capacitive deionization. *Carbon* **2016**, *101*, 1–8. [CrossRef]
36. Wang, Z.; Yan, T.; Fang, J.; Shi, L.; Zhang, D. Nitrogen-doped porous carbon derived from a bimetallic metal-organic framework as highly efficient electrodes for flow-through deionization capacitors. *J. Mater. Chem. A* **2016**, *4*, 10858–10868. [CrossRef]
37. Chang, L.; Hu, Y.H. 3D Channel-structured graphene as efficient electrodes for capacitive deionization. *J. Colloid Interface Sci.* **2019**, *538*, 420–425. [CrossRef]
38. Vinayan, B.P.; Nagar, R.; Raman, V.; Rajalakshmi, N.; Dhathathreyan, K.S.; Pamaprabhu, S. Synthesis of graphene-multiwalled carbon nanotubes hybrid nanostructure by strengthened electrostatic interaction and its lithium ion battery application. *J. Mater. Chem.* **2012**, *22*, 9949–9956. [CrossRef]
39. Liu, Y.; Chen, T.; Lu, T.; Sun, Z.; Chua, D.H.C.; Pan, L. Nitrogen-doped porous carbon spheres for highly efficient capacitive deionization. *Electrochim. Acta* **2015**, *158*, 403–409. [CrossRef]
40. Lin, Z.; Waller, G.; Liu, Y.; Liu, M.; Wong, C.-P. Facile synthesis of nitrogen-doped graphene via pyrolysis of graphene oxide and urea, and its electrocatalytic activity toward the oxygen-reduction reaction. *Adv. Energy Mater.* **2012**, *2*, 884–888. [CrossRef]
41. Tang, Y.; Zheng, S.; Cao, S.; Yang, F.; Guo, X.; Zhang, S.; Xue, H.; Pang, H. Hollow mesoporous carbon nanospheres space-confining ultrathin nanosheets superstructures for efficient capacitive deionization. *J. Colloid Interface Sci.* **2022**, *626*, 1062–1069. [CrossRef] [PubMed]
42. Zhang, W.; Liu, Y.; Jin, C.; Shi, Z.; Zhu, L.; Zhang, H.; Jiang, L.; Chen, L. Efficient capacitive deionization with 2D/3D heterostructured carbon electrode derived from chitosan and g-C3N4 nanosheets. *Desalination* **2022**, *538*, 115933. [CrossRef]
43. Li, Y.; Jiang, Y.; Wang, T.-J.; Zhang, C.; Wang, H. Performance of fluoride electrosorption using micropore-dominant activated carbon as an electrode. *Sep. Purif. Technol.* **2017**, *172*, 415–421. [CrossRef]
44. Xu, B.; Wang, R.; Fan, Y.; Li, B.; Zhang, J.; Peng, F.; Du, Y.; Yang, W. Flexible self-supporting electrode for high removal performance of arsenic by capacitive deionization. *Sep. Purif. Technol.* **2022**, *299*, 121732. [CrossRef]
45. Gaikwad, M.S.; Balomajumder, C. Removal of Cr(VI) and fluoride by membrane capacitive deionization with nanoporous and microporous *Limonia acidissima* (wood apple) shell activated carbon electrode. *Sep. Purif. Technol.* **2018**, *195*, 305–313. [CrossRef]

Disclaimer/Publisher's Note: The statements, opinions and data contained in all publications are solely those of the individual author(s) and contributor(s) and not of MDPI and/or the editor(s). MDPI and/or the editor(s) disclaim responsibility for any injury to people or property resulting from any ideas, methods, instructions or products referred to in the content.

Article

Montmorillonite-Based Natural Adsorbent from Colombia for the Removal of Organic Pollutants from Water: Isotherms, Kinetics, Nature of Pollutants, and Matrix Effects

Marcela Paredes-Laverde [1,2], Diego F. Montaño [3] and Ricardo A. Torres-Palma [1,*]

1. Grupo de Investigación en Remediación Ambiental y Biocatálisis (GIRAB), Instituto de Química, Facultad de Ciencias Exactas y Naturales, Universidad de Antioquia UdeA, Calle 70 No. 52-21, Medellín 50011, Colombia
2. Grupo de Investigación Cuidados de la Salud e Imágenes Diagnósticas, Facultad de Ciencias de la Salud, Fundación Universitaria Navarra—Uninavarra, Calle 10 No. 6-41, Neiva 31008, Colombia
3. Grupo de Investigación Ciencia y Diseño de Materiales (CiDiMat), Departamento de Química, Facultad de Ciencias Básicas, Universidad de Pamplona, Km 1 Vía Bucaramanga Ciudad Universitaria, Calle 5 No. 3-93, Pamplona 26909, Colombia
* Correspondence: ricardo.torres@udea.edu.co; Tel.: +57-315-314-98-76

Citation: Paredes-Laverde, M.; Montaño, D.F.; Torres-Palma, R.A. Montmorillonite-Based Natural Adsorbent from Colombia for the Removal of Organic Pollutants from Water: Isotherms, Kinetics, Nature of Pollutants, and Matrix Effects. *Water* **2023**, *15*, 1046. https://doi.org/10.3390/w15061046

Academic Editor: Hai Nguyen Tran

Received: 18 January 2023
Revised: 18 February 2023
Accepted: 19 February 2023
Published: 9 March 2023

Copyright: © 2023 by the authors. Licensee MDPI, Basel, Switzerland. This article is an open access article distributed under the terms and conditions of the Creative Commons Attribution (CC BY) license (https://creativecommons.org/licenses/by/4.0/).

Abstract: The presence of dyes and pharmaceuticals in natural waters is a growing concern worldwide. To address this issue, the potential of montmorillonite (MMT), an abundant clay in Colombia, was assessed for the first time for the removal of various dyes (indigo carmine (IC), congo red (CR), methylene blue (MB) and crystal violet (CV)) and pharmaceuticals (levofloxacin and diclofenac) from water. Initially, the MMT was characterized. TGA and FTIR showed OH groups and water adsorbed onto MMT. XRD showed an interlayer spacing of 11.09 Å and a BET surface area of 82.5 m^2g^{-1}. SEM/EDS revealed a typical flake surface composed mainly of Si and O. Subsequently, the adsorbent capacity of MMT was evaluated for the removal of the pollutants. Adsorption isotherms showed a fit to the Langmuir model, which was confirmed by the Redlich–Peterson isotherm, indicating a monolayer-type adsorption. Furthermore, adsorption kinetics were best described by the pseudo-second-order model. Adsorption capacity (for dyes CV > MB > CR > IC) was associated with the attractive forces between the contaminants and MMT (PZC 2.6). Moreover, the findings evidenced that MMT can remove MB, CR, CV, and levofloxacin by electrostatic attractions and hydrogen bonding, while for IC and diclofenac only hydrogen bonding takes place. It was shown that MMT was most cost-effective at removing CV. Additionally, the material was able to be reused. Finally, the MMT efficiently removed CV in textile wastewater and levofloxacin in urine due to the positive charge of the pollutants and the low PZC of MMT.

Keywords: pharmaceuticals; dyes; clays; adsorption; isotherm; kinetics; water treatment

1. Introduction

Textile industries consume large quantities of water in dyeing processes, generating considerable amounts of colored wastewater with anionic and cationic dyes. Both types of dyes in water, even at low concentration levels, can cause dramatic decreases in light transmittance, thus greatly affecting the photosynthesis of aquatic plants and seriously damaging the ecological environment [1]. In fact, the presence of anionic and cationic dyes in rivers, lakes, ponds, and wastewater treatment plant effluents have been reported at concentrations at or above 0.01 µg L^{-1} [2], while in textile wastewaters, the presence of anionic and cationic dyes can reach up to 45,000 µg L^{-1} [3].

The presence of pharmaceuticals in water resources is also of great concern. Pharmaceuticals are stable compounds created to improve human health and promote human well-being. There have been increasing concerns about the presence of pharmaceuticals in the environment due to the negative environmental impact of such compounds, such as their associated toxicity and their main role in the proliferation of antibacterial resistance, even at very low concentrations [4]. Concentrations of pharmaceuticals from 0.006 µg L^{-1} to 50 µg L^{-1} have been reported in the influent and effluent of wastewaters [5], while in river water it is possible to find concentrations in the range of 8.62–590.6 µg L^{-1} [6]. In aqueous matrices such as urine, concentrations of pharmaceuticals from 1 µg L^{-1} to 6400 µg L^{-1} have been reported [7].

According to the literature, the most cost-effective alternatives to deal with the presence of organic pollutants in waters include membrane separation, flocculation, ion exchange, and adsorption [8]. Adsorption systems using clays present numerous advantages, such as a relatively short time of operation, easy application, and a cheap cost. In addition, clays have stone properties, are resistant to aggressive media, do not require additional purification after secondary use, can be employed in large quantities, and contain highly dispersed hydroaluminosilicates, which have proven potential to effectively remove a variety of pollutants [9].

Clays such as bentonite, kaolin, and montmorillonite have been shown to be effective at the removal of dyes and pharmaceuticals [8]. Montmorillonite (MMT) is of special interest because it is one of the main components of clay minerals. MMT's structure is that of an octahedron and tetrahedron (TOT), with two layers of tetrahedral silica sheets sandwiching one octahedral aluminum sheet, where the interlayer charge can be neutralized by the presence of cations such as sodium or calcium [10]. Additionally, MMT has a specific surface area of around 74.2 m^2 g^{-1}, which is much higher than the surface areas of other clays such as illite (11.2 m^2 g^{-1}) and kaolinite (15.3 m^2 g^{-1}) [11]. Thus, MMT could be a key inorganic adsorbent for the adsorption of pollutants from water. However, in spite of the relative abundance of MMT in Colombia, few reports have evaluated the ability of MMT from Colombia to remove a variety of organic pollutants, particularly dyes and pharmaceuticals, from water.

Therefore, the objective of this study is to evaluate, for the first time, the use of MMT from Colombia as a versatile natural adsorbent to remove pollutants with different natures and chemical structures from water. Common anionic dyes—such as indigo carmine (IC) and congo red (CR)—and cationic dyes, such as methylene blue (MB) and crystal violet (CV)—were selected for this study. In addition, two highly consumed pharmaceuticals—the antibiotic levofloxacin (zwitterionic structure) and the analgesic diclofenac (anionic structure)—were considered. For these purposes, MMT from Bugalagrande, Colombia, was obtained and properly characterized. Subsequently, the ability of MMT to remove IC, CR, MB, and CV from distilled water was tested. The effects of the adsorbent dose and the type of pollutant were then addressed. To better understand the results, the experimental data were adjusted to Langmuir, Freundlich, and Redlich—Peterson adsorption isotherms, and pseudo-first order and pseudo-second order kinetics models were assessed. Moreover, the most probable interactions of dyes containing MMT were proposed using FTIR analysis before and after the adsorption process. Subsequently, the economic costs of the process and the reuse of the MMT were studied. Finally, to evaluate the effects of a matrix using MMT, the dye with the highest percentage of adsorption in distilled water was studied in textile wastewater. To further validate the potential and versatility of the technology, the adsorption of the pharmaceuticals levofloxacin (LEV) and diclofenac (DCF) in both distilled water and urine (a very complex matrix) was investigated.

2. Materials and Methods

2.1. Reagents

Congo red (CR) ($C_{32}H_{22}N_6Na_2O_6S_2$), indigo carmine (IC) ($C_{16}H_8N_2Na_2O_8S_2$), methylene blue (MB) ($C_{16}H_{18}ClN_3S$), crystal violet (CV) ($C_{24}H_{28}N_3Cl$), levofloxacin (LEV) ($C_{18}H_{20}FN_3O_4$), diclofenac (DCF) ($C_{14}H_{11}Cl_2NO_2$), formic acid ($CH_2O_2$), acetonitrile ($C_2H_3N$), sodium chloride (NaCl), urea (CH_4N_2O), sodium dihydrogen phosphate (NaH_2PO_4), sodium sulfate (Na_2SO_4), ammonium chloride (NH_4Cl), potassium chloride (KCl), magnesium chloride six hydrate ($MgCl_2 \bullet 6H_2O$), calcium chloride two hydrate

($CaCl_2 \cdot 2H_2O$), and sodium hydroxide (NaOH), all of analytical grade, were supplied by Merck S. A. (Darmstadt, Germany). Other chemical reagents, such as sodium carbonate (Na_2CO_3), sodium bicarbonate ($NaHCO_3$), sulfuric acid (H_2SO_4), and starch were provided by Sigma Aldrich (St. Louis, MO, USA). All solutions were prepared in distilled water. Additionally, the montmorillonite clay (MMT) with a general chemical formula of $(Si_4)^{IV}(Al_4)^{VI}O_{10}(OH)_8$ used in this study was obtained from Bentocol (Bugalagrande, Valle del Cauca, Colombia).

2.2. MMT Preparation

50 g of MMT was added to 500 mL of distilled water to eliminate the quartz fraction. It was then vigorously stirred for 2 h in a Labscient ultrasound instrument (power 360 W), after which the aqueous suspension was vacuum filtered using a Roker pump (power 125 W) for 15 min, thereby generating a solid. Later, the clay was dried at 110 °C for 12 h in a drying oven from Memmert (power 1700 W) and passed through a sieve with a particle size < 200 µm. Finally, 45 g of MMT was obtained, which was stored in a glass container for its posterior use as the adsorbent.

2.3. Characterization of MMT

The MMT was characterized by thermogravimetric analysis (TGA), Fourier transform infrared spectroscopy (FTIR) and scanning electronic microscopy (SEM). The TGA spectra was obtained using a Q500 thermogravimetric analyzer from TA Instruments (New Castle, DE, USA), and the heating rate was 10 °C min^{-1} from room temperature up to 800 °C in air. The FTIR was recorded in the range of 4000–400 cm^{-1} using attenuated total reflectance—ATR (PerkinElmer, Waltham, MA, USA). SEM analysis was carried out using a FEI Quanta 250 microscope (OR, USA), coupled with an energy dispersive spectrometer (EDS) to identify the relative content of elements.

An X-ray diffraction (XRD) experiment was performed using an XPert PANalytical Empyrean diffractometer (EA Almelo, The Netherlands) at 40 mA and 40 kV with monochromatized Cu Kα radiation (λ of 1.540598 Å and 2θ = 0–60°). Furthermore, the interlayer spacing of the clay was determined through the integration of the peaks reported by the MMT in the XRD spectrum. Additionally, the point of zero charge (PZC) of the MMT was calculated using the solid addition method [12]. Finally, the specific surface area was determined by nitrogen physisorption applying the Brunauer-Emmett-Teller (BET) equation to the isotherm [13] using a Micrometrics ASAP 2020 analyzer.

2.4. Adsorption Experiments

Initially, solutions at concentrations of 5.74 mg L^{-1} of IC, 8.57 mg L^{-1} of CR, 3.93 mg L^{-1} of MB and 4.85 mg L^{-1} of CV, which represent a dye concentration of 1.23×10^{-2} mmol L^{-1}, were prepared. 100 mL of each solution was placed in contact with 0.02 g of MMT and stirred at 200 rpm. Samples were taken at different intervals of time for 120 min. Subsequently, the adsorbed amounts of IC, CR, MB and CV were calculated using Equation (1) [12]. Likewise, to understand the effects of MMT dose and pollutant structure (anionic and cationic dyes), all dyes (at 1.23×10^{-2} mmol L^{-1}) were evaluated using adsorbent doses of 0.2, 0.6, 1.0 and 2.0 g L^{-1}. The amount of dye adsorbed onto MMT was then calculated using Equation (1). All the experiments were carried out at the natural pH of the solutions (IC: 5.7; CR: 5.7; MB: 5.6; CV: 6.0). Once these experiments were performed, the dose of MMT that was best at removing the pollutants was selected and the effect of the matrices was evaluated. Aqueous matrices such as textile wastewater [14] and urine [15] were prepared in the laboratory in accordance with previous reports (See Supplementary Materials Table S1). Both matrices were doped with a concentration of 1.23×10^{-2} mmol L^{-1} of contaminant (in the order of mg L^{-1}). The textile wastewater was doped with 4.85 mg L^{-1} of CV, while the urine was doped with 4.44 mg L^{-1} of LEV or with 3.64 mg L^{-1} of DCF. Consequently, 100 mL of each of the doped matrices under continuous stirring were put in contact with 0.1 g of MMT, and samples were taken at different time intervals over a 60 min period. The same procedure was performed using distilled water as a control test. The removal of pollutants was determined by calculating the adsorption percentage (%) using Equation (2) [12].

$$\text{Adsorbed Amount} = \frac{C_0 - C}{m} V \qquad (1)$$

$$\text{Adsorption (\%)} = \frac{C_0 - C}{C_0} * 100 \qquad (2)$$

where C_0 and C (mg L^{-1}) are the liquid phase concentrations of the pollutant at initial and equilibrium time, respectively, V (L) is the volume of the solution and m (g) is the mass of MMT. In some cases, the *Adsorbed Amount* (mg g^{-1}) was converted to mmol g^{-1} using the molecular weight of the pollutants.

To determine the cost of the process, the commercial value of 1 kg of MMT, and the energy and water required in the preparation process were considered. Thus, the cost of using MMT for the removal of dyes was determined from the ratio of the total production cost of the clay with a q_m of CV, IC, MB and CR.

To evaluate the reuse potential of the material, the MMT used was put in contact with 20 mL of HNO$_3$ (1 M) and 20 mL of HCl (1%) in ethanol for a period of 5 h. Subsequently, the material was washed with distilled water and dried at 105 °C for 24 h. After this time, the material was stored to be used in the next adsorption cycle.

2.5. Equilibrium Isotherms

The removal kinetics of 1.23×10^{-2} mmol L^{-1} of IC (5.74 mg L^{-1}), CR (8.57 mg L^{-1}), MB (3.93 mg L^{-1}), and CV (4.85 mg L^{-1}) in distilled water using 0.2–2.0 g L^{-1} of MMT were adjusted to the Langmuir, Freundlich and Redlich-Peterson isotherms. The Langmuir isotherm assumes monolayer sorption onto a surface containing a finite number of sorption sites of uniform sorption strategies with no transmigration of sorbate in the surface plane [1]. The Langmuir isotherm was calculated using the following equation (Equation (3))[9]:

$$q_e = \frac{q_m K_L C_e}{1 + K_L C_e} \qquad (3)$$

Equation (3) can be rearranged into a linear form:

$$\frac{C_e}{q_e} = \frac{C_e}{q_m} + \frac{1}{K_L q_m} \qquad (4)$$

where C_e (mg L^{-1}) and q_e (mg g^{-1}) are the equilibrium concentration and the amount of adsorbate adsorbed per unit mass of adsorbent, respectively, and q_m (mg g^{-1} or mmol g^{-1}) is the amount of dye adsorbed per unit mass of adsorbent equivalent to the formation of a complete monolayer. The affinity constant, K_L (L mg^{-1}), is the equilibrium constant of the adsorption process. Thus, from a plot of C_e/q_e vs. C_e, q_m can be obtained from the slope and K_L from the intercept. The essential characteristics of the Langmuir isotherm can be expressed by a dimensionless equilibrium parameter (known as the separation factor), which is defined as follows:

$$R_L = \frac{1}{1 + K_L C_0} \qquad (5)$$

where the R_L value indicates the type of adsorption: irreversible ($R_L = 0$), favorable ($0 < R_L < 1$), linear ($R_L = 1$) or unfavorable ($R_L > 1$); and C_0 (mg L^{-1}) is the initial concentration of the metal.

The Freundlich model is an empirical equation based on sorption on heterogeneous surfaces or surfaces supporting sites of varied affinities. This was studied using Equation (6) [16].

$$q_e = K_F C_e^{1/n} \qquad (6)$$

where K_F and n are indicators of the adsorption capacity and the adsorption intensity, respectively. If the n value lies in the range between 1 and 10, favorable adsorption takes place [12]. Rearranging Equation (6) gives the following:

$$ln\, q_e = ln\, K_F + \frac{1}{n} ln\, C_e \tag{7}$$

The Redlich–Peterson equation is widely used as a compromise between the Langmuir and Freundlich systems. This model has three parameters. It incorporates the advantageous significance of both models and can be represented as follows in Equation (8) [17].

$$q_e = \frac{K_{RP} C_e}{1 + a_R C_e^\beta} \tag{8}$$

where K_{RP} (L g^{-1}) and a_R (L mg^{-1})$^\beta$ are the Redlich-Peterson isotherm constants and β is an exponent that lies between 0 and 1. Equation (8) can be linearized by taking logarithms of both the sides, as shown in Equation (9):

$$ln\left(K_{RP}\frac{C_e}{q_e} - 1\right) = ln\, a_R + \beta\, ln\, C_e \tag{9}$$

2.6. Adsorption Kinetics

Pseudo-first order and pseudo-second order kinetic models were calculated from the results obtained when removing 1.23×10^{-2} mmol L^{-1} of IC (5.74 mg L^{-1}), CR (8.57 mg L^{-1}), MB (3.93 mg L^{-1}), and CV (4.85 mg L^{-1}) from distilled water with a dose variation between 0.2–2.0 g L^{-1} of MMT.

The pseudo-first order kinetic model assumes that one active sorption site on the clay is adsorbing one pollutant molecule [1]. Using this model, the rates of adsorption of IC, CR, MB and CV on MMT were determined by the following equation [12]:

$$ln\,(q_e - q_t) = ln\, q_e - k_1 t \tag{10}$$

where, q_t is the amount of dye adsorbed (mg g^{-1}) at the time t (min), q_e is the amount of pollutant adsorbed at the equilibrium (mg g^{-1}), and k_1 is the equilibrium rate constant of pseudo-first order adsorption (min^{-1}). Thus, from Equation (10), $ln\,(q_e - q_t)$ vs. t was plotted, obtaining q_e from the intercept and k_1 from the slope.

The pseudo-second order equation assumes that two active sorption sites on the MMT are adsorbing one dye molecule [1]. The pseudo-second order model can be expressed in the following form [17]:

$$\frac{t}{q_t} = \frac{1}{k_2 q_e^2} + \frac{t}{q_e} \tag{11}$$

where k_2 is the pseudo-second order model rate constant of adsorption (mg g^{-1} min^{-1}). Thus, the plots of t/q_t vs. t, allow q_e and k_2 to be determined from the slope and the intercept, respectively.

2.7. Evaluation of Applicability of the Isotherm Models and Kinetics

The comparisons of the applicability of each isotherm model and kinetics were based on the linear coefficient correlation (R^2), the average percentage error (APE) (Equation (12)) and the normalized standard deviation (Δq, %) (Equation (13)) [18].

$$APE\,(\%) = \frac{\sum_{i=1}^{N}\left|\left(\frac{q_{exp} - q_{cal}}{q_{exp}}\right)\right|}{N} * 100 \tag{12}$$

$$\Delta q\ (\%) = 100\ \sqrt{\frac{\sum \left(\frac{q_{exp}-q_{cal}}{q_{exp}}\right)^2}{N-1}} \qquad (13)$$

where q_{exp} and q_{cal} are the experimental and calculated amounts of dye adsorbed at equilibrium, respectively, and N the number of measurements. As can be seen, the APE and Δq indicates the fit between the experimental and predicted values of the adsorbed amounts.

2.8. Analytical Techniques

The quantitative determination of IC, CR, MB and CV were done by UV visible spectroscopy in a UV5 Mettler Toledo spectrophotometer set at 611 nm [19], 498 nm, 661 nm and 589.5 nm [20], respectively. Pharmaceutical evolutions were observed by liquid chromatography using a Thermo Scientific UHPLC (Dionex Ultimate 3000) equipped with a diode array detector (DAD) and an Acclaim™ 120 RP C18 column (5 µm, 4.6 × 150 mm). LEV removal was monitored at 290 nm, using mobile phase formic acid/acetonitrile 85/15 (% v/v) and a flow of 1.0 mL min^{-1} in isocratic mode. In addition, DCF removal was observed at 260 nm, with a mobile phase of formic acid/acetonitrile 30/70 (% v/v) and flow isocratic mode of 0.5 mL min^{-1}. In all cases, the injection volume was 25 µL. All the experiments were conducted in at least duplicate.

3. Results and Discussion

3.1. MMT Characterization

The composition and thermal stability of the MMT was studied by TGA (Figure 1a). The TGA showed three events. In the first and second event, the weight losses below 250 °C can be attributed to the moisture and water from the interlayer spacing of the clay, respectively. The subsequent event occurred gradually between 250 °C and 700 °C, which can be correlated with dehydroxylation of MMT [21].

In addition, the presence of water in the interlayer spacing and hydroxyl groups in the MMT was confirmed by FTIR analysis (Figure 1b). A band at 3698 cm^{-1} showed the presence of Si-OH, while the band at 3620 cm^{-1} was associated with the Al-OH group [22]. In turn, the bands around 3425 cm^{-1} and 1635 cm^{-1} are assigned to the stretching and bending of the hydroxyl group in the absorbed water, respectively. Likewise, bands related with the Si-O bonds were observed at 1038 cm^{-1} for Si-O-Si, 520 cm^{-1} for Si-O-Al, and 463 cm^{-1} for Si-O-Fe [23]. Other bands at 910 cm^{-1} and 820 cm^{-1} suggest the presence of Al-Al-OH and Al-Mg-OH, respectively in the clay [24]. This series of chemical compounds present in the MMT was confirmed by EDS analysis (Table 1). The MMT showed a high content of O, Si, Al and Fe. Other elements such as C, N, Mg and Na were also reported.

On the other hand, the crystallographic structure of MMT was determined by XRD (Figure 1c). The XRD pattern of the clay shows different peaks at 2 theta: 7.97, 20.08, 28.34, 35.99, 58.59, which were accordingly indexed to the (001), (100), (005), (110) and (210) diffractions of MMT [23]. In this study, 7.97 (2 theta) is a typical diffraction peak of MMT corresponding to an interlayer spacing (d001) of 11.09 Å. In fact, the interlayer spacing found in this study was greater than for other single (5.278 Å [25], 9.8 Å, 10.3 Å [26]) and modified (6.5 Å [27]) MMTs reported in the literature. Additionally, a peak at 26.81 (2 theta) characteristic of quartz was observed.

According to the SEM (Figure 1d), the surface of MMT featured a typical flake-layered morphology [28]. In addition, the clay presented interesting values for total pore volume and average pore width in terms of the adsorption process. It showed a specific surface area (BET) (Table 1) higher than that reported for other natural clays such as Kaolin [29], beidellite and montmorillonites [30] from other regions.

According to the aforementioned results, the MMT from Colombia showed good characteristics of surface and chemical composition, which is promising for adsorption processes in water treatment. Thus, in the following sections, the MMT will be tested for the removal of both anionic and cationic dyes and pharmaceuticals present in water.

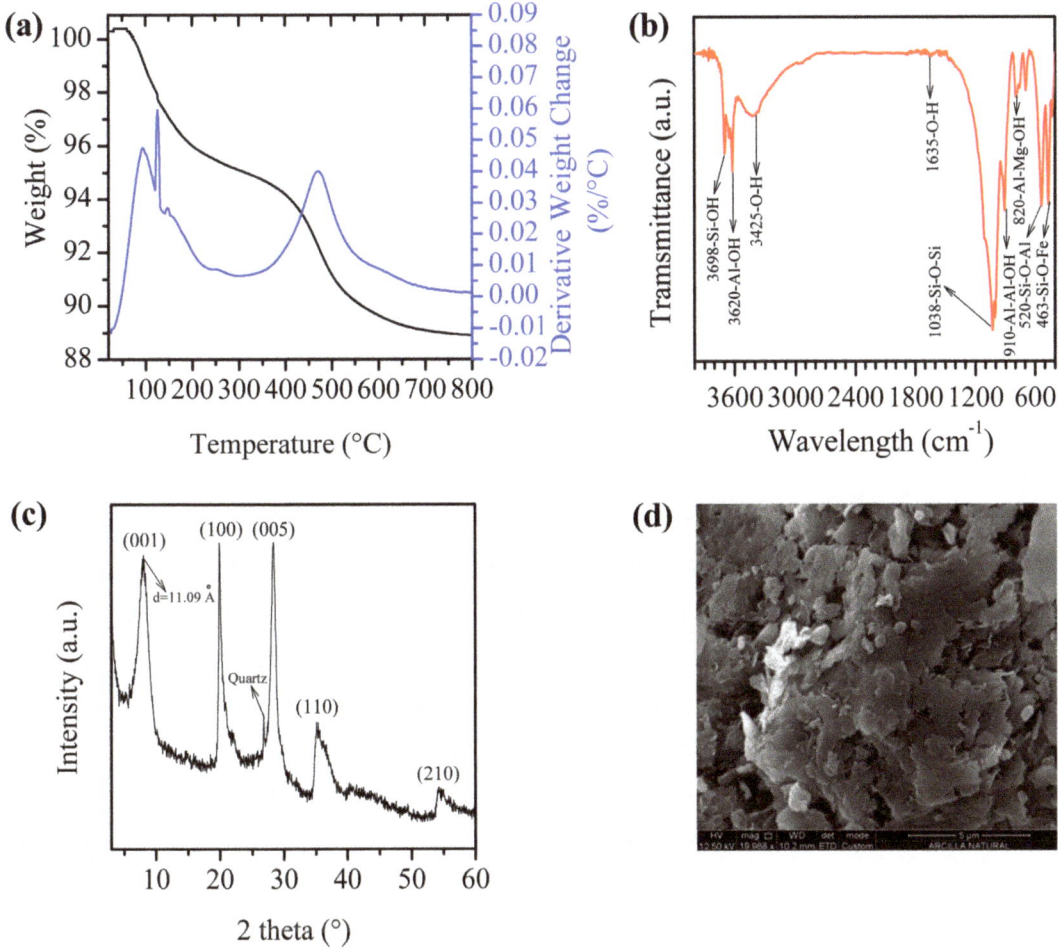

Figure 1. Physicochemical characterization of MMT. (**a**) TGA profile. Coordinates: Black color: Weight (%) vs. Temperature (°C); blue color: Derivative weight change (%/°C) vs. Temperature (°C); (**b**) FTIR spectra; (**c**) XRD analysis; (**d**) SEM micrograph.

Table 1. EDS analysis, specific surface area (BET) and PZC of MMT.

EDS Analysis			
Element	Wt (%)	Element	Wt (%)
C	3.25	Mg	2.19
N	0.76	Al	17.13
O	29.01	Si	36.50
Na	0.13	Fe	11.02
Nitrogen physisorption analysis			
Specific surface area (BET) 82.5 m^2 g^{-1} Total pore volume 0.004 cm^3 g^{-1} Average pore width 65.5 nm			
PZC			
2.6 ± 0.1			

3.2. Adsorption of Anionic and Cationic Dyes in Distilled Water Using MMT

The efficiency of MMT at absorbing both anionic (IC and CR) and cationic dyes (CV and MB) was evaluated, as shown in Figure 2. The results showed that MMT reached equilibrium in 60 min, following the order: CV > MB > CR > IC. The adsorption of the dyes using MMT can be attributed to the development of the surface area (BET), surface charge and the interlayer spacing of the natural clay [29]. The XRD analysis of MMT (Figure 1c) showed an inter-planar spacing of 11.09 Å. Therefore, due to the higher molecular size in terms of length, height or width of CV (11.37 Å × 8.70 Å × 10.25 Å) [31], MB (14.3 Å × 6.10 Å × 4.0 Å) [32], IC (17.71 Å × 8.03 Å × 6.16 Å) [33] and CR (26.2 Å × 7.40 Å × 4.30 Å) [32] none of the contaminants could enter the clay gaps. In this way, it can be assumed that there was no ion exchange between the pollutants and the Mg, Al, Si, Na and Fe ions present in the MMT as reported by the EDS analysis (Table 1). This suggests that the adsorption of the tested pollutants occurs via a different surface phenomenon.

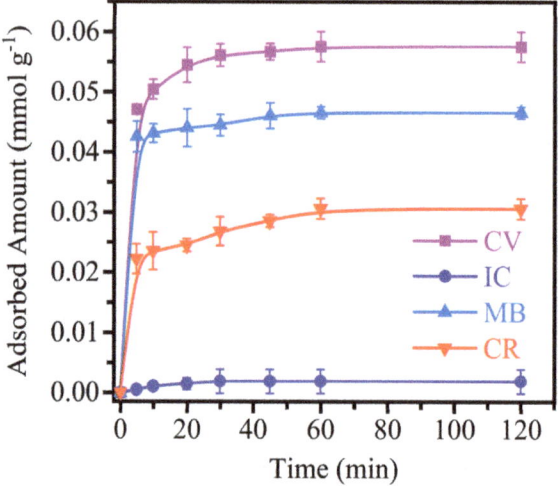

Figure 2. Amount of IC, CR, MB and CV in distilled water adsorbed onto MMT. Conditions: dye concentration 1.23×10^{-2} mmol L^{-1}, adsorbent dose 0.2 g L^{-1}, particle size <200 µm, pH: IC: 5.7; CR: 5.7; MB: 5.6; CV: 6.0, temperature 25 °C, stirring rate 200 rpm.

The PZC of the adsorbent was calculated (Figure S1) and is reported in Table 1. Considering the pH of the CV solution (6.0) and its pKa value (9.39) [34], the dye was positively charged (Figure S2), thus producing favorable electrostatic attraction with the negative charges of the material. In the case of MB, despite it also being a cationic dye, the MMT showed lower removal compared to the CV. This is because MB has a pKa of 5.6 [35] (Figure S3), which is similar to the pH of the solution. Therefore, 50% of the pollutant is positively charged, presenting an attraction with the MMT, while the other 50% of the molecule is in its neutral form, which limits the amount of MB adsorbed.

On the other hand, the adsorbent did not show significant adsorption of IC (Figure 2). This is probably due to the fact that the dye is negatively charged [36] (Figure S4) and electronic repulsions take place. Surprisingly, a significant amount of the other anionic dye (CR) was adsorbed onto the MMT. At a pH of 5.7, CR with a pKa of 4.1 [37] is found in its negative form with some protonated species (Figure S5). Therefore, a fraction of protonated CR molecules can experience attraction due to the negatively charged surface of the clay.

The adsorption results obtained for the four dyes are noteworthy. Nevertheless, the adsorbed amount of these pollutants can change depending on the type of dye and the variation of the adsorbent dose. The effect of these parameters is discussed below.

3.3. Effects of MMT Dose and Dye Type on the Amount of Adsorbed Organic Pollutants

The effect of the adsorbent dose on the amount of IC, CR, MB and CV adsorbed onto MMT is reported in Figure 3. The Figure shows that for each of the experiments, the amount of adsorbed dye per unit mass of adsorbent decreases with an increase in adsorbent dose. This is due to the concentration gradient between solute concentrations in the solution and the adsorbent surface. Thus, with increasing adsorbent dosage, the amount of dye adsorbed by unit weight of adsorbent is reduced, causing a decrease in adsorption capacity. However, there is an increase in the removal percentage of IC (from 3.2 to 16%), CR (from 50 to 88%), MB (from 76 to 91%), and CV (from 90 to 99%), with an increase in MMT dosage from 0.2 to 1 g L^{-1} (Figure S6). This can be attributed to the increased surface area and the availability of more adsorption sites [12]. However, at doses of MMT above 1 g L^{-1} there is an excess of clay, and the adsorption of pollutants does not increase. Therefore, under the conditions studied, a dose of 1 g L^{-1} will avoid incurring additional process costs.

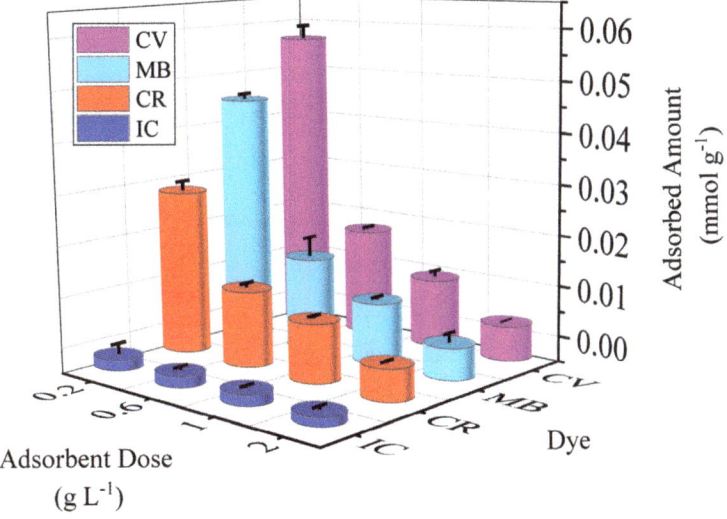

Figure 3. Effect of adsorbent dose on the adsorbed amount of IC, CR, MB and CV by MMT. Conditions: dye concentration 1.23 × 10^{-2} mmol L^{-1}, adsorbent dose 0.2–2 g L^{-1}, particle size < 200 μm, pH: IC: 5.7; CR: 6.5; MB: 5.6; CV: 7.3, temperature 25 °C, time 60 min, stirring rate 200 rpm.

Moreover, Figure 3 shows the effect of the dye type, where the adsorption of the cationic dyes was greater than that experienced by the anionic dyes, following the order of CV > MB > CR > IC. This is due to the affinity for the charges analyzed, considering the PZC of the MMT, the pKa of the pollutant and the pH of the dyes in solution, as detailed in Section 3.2.

Consequently, the amounts of dye removed by the different doses of clay indicate the type of adsorption onto MMT experienced by IC, CR, MB, and CV. To evaluate this phenomenon, the adsorption isotherms of the process are analyzed in the next session.

3.4. Study of Adsorption Isotherms

Several adsorption isotherm models (Langmuir, Freundlich and Redlich–Peterson) were used to analyze the equilibrium data obtained in the present study (Figure S7). Mathematical adjustments were made for the Langmuir, Freundlich and Redlich–Peterson models during the IC, CR, MB, CV adsorption (Figures S8, S9, S10 and S11, respectively) and the parameters are shown in Table 2. According to the results of the Langmuir model, the dye adsorption presented affinity with the binding sites (K_L), and showed a maximum capacity, q_m, of 18.2, 16.2, 14.1 and 1.07 mg g^{-1} for CV, CR, MB and IC, respectively. However, the q_m in mmol g^{-1} indicated that adsorption onto MMT followed the order: CV > MB > CR > IC, thus confirming the affinity of MMT for positives charges. Additionally, a separation factor R_L in a range between 0 and 1 was obtained for dye adsorption onto MMT, indicating that in all cases the adsorption process was favorable.

Table 2. Adsorption parameters for Langmuir, Freundlich and Redlich-Peterson isotherms for the removal of IC, CR, MB and CV from distilled water using MMT. Conditions: dye concentration 1.23×10^{-5} mmol L^{-1} (IC: 5.74 mg L^{-1}, CR: 8.57 mg L^{-1}, MB: 3.93 mg L^{-1} and CV: 4.85 mg L^{-1}), adsorbent dose 0.2–2 g L^{-1}, particle size < 200 μm, pH: IC: 5.7; CR: 6.5; MB: 5.6; CV: 7.3, time 60 min, temperature 25 °C, stirring rate 200 rpm.

Isotherm Models	Parameters	Dye			
		IC	CR	MB	CV
Langmuir	q_m (mg g^{-1})	1.07	16.2	14.1	18.2
	q_m (mmol g^{-1})	2.29×10^{-3}	2.33×10^{-2}	4.41×10^{-2}	4.62×10^{-2}
	K_L (L mg^{-1})	1.18	0.287	0.496	0.642
	K_L (L mmol^{-1})	550	200	159	253
	R_L	4.23×10^{-3}–4.25×10^{-4}	1.72×10^{-2}–1.74×10^{-3}	9.98×10^{-3}–1.01×10^{-3}	7.73×10^{-3}–7.78×10^{-4}
	R^2	0.9992	0.9909	0.9914	0.9999
	APE (%)	0.913	5.46	1.96	0.981
	Δq (%)	1.01	7.74	3.40	1.47
Freundlich	K_F (mg g^{-1})(L mg^{-1})$^{1/n}$	1.58	13.9	12.9	26.1
	n	4.44	1.55	0.719	0.688
	R^2	0.9885	0.9899	0.9870	0.9884
	APE (%)	3.51	7.44	4.70	5.21
	Δq (%)	5.43	11.48	6.34	7.37
Redlich-Peterson	K_{RP} (L g^{-1})	133	360	361	187
	a_R (L mg^{-1})$^\beta$	155	47.7	41.9	19.1
	β	0.95	0.91	0.89	0.95
	R^2	0.9990	0.9975	0.9997	0.9999
	APE (%)	0.715	0.825	0.426	0.261
	Δq (%)	0.984	1.35	0.640	0.554

Moreover, the adjustment of data to the Freundlich model is reported in Table 2, which shows adsorption capacity (K_F) values of 26.1, 13.9, 12.9 and 1.58 mg$^{1-1/n}$ L$^{1/n}$ g^{-1} for CV, CR, MB and IC, respectively. Only CV presents a n value in the range of 1–10, indicating favorable sorption for this pollutant.

However, the Langmuir isotherm model has a higher regression coefficient R^2 and a lower value of APE and Δq than the Freundlich model (Table 2), indicating that the Langmuir model has a better adjustment for the adsorption process of MMT with CV, CR, MB and IC. The results suggest monolayer adsorption for all the dyes onto the MMT surface.

Consequently, the Redlich-Peterson model was used to confirm the adjustment to the Langmuir model. The results of the Redlich-Peterson model are reported in Table 2. The model showed high values of R^2, and low values of APE and Δq for adsorption. Likewise, the values of β tend to 1 in all the experiments with dyes, which confirmed the best fit of the data with the Langmuir model and suggests a monolayer-type adsorption.

Table 3 shows the q_m values for the removal of IC, CR, MB and CV using MMT. In some cases, the clay presented a q_m greater than that reported by other adsorbents. However, there are also materials that presented a q_m greater than the value found in this investigation, possibly because these materials were rigorously washed with NaOH and/or subjected to high temperatures or doped with a metal, thus providing an improvement in the surface of the material, and consequently, an increase in its adsorbent capacity.

Table 3. Comparison of maximum adsorption capacity (q_m) of MMT for the removal of IC, CR, MB and CV against other adsorbents reported in the literature.

Adsorbed Dye	Adsorbent	q_m (mg g^{-1})	q_m (mmol g^{-1})	Reference
IC	Abrasive spherical materials made of rice husk ash	0.4	8.58×10^{-4}	[17]
	MMT	1.07	2.29×10^{-3}	Present Study
	Fe-zeolitic	32.8	7.04×10^{-2}	[38]
	Activated carbon with ZnCl$_2$ from rice husk	36.6	7.85×10^{-2}	[18]
CR	Algerian kaolin	5.94	8.53×10^{-3}	[39]
	MMT	16.2	2.33×10^{-2}	Present Study
	Copolymer of luffa cylindrica fibre	19.2	2.76×10^{-2}	[40]
	Palygorskite clay	51.2	7.35×10^{-2}	[41]
MB	Sawdust	7.84	2.45×10^{-2}	[16]
	Kaolinite	8.88	2.78×10^{-2}	[42]
	MMT	14.1	4.41×10^{-2}	Present Study
	Mesoporous Iraqi red kaolin clay	240	7.52×10^{-1}	[43]
CV	Composite of typha latifolia activated carbon	2.37	6.02×10^{-3}	[44]
	Nano mesocellular foam silica	6.60	1.68×10^{-2}	[45]
	MMT	18.2	4.62×10^{-2}	Present Study
	Tunisian Smectite Clay	86.5	2.20×10^{-1}	[46]

The MMT showed interesting q_m values for CV, CR, MB and IC. However, the speed with which each of these processes occur is still unknown. Therefore, in the following section, the adsorption kinetics of MMT for each of the dyes will be studied.

3.5. Study of Adsorption Kinetics

The kinetics of the process were determined using different doses of MMT (Figure S7). The adsorption kinetics of IC, CR, MB and CV were then modeled with the pseudo-first order and pseudo-second order kinetics (Figures S12, S13, S14 and S15, respectively) and the results are shown in Table 4. The pseudo-first order and pseudo-second order models showed that $q_{eq, cal}$ is similar to $q_{eq, exp}$. However, the pseudo-second order model presented R^2 values >0.99 and percentages of *APE* and Δq lower than those reported by the pseudo-first order model. This indicates that the pseudo-second order model is suitable to explain the results obtained. Thus, it is possible to assume that two active adsorption sites on MMT are able to adsorb one molecule of pollutant. In addition, Table 4 shows that increasing the initial MMT dose also increases the pseudo-second order kinetic rate constant, k_2. This confirms what was mentioned in Section 3.3, that at higher doses of clay, the active adsorption sites increase, and consequently, the equilibrium in the adsorption of all dyes is obtained more quickly.

The high correlation of the pseudo-second order kinetic equation with the kinetic data is consistent with previous results from the literature which deal with the adsorption of dyes onto clays, e.g., IC adsorption onto CdSe-Montmorillonite nanocomposites [47], CR adsorption onto carboxymethyl chitosan hybrid Montmorillonite [48], MB adsorption onto modified montmorillonite with magnetic nanoparticles and surfactants, and CV adsorption onto Na-Montmorillonite Nano Clay [49].

From the results obtained throughout this study, it can be inferred that even if MMT is able to remove negatively charged dyes, the material has a greater affinity, and is therefore more efficient, with positively charged dyes. In spite of this, the functional groups that participate in the adsorption process are unknown, and accordingly, this topic will be explored in next section.

Table 4. Kinetics of the removal of IC, CR, MB and CV from distilled water using MMT. Conditions: dye concentration 1.23×10^{-2} mmol L^{-1} (IC: 5.74 mg L^{-1}, CR: 8.57 mg L^{-1}, MB: 3.93 mg L^{-1} and CV: 4.85 mg L^{-1}), adsorbent dose 0.2–2 g L^{-1}, particle size < 200 μm, pH: IC: 5.7; CR: 6.5; MB: 5.6; CV: 7.3, temperature 25 °C, time 60 min, stirring rate 200 rpm.

Dye	Adsorbent Dose (g L^{-1})	Pseudo-First Order Model						
		$q_{eq, exp}$ (mg g^{-1})	$q_{eq, exp}$ (mmol g^{-1})	$q_{eq, cal}$ (mg g^{-1})	$q_{eq, cal}$ (mmol g^{-1})	k_1 (min^{-1})	R^2	APE (%)	Δq (%)
IC	0.2	0.919	1.97×10^{-3}	0.957	2.05×10^{-3}	0.0758	0.9912	6.05	9.61
	0.6	0.911	1.95×10^{-3}	0.942	2.02×10^{-3}	0.0407	0.9887	4.65	7.59
	1	0.909	1.95×10^{-3}	0.856	1.84×10^{-3}	0.146	0.9701	6.83	10.3
	2	0.722	1.55×10^{-3}	0.684	1.47×10^{-3}	0.160	0.9803	7.05	10.0
CR	0.2	21.4	3.07×10^{-2}	18.2	2.61×10^{-2}	6.01	0.9191	14.8	19.0
	0.6	9.81	1.41×10^{-2}	8.49	1.22×10^{-2}	17.8	0.9404	13.4	18.9
	1	7.56	1.09×10^{-2}	7.81	1.12×10^{-2}	0.209	0.9802	3.30	4.93
	2	4.06	5.83×10^{-3}	3.83	5.50×10^{-3}	7.75	0.9618	5.72	8.20
MB	0.2	14.9	4.66×10^{-2}	14.2	4.44×10^{-2}	4.15	0.9832	4.76	6.76
	0.6	5.50	1.72×10^{-2}	5.46	1.71×10^{-2}	0.809	0.9891	0.789	1.13
	1	3.59	1.12×10^{-2}	3.52	1.10×10^{-2}	18.8	0.9999	1.87	2.68
	2	1.87	5.85×10^{-3}	1.82	5.69×10^{-3}	4.14	0.9873	2.61	3.70
CV	0.2	22.7	5.76×10^{-2}	21.2	5.38×10^{-2}	18.3	0.9669	6.60	9.35
	0.6	7.87	2.00×10^{-2}	7.69	1.95×10^{-2}	0.371	0.9767	2.29	3.24
	1	4.96	1.26×10^{-2}	4.89	1.24×10^{-2}	17.96	0.9864	1.35	1.92
	2	2.51	6.37×10^{-3}	2.50	6.35×10^{-3}	0.741	0.9877	0.398	0.563

Table 4. Cont.

| Dye | Adsorbent Dose (g L^{-1}) | $q_{eq,exp}$ (mg g^{-1}) | $q_{eq,exp}$ (mmol g^{-1}) | Pseudo-Second Order Model |||||||
				$q_{eq,cal}$ (mg g^{-1})	$q_{eq,cal}$ (mmol g^{-1})	k_2 (mg g^{-1} min^{-1})	k_2 (mmol g^{-1} min^{-1})	R^2	APE (%)	Δq (%)
IC	0.2	0.919	1.97×10^{-3}	0.983	2.11×10^{-3}	0.0636	1.36×10^{-4}	0.9967	3.50	6.67
	0.6	0.911	1.95×10^{-3}	0.934	2.00×10^{-3}	0.0674	1.45×10^{-4}	0.9981	4.62	7.01
	1	0.909	1.95×10^{-3}	0.906	1.94×10^{-3}	0.207	4.44×10^{-4}	0.9919	2.59	4.16
	2	0.722	1.55×10^{-3}	0.767	1.64×10^{-3}	0.289	6.20×10^{-4}	0.9963	6.47	9.1
CR	0.2	21.4	3.07×10^{-2}	20.6	2.96×10^{-2}	0.0239	3.43×10^{-5}	0.9901	3.61	5.36
	0.6	9.81	1.41×10^{-2}	9.41	1.35×10^{-2}	0.0634	9.10×10^{-5}	0.9974	3.99	5.89
	1	7.56	1.09×10^{-2}	7.66	1.10×10^{-2}	0.0815	1.17×10^{-4}	0.9956	1.31	2.43
	2	4.06	5.83×10^{-3}	4.15	5.96×10^{-3}	0.180	2.58×10^{-4}	0.9968	2.15	3.35
MB	0.2	14.9	4.66×10^{-2}	14.7	4.60×10^{-2}	0.147	4.60×10^{-4}	0.9981	1.41	2.08
	0.6	5.50	1.72×10^{-2}	5.49	1.72×10^{-2}	1.39	4.35×10^{-3}	0.9999	0.244	0.403
	1	3.59	1.12×10^{-2}	3.52	1.10×10^{-2}	1.76	5.50×10^{-3}	0.9999	1.87	2.68
	2	1.87	5.85×10^{-3}	1.85	5.78×10^{-3}	1.80	5.63×10^{-3}	0.9999	1	1.45
CV	0.2	22.7	5.76×10^{-2}	21.2	5.38×10^{-2}	0.0515	1.31×10^{-4}	0.9973	6.60	9.35
	0.6	7.87	2.00×10^{-2}	7.99	2.03×10^{-2}	0.117	2.97×10^{-4}	0.9999	1.52	2.16
	1	4.96	1.26×10^{-2}	4.95	1.26×10^{-2}	1.28	3.25×10^{-3}	0.9999	0.145	0.223
	2	2.51	6.37×10^{-3}	2.52	6.40×10^{-3}	2.27	5.76×10^{-3}	0.9999	0.398	0.563

3.6. Identification of the Adsorption Sites on MMT for Removal of Dyes

Figure 4 shows the FTIR spectrum of MMT before and after the adsorption of each dye. For the case of IC (Figure 4a), after the removal of the pollutant, the FTIR spectrum of the clay shows changes in the bands at 3425 and 1635 cm^{-1}. These changes suggest that the adsorption reported for IC takes place via hydrogen bonds formed between water present in the clay and the hydrogen of the amine (Figure 5a). Moreover, the water on the MMT can form hydrogen bonds with the oxygens of the sulfonate group or with those of the cyclopentane in the dye (Figure 5a).

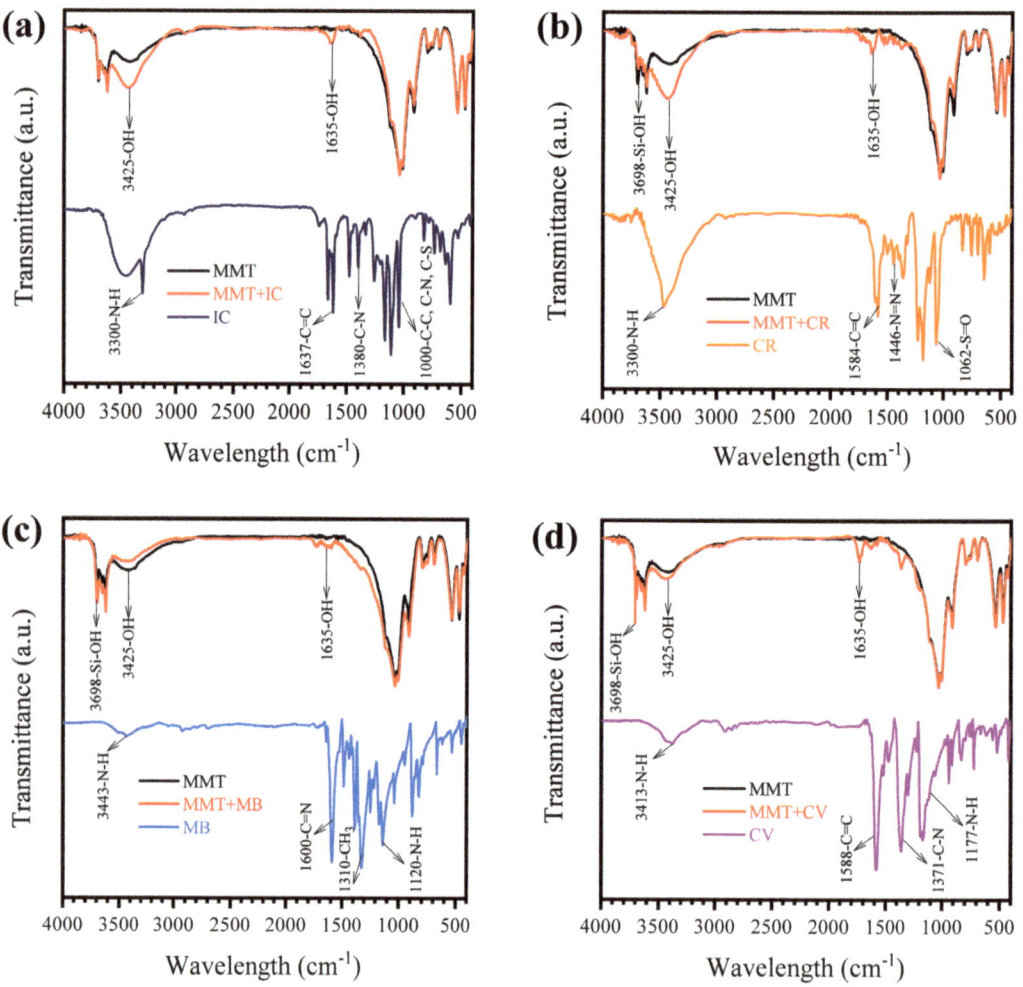

Figure 4. FTIR spectrum before and after the adsorption of the dyes in distilled water using MMT. (**a**) IC; (**b**) CR; (**c**) MB and (**d**) CV. Conditions: dye concentration 1.23×10^{-5} mol L^{-1}, adsorbent dose 1 g L^{-1}, particle size <200 µm. pH: IC: 5.7; CR: 6.5; MB: 5.6; CV: 7.3, temperature 25 °C, stirring rate 200 rpm.

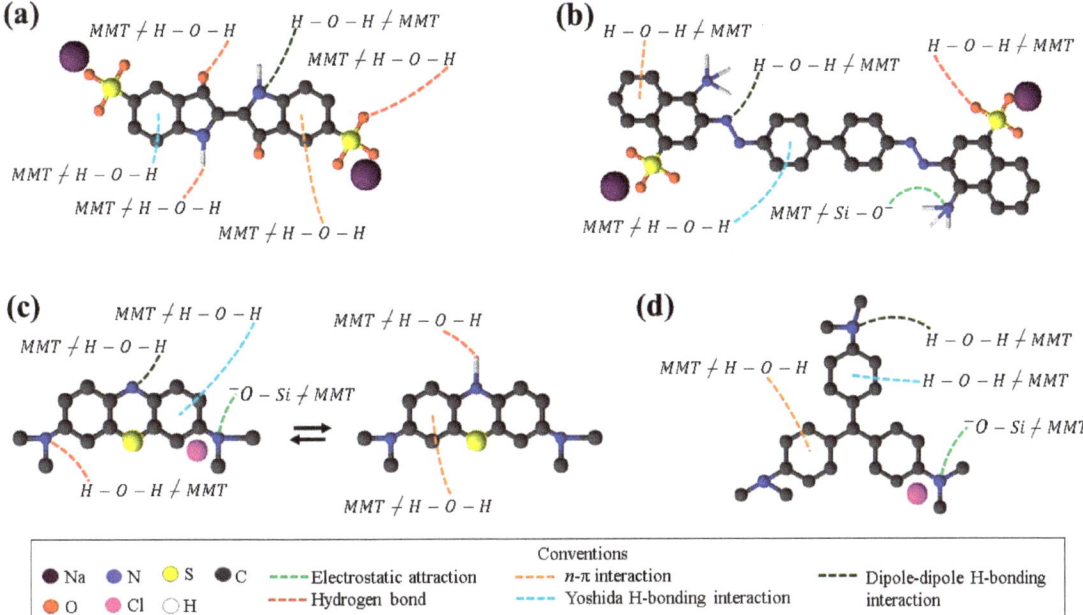

Figure 5. Most favorable interactions between MMT and the dyes during the adsorption process. (**a**) IC; (**b**) CR; (**c**) MB; (**d**) CV.

Figure 4b–d shows the FTIR spectra for CR, MB and CV, respectively, and for MMT before and after the adsorption process. For the three spectra of MMT after the adsorption process, changes are observed in the bands of the Si-OH groups at 3698 cm^{-1}. Similar to what was found for IC, there are variations in the bands at 3425 and 1635 cm^{-1} of the water present in the MMT. These changes allow the most probable adsorption sites for each of the three dyes (CR, MB and CV) to be proposed.

The most probable interaction sites between MMT and CR removal are shown in Figure 5b. Thus, CR can be adsorbed by electrostatic attractions between the positively charged amine present in its aromatic ring and the Si-O$^-$ group of the MMT, which is deprotonated at the pH of the experiment (pKa of 4.8) [50]. In addition, there may be hydrogen bonds between the oxygens of the sulfonate group of CR and H within the interlaminar water of the clay.

For MB, Figure 5c illustrates electrostatic attractions between the positively charged amine of the dye and the Si-O$^-$ group of the MMT. Even hydrogen bonds can occur between the N of the neutral amine and the H of the water in the clay. Furthermore, MB can be in its neutral form, thus it can form hydrogen bonds between the hydrogen of the amine of the aromatic ring and the water present in the MMT.

Figure 5d shows the CV adsorption, which may occur by electrostatic attractions between the Si-O$^-$ of the clay and the positively charged amine of the dye.

Finally, Figure 5a–d also shows additional favorable interactions between MMT and the dyes (CV, MB, CR and IC, respectively). In this way, the aromatic rings of the contaminants with O or H of the water in the clay can form n-π interactions and Yoshida H-bonding interactions, respectively. Other possible interactions such as dipole-dipole H-bonding may also occur between the N of the dyes and H in the MMT.

3.7. Cost and Reuse of MMT

To further evaluate the feasibility of the technology, an economic analysis of the performance of the system was carried out and the results are shown in Table 5. A low cost was reported for the production of MMT (5.76 USD kg^{-1}), which is associated with its abundance, easy acquisition and simple preparation process. The cost of using MMT to remove IC, CR, MB and CV was then calculated. The results indicate that MMT can remove more grams of CV than CR, MB and IC per USD. Therefore, MMT can be considered a low-cost and efficient alternative for CV removal.

Table 5. Cost of using MMT for the removal of dyes.

	Cost Production of MMT		
Items	Unit Price	Consumption	Cost (USD)
Power	0.160 USD kWh^{-1}	21.2 kW h^{-1}	3.39
MMT	2.13 USD kg^{-1}	1.11 kg	2.36
Water	0.950 USD m^{-3}	0.0110 m^3	0.0105
	Total Cost production MMT (USD kg^{-1})		5.76
	Cost of MMT for removal of dyes		
Dye	IC	CR	MB	CV
q_m (g kg^{-1})	0.922	16.2	14.1	18.2
(g dye/USD)	0.160	2.81	2.45	3.16

The positive results obtained for removing CV using MMT are a good reason to study other properties of the material, such as its reusability. Therefore, the reuse of MMT after CV removal was explored. Figure 6 shows that the adsorption percentage of the dye decreases when the MMT is reused. However, clay can be effective for up to 4 cycles, reaching 76% of pollutant removal in the last cycle. This additional advantage of MMT favors its use in complex water treatment.

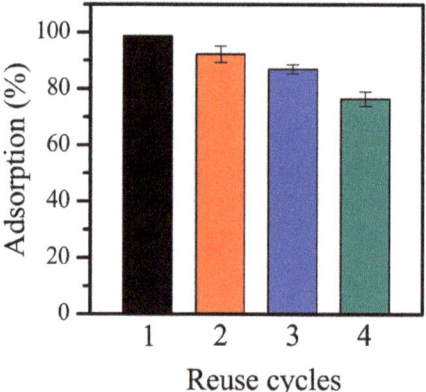

Figure 6. Reuse of MMT to remove CV from distilled water after 60 min. Conditions: CV: 4.85 mg L^{-1}, adsorbent dose 1 g L^{-1}, particle size < 200 μm, temperature 25 °C, stirring rate 200 rpm and pH of 7.3.

3.8. Complex Matrix Effect

The success of MMT in this study at removing organic contaminants present in complex waters would help persuade the industrial sector to use this material on a real scale. To further investigate this, the adsorption of CV (the dye with the best adsorption properties in this study) onto MMT in synthetic textile wastewater was explored. Considering the high levels of pharmaceutical contamination and that fact that urine represents a primary source of wastewater contamination, the ability of MMT to remove an antibiotic and an analgesic in simulated urine was also studied.

3.8.1. CV Adsorption in Textile Wastewater

Figure 7a shows the use of MMT to remove CV from both distilled water and textile wastewater. In distilled water, it is possible to remove practically 100% of the dye in less than 60 min because at the study pH (7.3) the contaminant is favored by electrostatic attractions and hydrogen bonds, as demonstrated in Section 3.6. However, in textile wastewater the adsorption decreased by around 14%, possibly due to the increase in pH (value of 10). Thus, CV was in its neutral form (Figure S2) and the pollutant adsorption took place via hydrogen bonds that formed between the N groups of the dye and the water in the interlayer spacing of the MMT. Furthermore, this small detrimental matrix effect may be associated with the components present in the textile wastewater [14]. For example, starch is an organic polymer with a chemical structure rich in H and O. It can therefore also adsorb onto the clay via hydrogen bonds. Consequently, competition occurs with the MMT, thus reducing the removal efficiency of the dye in this complex matrix.

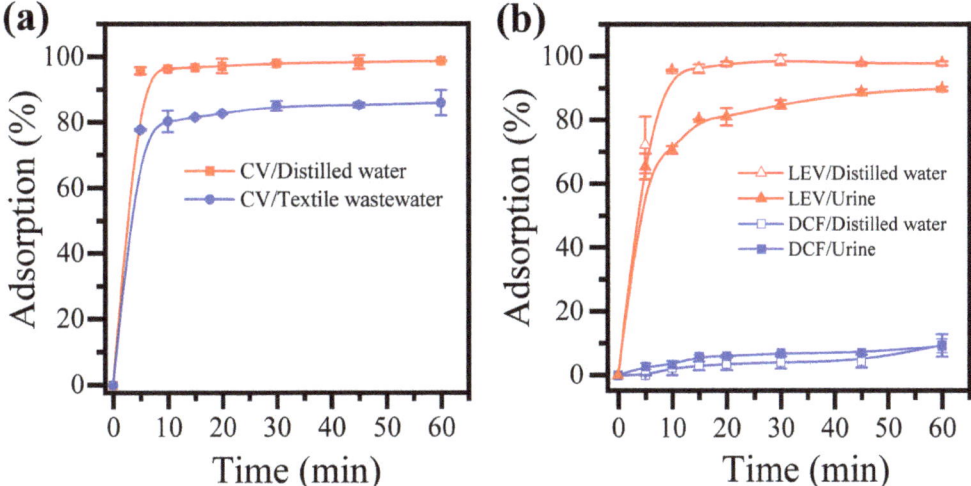

Figure 7. Complex matrix effect using MMT. (**a**) Adsorption of CV in distilled water and textile wastewater; (**b**) LEV and DCF adsorption in distilled water and urine. Conditions: pollutant concentration 1.23×10^{-2} mmol L^{-1} (CV: 4.85 mg L^{-1}, LEV: 4.44 mg L^{-1}, DCF: 3.64 mg L^{-1}), adsorbent dose 1 g L^{-1}, particle size <200 µm, temperature 25 °C, stirring rate 200 rpm. Experiments were performed at natural pH, for example 7.3 for CV in distilled water and 10 for CV in textile wastewater. For LEV and DCF in distilled water, the pH was 6.3 and 6.0 respectively, while in urine the pH was 6.0 for both pharmaceuticals.

3.8.2. Pharmaceutical Adsorption in Urine

The efficiency of MMT was evaluated for the adsorption of LEV and DCF present in urine and the results obtained are shown in Figure 7b. Even after 60 min, the adsorption of DCF in distilled water and in urine is less than 10%. DCF has a pKa of 4.15 [51]. Therefore, at the pH of the solution, in both matrices the pollutant will have most of the molecules negatively charged causing an electrostatic repulsion within the surface of the clay, which is negatively charged due to its low PZC value (Table 1). In contrast, a high and efficient removal of LEV was observed in distilled water after only 5 min (95%). The LEV had pKa values of 6.0 and 8.1 [52]. Therefore, at the experimental pH, LEV is in its zwitterionic form, which enables electrostatic attractions between its positive charge and the negatively charged surface of the material. Additionally, LEV adsorption in urine showed a decrease of 8%, possibly because this type of matrix is characterized by a high content of inorganic salts rich in ions [53] such as Na$^+$, K$^+$, Ca^{2+} and Mg^{2+} that can compete with the pollutant for the active sites present in the MMT. Thus, these results show the affinity of MMT in the removal of pharmaceuticals with a positive charge, even in a highly complex matrix such as urine.

4. Conclusions

This study analyzed the potential of MMT, a natural clay from Colombia, to remove dyes and pharmaceuticals from wastewater. The MMT presented a high content of hydroxyl groups and bounded water. Additionally, it had a high BET surface area (82.5 m^2 g^{-1}) and wide interlaminar spacing (11.09 Å), meaning that it had greater adsorption characteristics than other Montmorillonite clays.

At the natural pH of the experiments, MMT efficiently removed pollutants that had positive charges, such as the CV and MB dyes and the levofloxacin antibiotic. This was due to the favorable attraction between the positive charges of the pollutants and the negative surface of the clay (PZC 2.6).

The evaluation of the adsorbent dose showed that the increase in the adsorbent concentration increased the MMT sites and improved the removal of the pollutants. Furthermore, the equilibrium data for the adsorption of IC, CR, MB and CV onto MMT adjusted best to the Langmuir model, indicating that the pollutants form a monolayer on the surface of the clay. It is worth noting that the q_m for CV elimination (18.2 mg g^{-1}) was even higher than that found for other materials such as composites of typha latifolia activated carbon (2.37 mg g^{-1}) and nano mesocellular foam silica (6.60 mg g^{-1}).

In addition, the adsorption kinetics fitted well with the pseudo-second order kinetic model, indicating that one pollutant molecule is adsorbed onto two active adsorption sites of MMT.

The removal of CV from textile wastewater using MMT was negatively affected to some extent, probably due to the presence of starch, while LEV adsorption in urine decreased slightly, maybe due to the presence of cations in the matrix. In addition, DCF removal was not possible due to the anionic nature of this pharmaceutical. According to the aforementioned results, MMT proved to be an effective and low-cost natural adsorbent with good reuse characteristics for the efficient removal of cationic organic pollutants, which suggests it is suitable for full-scale applications.

Supplementary Materials: The following supporting information can be downloaded at: https://www.mdpi.com/article/10.3390/w15061046/s1, Table S1: Chemical composition of fresh urine and textile wastewater; Figure S1: Determination of PZC value of MMT; Figure S2: CV Structure; Figure S3: MB Structure; Figure S4: IC Structure; Figure S5: CR Structure; Figure S6: Effect of MMT dose on the adsorption percentage of CV, MB, CR and IC after 60 min; Figure S7: IC, CR, MB and CV removal in distilled water using different doses of MMT; Figure S8: Adsorption isotherms for IC removal in distilled water using MMT as an adsorbent. (a) Langmuir model; (b) Freundlich model; (c) Redlich-Peterson model; Figure S9: Adsorption isotherms for CR removal in distilled water using MMT as an adsorbent. (a) Langmuir model; (b) Freundlich model; (c) Redlich-Peterson model; Figure S10: Adsorption isotherms for MB removal in distilled water using MMT as an adsorbent. (a) Langmuir model; (b) Freundlich model; (c) Redlich-Peterson model; Figure S11: Adsorption isotherms for CV removal in distilled water using MMT as an adsorbent. (a) Langmuir model; (b) Freundlich model; (c) Redlich-Peterson model; Figure S12: Kinetics of pseudo-first order model and pseudo-second order model for IC removal in distilled water using MMT as an adsorbent. (a) 0.2 g L^{-1}; (b) 0.6 g L^{-1}; (c) 1 g L^{-1}; (d) 2 g L^{-1}; Figure S13: Kinetics of pseudo-first order model and pseudo-second order model for CR removal in distilled water using MMT as an adsorbent. (a) 0.2 g L^{-1}; (b) 0.6 g L^{-1}; (c) 1 g L^{-1}; (d) 2 g L^{-1}; Figure S14: Kinetics of pseudo-first order model and pseudo-second order model for MB removal in distilled water using MMT as an adsorbent. (a) 0.2 g L^{-1}; (b) 0.6 g L^{-1}; (c) 1 g L^{-1}; (d) 2 g L^{-1}; Figure S15: Kinetics of pseudo-first order model and pseudo-second order model for CV removal in distilled water using MMT as an adsorbent. (a) 0.2 g L^{-1}; (b) 0.6 g L^{-1}; (c) 1 g L^{-1}; (d) 2 g L^{-1}.

Author Contributions: Conceptualization, M.P.-L., D.F.M. and R.A.T.-P.; methodology, M.P.-L. and D.F.M.; validation, M.P.-L.; formal analysis, M.P.-L.; investigation, M.P.-L.; resources, D.F.M. and R.A.T.-P.; writing—original draft preparation, M.P.-L.; writing—review and editing, M.P.-L., D.F.M.

and R.A.T.-P.; visualization, M.P.-L., D.F.M. and R.A.T.-P.; supervision, R.A.T.-P.; project administration, M.P.-L. and R.A.T.-P.; funding acquisition, R.A.T.-P. All authors have read and agreed to the published version of the manuscript.

Funding: This research received no external funding.

Data Availability Statement: Not applicable.

Acknowledgments: The authors wish to thank MINCIENCIAS for project No. 1115-852-69594 "Program to improve the quality of irrigation water through the elimination of emerging contaminants by advanced oxidation processes (PRO-CEC-AGUA)". M. Paredes-Laverde would like to thank the Gobernación del Departamento del Huila and MINCIENCIAS for her scholarship through the program "Becas de Excelencia Doctoral del Bicentenario—Primer Corte".

Conflicts of Interest: The authors declare no conflict of interest.

References

1. Salleh, M.A.M.; Mahmoud, D.K.; Karim, W.A.W.A.; Idris, A. Cationic and Anionic Dye Adsorption by Agricultural Solid Wastes: A Comprehensive Review. *Desalination* **2011**, *280*, 1–13. [CrossRef]
2. Tkaczyk-Wlizło, A.; Mitrowska, K.; Błądek, T. Quantification of Twenty Pharmacologically Active Dyes in Water Samples Using UPLC-MS/MS. *Heliyon* **2022**, *8*, e09331. [CrossRef]
3. Sivakumar, D. Role of Lemna Minor Lin. in Treating the Textile Industry Wastewater. *Int. J. Fash. Text. Eng.* **2014**, *8*, 208–212. [CrossRef]
4. Luján-Facundo, M.J.; Iborra-Clar, M.I.; Mendoza-Roca, J.A.; Alcaina-Miranda, M.I. Pharmaceutical Compounds Removal by Adsorption with Commercial and Reused Carbon Coming from a Drinking Water Treatment Plant. *J. Clean. Prod.* **2019**, *238*, 117866. [CrossRef]
5. Botero-Coy, A.M.; Martínez-Pachón, D.; Boix, C.; Rincón, R.J.; Castillo, N.; Arias-marín, L.P.; Losada, L.M.-; Torres-Palma, R.; Moncayo-Lasso, A.; Hérnandez, F. An Investigation into the Occurrence and Removal of Pharmaceuticals in Colombian Wastewater. *Sci. Total Environ.* **2018**, *642*, 842–853. [CrossRef]
6. Omotola, E.O.; Olatunji, O.S. Quantification of Selected Pharmaceutical Compounds in Water Using Liquid Chromatography-Electrospray Ionisation Mass Spectrometry (LC-ESI-MS). *Heliyon* **2020**, *6*, e05787. [CrossRef]
7. Bischel, H.N.; Özel Duygan, B.D.; Strande, L.; McArdell, C.S.; Udert, K.M.; Kohn, T. Pathogens and Pharmaceuticals in Source-Separated Urine in EThekwini, South Africa. *Water Res.* **2015**, *85*, 57–65. [CrossRef]
8. Ewis, D.; Ba-Abbad, M.M.; Benamor, A.; El-Naas, M.H. Adsorption of Organic Water Pollutants by Clays and Clay Minerals Composites: A Comprehensive Review. *Appl. Clay Sci.* **2022**, *229*, 106686. [CrossRef]
9. Rakhym, A.B.; Seilkhanova, G.A.; Kurmanbayeva, T.S. Adsorption of Lead (II) Ions from Water Solutions with Natural Zeolite and Chamotte Clay. *Mater. Sci. Eng.* **2020**, *31*, 482–485. [CrossRef]
10. Vidal, C.B.; dos Santos, A.B.; do Nascimento, R.F.; Bandosz, T.J. Reactive Adsorption of Pharmaceuticals on Tin Oxide Pillared Montmorillonite: Effect of Visible Light Exposure. *Chem. Eng. J.* **2015**, *259*, 865–875. [CrossRef]
11. Wang, L.; Wang, D.; Cai, C.; Li, N.; Zhang, L.; Yang, M. Effect of Water Occupancy on the Excess Adsorption of Methane in Montmorillonites. *J. Nat. Gas Sci. Eng.* **2020**, *80*, 103393. [CrossRef]
12. Guechi, E.-K.; Hamdaoui, O. Biosorption of Methylene Blue from Aqueous Solution by Potato (*Solanum tuberosum*) Peel: Equilibrium Modelling, Kinetic, and Thermodynamic Studies. *Desalin. Water Treat.* **2015**, *57*, 10270–10285. [CrossRef]
13. Legras, A.; Kondor, A.; Heitzmann, M.T.; Truss, R.W. Inverse Gas Chromatography for Natural Fibre Characterisation: Identification of the Critical Parameters to Determine the Brunauer—Emmett—Teller Specific Surface Area. *J. Chromatogr. A* **2015**, *1425*, 273–279. [CrossRef]
14. Pillai, I.M.S.; Gupta, A.K. Performance Analysis of a Continuous Serpentine Flow Reactor for Electrochemical Oxidation of Synthetic and Real Textile Wastewater: Energy Consumption, Mass Transfer Coefficient and Economic Analysis. *J. Environ. Manag.* **2017**, *193*, 524–531. [CrossRef]
15. Jojoa-Sierra, D.; Silva-Agredo, J.; Herrera-Calderon, E.; Torres-Palma, R. Elimination of the Antibiotic Norfloxacin in Municipal Wastewater, Urine and Seawater by Electrochemical Oxidation on IrO_2 Anodes. *Sci. Total Environ.* **2016**, *575*, 1228–1238. [CrossRef]
16. Bouyahia, C.; Rahmani, M.; Bensemlali, M.; El Hajjaji, S.; Slaoui, M.; Bencheikh, I.; Azoulay, K.; Labjar, N. Influence of Extraction Techniques on the Adsorption Capacity of Methylene Blue on Sawdust: Optimization by Full Factorial Design. *Mater. Sci. Energy Technol.* **2023**, *6*, 114–123. [CrossRef]
17. Arenas, C.N.; Vasco, A.; Betancur, M.; Martínez, J.D. Removal of Indigo Carmine (IC) from Aqueous Solution by Adsorption through Abrasive Spherical Materials Made of Rice Husk Ash (RHA). *Process. Saf. Environ. Prot.* **2017**, *106*, 224–238. [CrossRef]
18. Paredes-Laverde, M.; Salamanca, M.; Diaz-Corrales, J.D.; Flórez, E.; Silva-Agredo, J.; Torres-Palma, R.A. Understanding the Removal of an Anionic Dye in Textile Wastewaters by Adsorption on $ZnCl_2$ Activated Carbons from Rice and Coffee Husk Wastes: A Combined Experimental and Theoretical Study. *J. Environ. Chem. Eng.* **2021**, *9*, 105685. [CrossRef]
19. Pan, X.; Qin, X.; Zhang, Q.; Ge, Y.; Ke, H.; Cheng, G. N- and S-Rich Covalent Organic Framework for Highly Efficient Removal of Indigo Carmine and Reversible Iodine Capture. *Microporous Mesoporous Mater.* **2020**, *296*, 109990. [CrossRef]

20. Bentahar, S.; Dbik, A.; El Khomri, M.; El Messaoudi, N.; Lacherai, A. Adsorption of Methylene Blue, Crystal Violet and Congo Red from Binary and Ternary Systems with Natural Clay: Kinetic, Isotherm, and Thermodynamic. *J. Environ. Chem. Eng.* **2017**, *5*, 5921–5932. [CrossRef]
21. Lang, W.; Yang, Q.; Song, X.; Yin, M.; Zhou, L. Cu Nanoparticles Immobilized on Montmorillonite by Biquaternary Ammonium Salts: A Highly Active and Stable Heterogeneous Catalyst for Cascade Sequence to Indole-2-Carboxylic Esters. *RSC Adv.* **2017**, *7*, 13754–13759. [CrossRef]
22. Patel, H.A.; Somani, R.S.; Bajaj, H.C.; Jasra, R.V. Nanoclays for Polymer Nanocomposites, Paints, Inks, Greases and Cosmetics Formulations, Drug Delivery Vehicle and Waste Water Treatment. *Bull. Mater. Sci.* **2006**, *29*, 133–145. [CrossRef]
23. Zhang, B.; Zhang, T.; Zhang, Z.; Xie, M. Hydrothermal Synthesis of a Graphene/Magnetite/Montmorillonite Nanocomposite and Its Ultrasonically Assisted Methylene Blue Adsorption. *J. Mater. Sci.* **2019**, *54*, 11037–11055. [CrossRef]
24. Samudrala, S.P.; Kandasamy, S.; Bhattacharya, S. Turning Biodiesel Waste Glycerol into 1,3-Propanediol: Catalytic Performance of Sulphuric Acid-Activated Montmorillonite Supported Platinum Catalysts in Glycerol Hydrogenolysis. *Sci. Rep.* **2018**, *8*, 7484. [CrossRef]
25. Martins, M.G.; Martins, D.O.T.A.; De Carvalho, B.L.C.; Mercante, L.A.; Soriano, S.; Andruh, M.; Vieira, M.D.; Vaz, M.G.F. Synthesis and Characterization of Montmorillonite Clay Intercalated with Molecular Magnetic Compounds. *J. Solid State Chem.* **2015**, *228*, 99–104. [CrossRef]
26. Yi, D.; Yang, H.; Zhao, M.; Huang, L.; Camino, G.; Frache, A.; Yang, R. A Novel, Low Surface Charge Density, Anionically Modified Montmorillonite for Polymer Nanocomposites. *RSC Adv.* **2017**, *7*, 5980–5988. [CrossRef]
27. Li, X.; Peng, K. $MoSe_2$/Montmorillonite Composite Nanosheets: Hydrothermal Synthesis, Structural Characteristics, and Enhanced Photocatalytic Activity. *Minerals* **2018**, *8*, 268. [CrossRef]
28. Zhang, Q.; Zhang, Y.; Liu, S.; Wu, Y.; Zhou, Q.; Zhang, Y.; Zheng, X.; Han, Y.; Xie, C.; Liu, N. Adsorption of Deoxynivalenol by Pillared Montmorillonite. *Food Chem.* **2021**, *343*, 128391. [CrossRef]
29. Panda, A.K.; Mishra, B.G.; Mishra, D.K.; Singh, R.K. Effect of Sulphuric Acid Treatment on the Physico-Chemical Characteristics of Kaolin Clay. *Colloids Surf. A Physicochem. Eng. Asp.* **2010**, *363*, 98–104. [CrossRef]
30. Dogan, M.; Umran Dogan, A.; Irem Yesilyurt, F.; Alaygut, D.; Buckner, I.; Wurster, D.E. Baseline Studies of the Clay Minerals Society Special Clays: Specific Suerface Area by the Brunauer Emmett Teller (BET) Method. *Clays Clay Miner.* **2007**, *55*, 534–541. [CrossRef]
31. Abbasi, A.R.; Karimi, M.; Daasbjerg, K. Efficient Removal of Crystal Violet and Methylene Blue from Wastewater by Ultrasound Nanoparticles Cu-MOF in Comparison with Mechanosynthesis Method. *Ultrason. Sonochem.* **2017**, *37*, 182–191. [CrossRef]
32. Szlachta, M.; Wójtowicz, P. Adsorption of Methylene Blue and Congo Red from Aqueous Solution by Activated Carbon and Carbon Nanotubes. *Water Sci. Technol.* **2013**, *68*, 2240–2248. [CrossRef]
33. Yao, M.; Kuratani, K.; Kojima, T.; Takeichi, N.; Senoh, H.; Kiyobayashi, T. Indigo Carmine: An Organic Crystal as a Positive-Electrode Material for Rechargeable Sodium Batteries. *Sci. Rep.* **2014**, *4*, 3650. [CrossRef]
34. Adams, E. The Antibacterial Action of Crystal Violet. *J. Pharm. Pharmacol.* **1967**, *19*, 821–826. [CrossRef]
35. Dotto, G.L.; Santos, J.M.N.; Rodrigues, I.L.; Rosa, R.; Pavan, F.A.; Lima, E.C. Adsorption of Methylene Blue by Ultrasonic Surface Modified Chitin. *J. Colloid Interface Sci.* **2015**, *446*, 133–140. [CrossRef]
36. Adel, M.; Ahmed, M.A.; Mohamed, A.A. Effective Removal of Indigo Carmine Dye from Wastewaters by Adsorption onto Mesoporous Magnesium Ferrite Nanoparticles. *Environ. Nanotechnol. Monit. Manag.* **2021**, *16*, 100550. [CrossRef]
37. Wang, H.; Luo, W.; Guo, R.; Li, D.; Xue, B. Effective Adsorption of Congo Red Dye by Magnetic Chitosan Prepared by Solvent-Free Ball Milling. *Mater. Chem. Phys.* **2022**, *292*, 126857. [CrossRef]
38. Gutiérrez-Segura, E.; Solache-Ríos, M.; Colín-Cruz, A. Sorption of Indigo Carmine by a Fe-Zeolitic Tuff and Carbonaceous Material from Pyrolyzed Sewage Sludge. *J. Hazard. Mater.* **2009**, *170*, 1227–1235. [CrossRef]
39. Meroufel, B.; Benali, O.; Benyahia, M.; Benmoussa, Y.; Zenasni, M.A. Adsorptive Removal of Anionic Dye from Aqueous Solutions by Algerian Kaolin: Characteristics, Isotherm, Kinetic and Thermodynamic Studies. *J. Mater. Environ. Sci.* **2013**, *4*, 482–491.
40. Pathania, D.; Sharma, A.; Sethi, V. Microwave Induced Graft Copolymerization of Binary Monomers onto Luffa Cylindrica Fiber: Removal of Congo Red. *Procedia Eng.* **2017**, *200*, 408–415. [CrossRef]
41. Dong, W.; Lu, Y.; Wang, W.; Zong, L.; Zhu, Y.; Kang, Y.; Wang, A. A New Route to Fabricate High-Efficient Porous Silicate Adsorbents by Simultaneous Inorganic-Organic Functionalization of Low-Grade Palygorskite Clay for Removal of Congo Red. *Microporous Mesoporous Mater.* **2019**, *277*, 267–276. [CrossRef]
42. Ghosh, D.; Bhattacharyya, K.G. Adsorption of Methylene Blue on Kaolinite. *Appl. Clay Sci.* **2002**, *20*, 295–300. [CrossRef]
43. Jawad, A.H.; Abdulhameed, A.S. Mesoporous Iraqi Red Kaolin Clay as an Efficient Adsorbent for Methylene Blue Dye: Adsorption Kinetic, Isotherm and Mechanism Study. *Surf. Interfaces* **2020**, *18*, 100422. [CrossRef]
44. Jayasantha Kumari, H.; Krishnamoorthy, P.; Arumugam, T.K.; Radhakrishnan, S.; Vasudevan, D. An Efficient Removal of Crystal Violet Dye from Waste Water by Adsorption onto TLAC/Chitosan Composite: A Novel Low Cost Adsorbent. *Int. J. Biol. Macromol.* **2017**, *96*, 324–333. [CrossRef]
45. Zhai, Q. Studies of Adsorption of Crystal Violet from Aqueous Solution by Nano Mesocellular Foam Silica: Process Equilibrium, Kinetic, Isotherm, and Thermodynamic Studies. *Water Sci. Technol.* **2020**, *81*, 2092–2108. [CrossRef]

46. Hamza, W.; Dammak, N.; Hadjltaief, H.B.; Eloussaief, M.; Benzina, M. Sono-Assisted Adsorption of Cristal Violet Dye onto Tunisian Smectite Clay: Characterization, Kinetics and Adsorption Isotherms. *Ecotoxicol. Environ. Saf.* **2018**, *163*, 365–371. [CrossRef]
47. Chikate, R.C.; Kadu, B.S. Improved Photocatalytic Activity of CdSe-Nanocomposites: Effect of Montmorillonite Support towards Efficient Removal of Indigo Carmine. *Spectrochim. Acta—Part A Mol. Biomol. Spectrosc.* **2014**, *124*, 138–147. [CrossRef]
48. Zhang, H.; Ma, J.; Wang, F.; Chu, Y.; Yang, L.; Xia, M. Mechanism of Carboxymethyl Chitosan Hybrid Montmorillonite and Adsorption of Pb (II) and Congo Red by CMC-MMT Organic-Inorganic Hybrid Composite. *Int. J. Biol. Macromol.* **2020**, *149*, 1161–1169. [CrossRef]
49. El Haouti, R.; Ouachtak, H.; El Guerdaoui, A.; Amedlous, A.; Amaterz, E.; Haounati, R.; Addid, A.A.; Akbal, F.; El Alem, N.; Taha, M.L. Cationic Dyes Adsorption by Na-Montmorillonite Nano Clay: Experimental Study Combined with a Theoretical Investigation Using DFT- Based Descriptors and Molecular Dynamics Simulations. *J. Mol. Liq.* **2019**, *290*, 111139. [CrossRef]
50. Liu, X.; Cheng, J.; Lu, X.; Wang, R. Surface Acidity of Quartz: Understanding the Crystallographic Control. *Phys. Chem. Chem. Phys.* **2014**, *16*, 26909–26916. [CrossRef]
51. Quesada, H.B.; Baptista, A.T.A.; Cusioli, L.F.; Seibert, D.; Bezerra, C.; de Oliveira Bezerra, C.; Bergamasco, R. Surface Water Pollution by Pharmaceuticals and an Alternative of Removal by Low-Cost Adsorbents: A Review. *Chemosphere* **2019**, *222*, 766–780. [CrossRef]
52. Mpelane, S.; Mketo, N.; Bingwa, N.; Nomngongo, P.N. Synthesis of Mesoporous Iron Oxide Nanoparticles for Adsorptive Removal of Levofloxacin from Aqueous Solutions: Kinetics, Isotherms, Thermodynamics and Mechanism. *Alex. Eng. J.* **2022**, *61*, 8457–8468. [CrossRef]
53. Paredes-Laverde, M.; Salamanca, M.; Silva-Agredo, J.; Manrique-Losada, L.; Torres-Palma, R.A. Selective Removal of Acetaminophen in Urine with Activated Carbons from Rice (*Oryza sativa*) and Coffee (*Co Ffea arabica*) Husk: Effect of Activating Agent, Activation Temperature and Analysis of Physical-Chemical Interactions. *J. Environ. Chem. Eng.* **2019**, *7*, 103318. [CrossRef]

Disclaimer/Publisher's Note: The statements, opinions and data contained in all publications are solely those of the individual author(s) and contributor(s) and not of MDPI and/or the editor(s). MDPI and/or the editor(s) disclaim responsibility for any injury to people or property resulting from any ideas, methods, instructions or products referred to in the content.

Article

Mesoporous Zr-G-C₃N₄ Sorbent as an Exceptional Cu (II) Ion Adsorbent in Aquatic Solution: Equilibrium, Kinetics, and Mechanisms Study

Lotfi Khezami [1,*], Abueliz Modwi [2], Kamal K. Taha [3], Mohamed Bououdina [4], Naoufel Ben Hamadi [1] and Aymen Amine Assadi [5,*]

[1] Chemistry Department, College of Science, Imam Mohammad Ibn Saud Islamic University (IMSIU), P.O. Box 5701, Riyadh 11432, Saudi Arabia
[2] Department of Chemistry, College of Science and Arts at Al-Rass, Qassim University, Buraydah 52571, Saudi Arabia
[3] Chemistry & Industrial Chemistry Department, College of Applied & Industrial Sciences, Bahri University, Khartoum 11111, Sudan
[4] Department of Mathematics and Sciences, College of Humanities and Sciences, Prince Sultan University, Riyadh 11586, Saudi Arabia
[5] École Nationale Supérieure de Chimie de Rennes (ENSCR), Université de Rennes, UMR CNRS 6226, 11 Allée de Beaulieu, 35700 Rennes, France
* Correspondence: lhmkhezami@imamu.edu.sa (L.K.); aymen.assadi@ensc-rennes.fr (A.A.A.)

Abstract: A mesoporous Zr-G-C₃N₄ nanomaterial was synthesized by a succinct-step ultrasonication technique and used for Cu^{2+} ion uptake in the aqueous phase. The adsorption of Cu^{2+} was examined by varying the operating parameters, including the initial metal concentration, contact time, and pH value. Zr-G-C₃N₄ nanosorbent displays graphitic carbon nitride (g-C₃N₄) and ZrO₂ peaks with a crystalline size of ~14 nm, as determined by XRD analysis. The Zr-G-C₃N₄ sorbent demonstrated a BET-specific surface area of 95.685 m^2/g and a pore volume of 2.16×10^{-7} m$^3 \cdot$g^{-1}. Batch mode tests revealed that removing Cu (II) ions by the mesoporous Zr-G-C₃N₄ was pH-dependent, with maximal removal achieved at pH = 5. The adsorptive Cu^{2+} ion process by the mesoporous nanomaterial surface is well described by the Langmuir isotherm and pseudo-second-order kinetics model. The maximum adsorption capacity of the nanocomposite was determined to be 2.262 mol·kg^{-1} for a contact time of 48 min. The results confirmed that the fabricated mesoporous Zr-G-C₃N₄ nanomaterial is effective and regenerable for removing Cu^{2+} and could be a potent adsorbent of heavy metals from aqueous systems.

Keywords: Zr-doped G-C₃N₄; wastewater treatment; mineral pollutant; catalysts regeneration

1. Introduction

Numerous harmful heavy metals are released to the environment due to industrial wastewater discharges. Even at low concentrations, they contribute significantly to pollution and endanger human health. Simultaneously, some of them, such as silver and copper, are valuable and may be recycled and used in various applications [1,2]. Toxic heavy metals found in liquid effluents, such as copper (II), are persistent, nonbiodegradable, and bioaccumulated, hence posing a severe threat to natural human health and the environment [3–5]. Many adsorbents were modified or functionalized to adsorb copper ions, for instance, 48.6, 25, 105.3, 86.95, and 31 mg/g were eliminated using polyethylenimine modified wheat straw [6], chitosan/poly (vinyl alcohol) beads functionalized with poly (ethylene glycol) [7], hematite (α-Fe₂O₃) iron oxide-coated sand [8], chitosan–montmorillonite composite [9] and chitosan@TiO₂ composites [10], respectively.

Mining and electroplating industries, for example, discharge aqueous effluents containing high amounts of heavy metals such as uranium, mercury, cadmium, lead, and copper into the environment [11,12]. Cu (II) concentrations in wastewater produced by industrial activities are frequently high [13]. The heavy metal Cu (II) can settle down in the human body and cause significant health problems in the skin, liver, heart, and brain. Cu (II) was found to be carcinogenic [13,14], especially when its concentration exceeded the (~20 µM) limit set by the U.S. Environment Protection Agency (EPA) for potable water [15]. As a result, extensive efforts have been devoted to removing Cu (II) from the aquatic environment.

Recently researchers are showing great fascination with graphitic carbon nitride (g-C_3N_4) due to its simple synthesis from the most earth-abundant elements (carbon and nitrogen) to make strong covalent bonds in its conjugated layer structure [16]. It is appealing, by virtue of the high chemical and thermal stability narrow-bandgap energy 2.7 eV (460) nm and efficient visible-light absorption. Additionally, the exceptional delocalized conjugated system of the stacked graphitic C_3N_4 layers, which are interconnected with nitrogen-containing functional groups, are capable of bonding to pollutants during the adsorption of photocatalysis [17]. Nevertheless, C_3N_4 suffers some drawbacks, including high electron–hole recombination, a low absorption coefficient, and a low specific surface area, which are mainly overcome by doping with metal and metal oxides [18].

Chemical precipitation, ion exchange, reverse osmosis, nanofiltration, and adsorption are some of the heavy metal removal processes that are commonly used [19]. Most of these approaches are not suited for small-scale companies due to their large capital expenditures. The adsorption process is the most effective method of eliminating ecologically harmful and hazardous organic–inorganic compounds from the environment [20–22]. Further, the development of cost-effective adsorbents for heavy metal removal from wastewater relies on the fabrication of nanostructured materials with enhanced properties—primarily high surface area—that play an essential role in the adsorption process. Numerous studies have focused particularly on metal oxides and composites based on metal oxides to remove heavy metals from aqueous solutions [23–26].

Interestingly, both precursors used for the Zr-G-C_3N_4 fabrication are biocompatible where ZrO_2 finds medical application in dentistry and biomedical implants without adverse reactions [27,28], while G-C_3N_4 is employed as an anode in microbial fuel cells [29] and human-safe antimicrobial agents [30]. Therefore, this paper aims to prepare a harmless Zr-G-C_3N_4 nanosorbent to treat wastewater contaminated with copper ions, one of the most prevalent hazardous heavy metals. Using thermal pyrolysis, Zr-G-C_3N_4 sorbent was fabricated. XRD, SEM/EDX, and FTIR analyses were carried out to examine nanomaterials' structure, morphology and chemical composition, specific surface and pore size, and functional groups. The effect of operational parameters such as contact time and pH on the adsorption capacity of Cu ions was investigated. Moreover, kinetic and isothermal adsorption tests were conducted. A plausible mechanism of Cu^{2+} ion adsorption onto Zr-G-C_3N_4 sorbent was discussed and proposed.

2. Experimental

2.1. Nanomaterial Fabrication, Morphological, and Structural Characterization

Mesoporous Zr-G-C_3N_4 sorbent was already synthesized [18] using a straightforward one-step ultrasonication technique.

Using a Bruker D8 Advance X-ray diffractometer (Bruker AXS, Karlsruhe, Germany) with Cu-Kα1 radiation and k = 1.5406 at a scan speed of 0.002/per second, the structural characteristics of the Zr-G-C_3N_4 sorbent were investigated. The surface texture of the manufactured sorbent was estimated via N_2 adsorption at 77 K using ASAP 2020 HD 88 equipment. Prior to each investigation, the ZnO sample was outgassed at 250 °C for 6 h by a constant helium current flow. On a scale of 400–4000 cm^{-1}, the FT-IR spectra of the ZOCN mesosorbent before and after MG dye elimination were monitored with a Nicolet 5700 FT-IR spectrophotometer. The Cu^{2+} ion content was measured on a Shimadzu 680 A or a Perkin-Elmer 603 atomic absorption spectrometer with a hallow cathode lamp and a deuterium background corrector, at respective resonance line using an air–acetylene flame.

On the other hand, a Hitachi S4700 Scanning electron microscopy (SEM) (Anhui, China) operating at a 25 kV and a Tecnai G20 TEM (Houston, TX, USA) operating at a 200 kV were used for morphological observations and elemental chemical composition.

2.2. Cu^{2+} Ion Removal Experiments

The efficiency of the produced nanomaterial for removing Cu metal ions was realized in a batch mode reactor at room temperature. A total of 10 mg of Zr-G-C_3N_4 sorbent was added to a series of Erlenmeyer flasks containing 25 mL synthetic solutions with different starting concentrations, C_0 (5 to 200 ppm), 27 °C, and a pH ranging from 1 to 8. The flasks were firmly sealed to prevent evaporation, and the suspension was constantly agitated at 420 rpm for 24 hrs. The supernatant solution of Cu ions and Zr-G-C_3N_4 sorbent was separated by 10 min centrifugation at 2500 rpm. In the meantime, the remaining Cu^{2+} ion concentrations, C_e, were analyzed by atomic absorption spectroscopy (AAS). The ultimate adsorptive capacity (q_e) relative to the starting and equilibrium concentrations was calculated using the following equation:

$$q_e = \frac{V}{m}(C_0 - C_e) \quad (1)$$

where m (in g) is the Zr-G-C_3N_4 adsorbent mass, V (in L) is the solution volume.

To specify the period of time needed to achieve sorption equivalency for Cu^{2+} metal ions onto Zr-G-C_3N_4 sorbent, the removal process was examined at various predetermined contact times from 5 to 1440 min. At room temperature, the Cu^{2+} metal ion kinetics rate was provided at a constant Zr-G-C_3N_4 sorbent mass of 60 mg and 150 mL initial copper metal solution concentration of about 60 ppm. The remaining Cu ion concentration (C_t) was measured using AAS, and the adsorption kinetic models are mentioned in the Results section.

The Cu^{2+} ion concentration was determined and utilized to specify the adsorption capacity at time t (q_t) and the percentage removal (%R), which were estimated using the following expressions:

$$q_t = \frac{V}{m}(C_0 - C_t) \quad (2)$$

$$\%R = (C_0 - C_t)\frac{100\%}{C_0} \quad (3)$$

At pH > 6, the filtrate was acidified with dilute HNO_3 prior to Cu^{2+} ion concentration by AAS.

3. Results and Discussion

3.1. Composition Phase and Surface Characteristics of Mesoporous Zr-G-C_3N_4 Sorbent

The recorded XRD pattern of the Zr-G-C_3N_4 sorbent is shown in Figure 1a. The peaks at 24.00° (110), 28.22° (−111), 31.35° (111), 34.07° (200), 35.27° (002), 40.64° (120), 44.80° (211), 49.24° (220), and 50.09° (022) are indexed within a monoclinic crystal structure of the ZrO_2 phase (JCPDS card No. 00-37-1484) [31,32]. The remaining peaks located at 13.0° (100) and 27.58° (002) correspond to the layered two-dimensional structure of the G-C_3N_4 phase (JCPDS card No 87-1526 [33]. No additional peaks are detected, except ZrO_2 and g-C_3N_4, indicating the formation of a pure biphasic composite composed of the ZrO_2 phase with a monoclinic crystal structure and the g-C_3N_4 phase with a layered two-dimensional structure. The crystallite sizes determined by the Scherrer formula [34] using the most intense reflection (002) for G-C_3N_4 and (−111) for ZrO_2 are found to be 7.94 and 13.95 nm, respectively. Figure 1b illustrates the N_2 sorption–desorption isotherm of the Zr-G-C_3N_4 sorbent. The nanostructure's isotherm is of type IV, suggesting the existence of mesopores throughout the material resulting from the narrow hysteresis loop generation [35,36]. The Zr-G-C_3N_4 sorbent exhibits a specific surface area of 95.7 $m^2 \cdot g^{-1}$ and a pore volume of

0.000000216 m^3·g^{-1}. The relatively high surface area is expected to improve the sorption capability of the as-fabricated mesoporous Zr-G-C$_3$N$_4$ nanocomposite [18].

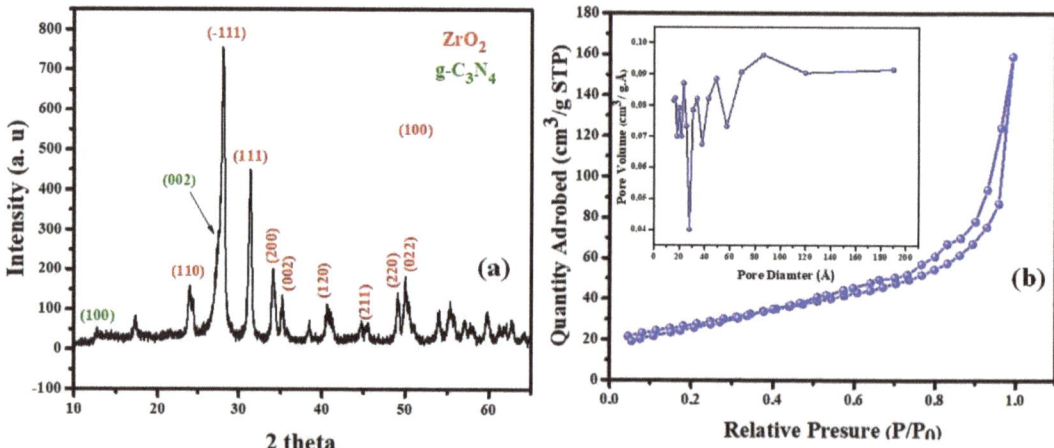

Figure 1. (**a**) XRD pattern; (**b**) N$_2$ sorption–desorption isotherm of Zr-G-C$_3$N$_4$ sorbent. The inset in (**b**) presents the pore size distribution.

Figure 2 illustrated SEM morphological observations with EDX chemical analysis and elemental mapping of the nanocomposite before the adsorption process.

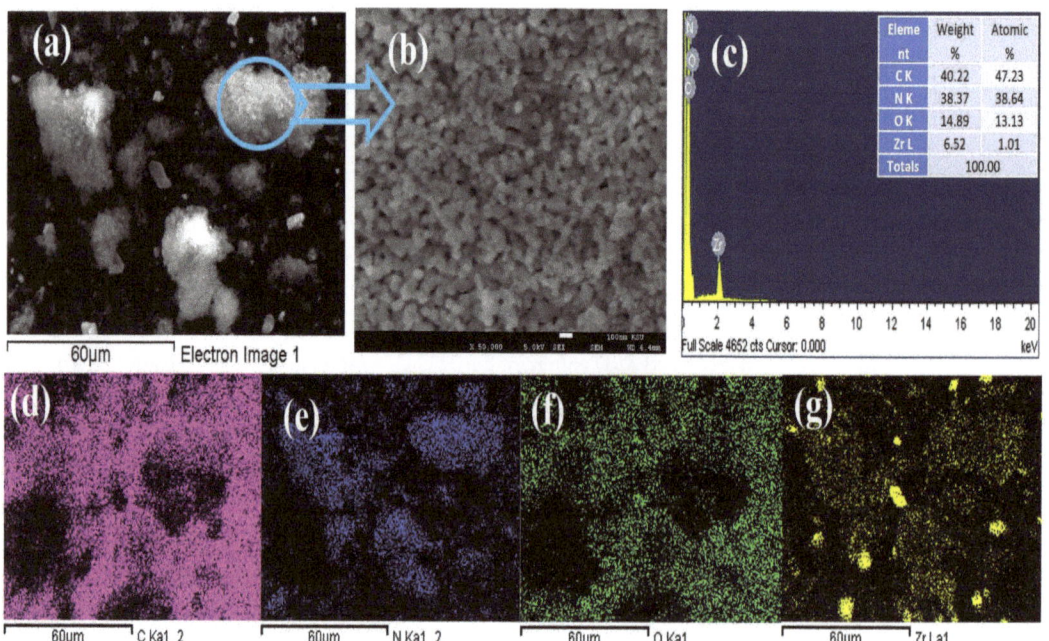

Figure 2. (**a**) SEM image and (**c**) EDX spectrum for element composition; (**b**,**d**–**g**) elemental mapping of C, N, O, and Zr, respectively, in Zr-G-C$_3$N$_4$ nanocomposite.

Figure 2a reveals smaller elongated particles with smooth surfaces alongside larger aggregates formed by spherical-shaped particles in the nanoscale with homogenous size distribution (Figure 2b). The nanoparticles are interconnected, forming a mesoporous network nanostructure all over the material, generated from the thin hysteresis loop spawning, which is convenient with the BET analysis manifesting a type II isotherm [18].

The EDS spectrum of Zr-G-C_3N_4 (Figure 2c) contains the peaks related to C, N, O, and Zr, confirming once again that the purity of the as-fabricated nanocomposite material made of graphitic carbon nitride and zirconium oxide corroborates the literature [18,19], and EDS elemental mapping indicates homogenous distribution Figure 2d–g.

3.2. Adsorption Measurements of Zr-G-C_3N_4 sorbent

3.2.1. Impact of Adsorption Time

Figure 3a depicts the effect of contact time on the % removal of Cu^{2+} ions at room temperature. The adsorption of copper ions onto the Zr-G-C_3N_4 sorbent has been examined during agitating durations varying from 5 to 1440 min for an initial concentration of 0.708 mmol/L and 27 °C. The half-amount of metal ions eliminated was attained in 22 min, and 97% of the adsorbate was completely removed in the process. Owing to the multiple active sites on the Zr-G-C_3N_4 nanomaterial surface, the initial adsorption rate is significantly high, reaching a value of h_0 = 4.45 mg/(g.min).

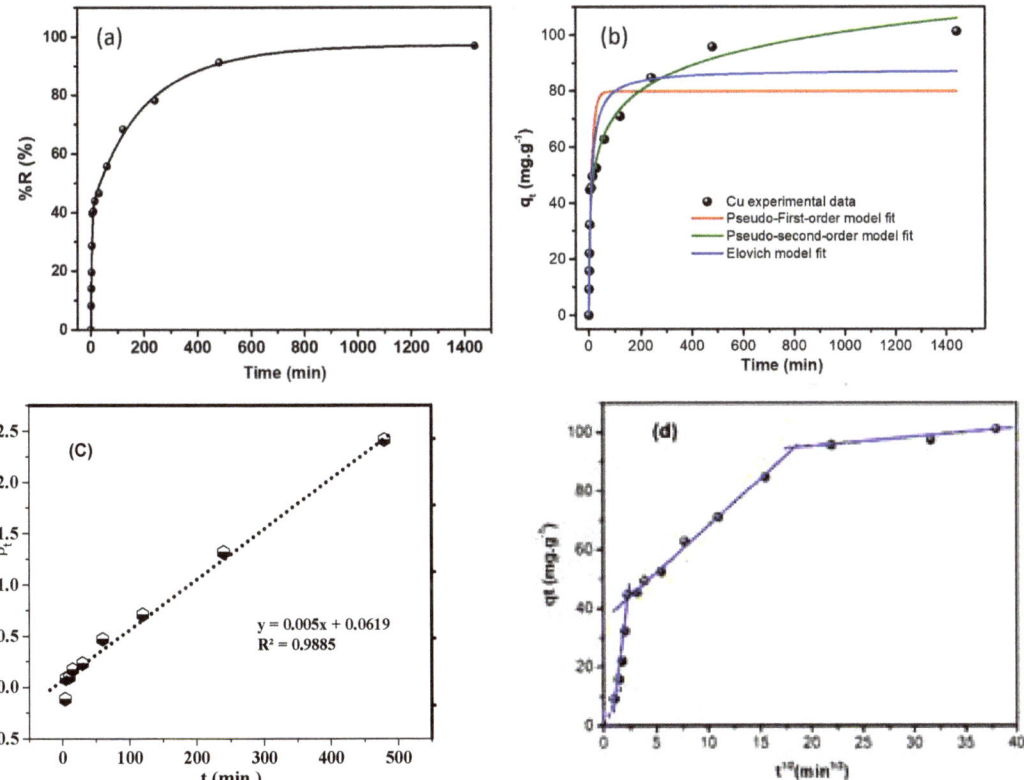

Figure 3. (a) Impact of adsorption time and (b) fitting experimental data with different nonlinear kinetics models; (c) Boyd model and (d) IPDT model for Cu^{2+} metal ion uptake onto the sorbent Zr-G-C_3N_4, using a nonlinear trend.

3.2.2. Kinetic Study

In order to obtain an insight into the process that governs the adsorption of Cu ions, the kinetic measurement is often conducted using various kinetic models (Table 1). The obtained experimental data may be well fitted by the pseudo-first-order (PFO), (the pseudo-second-order (PSO), the Elovich, and the intraparticle diffusion (IPD) models as depicted by the nonlinear model fittings in Table 1 and Figure 4.

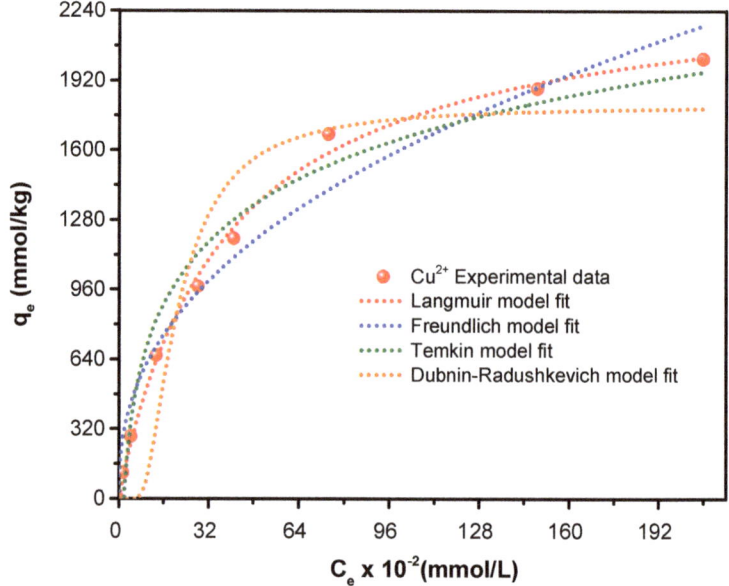

Figure 4. The nonlinear illustration of the equilibrium experimental data.

The PFO kinetic model (Equation (4)) demonstrates the presence of proportionality between the binding capability and the number of available active sites [20]. From the PFO graph (Figure 3), the rate constant k_1 and the adsorbed amount during the contact time were $1.4 \times 10^{-3} \cdot min^{-1}$ and 51.5 mg·g^{-1} in turn. The difference in the adsorbed amount compared to the experimental data and the lesser correlation coefficient value (Table 1) indicates that the PFO kinetic model is inappropriate to describe this process. The PSO kinetics model (Figure 3 and Equation (5)) assumes the chemisorption process as the rate-limiting phase accompanying the valence electrons' involvement between the nanomaterial and the pollutant [21].

The PSO model rate constant (k_2) and ultimate adsorbed amount of Cu^{2+} ions were 5.6×10^{-3} g·mg^{-1}·min^{-1} and 90 g·mg^{-1}, respectively. The almost-equal values of the calculated (90 g·mg^{-1}) and experimentally (87 g·mg^{-1}) adsorbed amount and the large rate constant (k_2), as well as the strong correlation coefficient close to the unit (i.e., 0.9984), demonstrate that adsorption follows the PSO kinetic model (e.g., Table 1 and Figure 3) [20].

Table 1. Kinetics models' parameters for the removal of Cu (II) by Zr-G-C$_3$N$_4$.

Kinetics Model	Linear and Nonlinear Kinetic Equations	Equation No.	Refs.	Parameters	Values
Pseudo-first-order	$q_t = q_e(1-e^{-k_1 t})$	(4)	[37,38]	q_m (mg·g^{-1})	51.5
				k_1 (min^{-1})	1.4×10^{-3}
				R^2	0.8207
Pseudo-second-order	$q_t = \frac{k_2 q_e^2 t}{1+k_2 q_e t}$	(5)	[37]	$q_{m(exp.)}$ (mg·g^{-1})	87
				$q_{m(cal.)}$ (mg·g^{-1})	90
				k_2 (g·mg^{-1}·min^{-1})	5.6×10^{-3}
				h_0 (mg·(g^{-1}·min^{-1}))	4.65
				$t_{1/2}$ (min^{-1})	21.95
				R^2	0.9984
Elovich	$q_t = \frac{1}{\beta}\ln(1+\alpha\beta t)$	(6)	[39]	β (g·mg^{-1})	0.0863
				α (mg·g^{-1}·min^{-1})	57.37
				R^2	0.9687
Intraparticle diffusion	$q_t = k_{dif}t^{1/2} + C$	(7)	[39]	k_{dif1} (mg·g^{-1}·min$^{1/2}$)	70.89
				C_1 (mg·g^{-1})	41.76
				R^2	0.9737
				k_{dif2} (mg·(g^{-1}·min$^{-1/2}$))	1.14
				C_2 (mg·g^{-1})	81.88
				R^2	0.9527
				k_{dif3} (mg·(g^{-1}·min$^{-1/2}$))	0.046
				C_3 (mg·g^{-1})	115.32
				R^2	0.9612
Film diffusion	$-0.4977 - \ln(1-F)$ $F = \frac{q_t}{q_e}$	(8)	[40]	k (min^{-1})	5.00×10^{-3}
				D_i (cm^{-2}·g^{-1}·min^{-1})	9.166×10^{-9}
				R^2	0.9885

Notes: q_t: adsorption capacity at time (t). q_e: adsorption capacity at equilibrium. C_0: initial concentration. C_e: equilibrium concentration. k_1: PFO rate constant. k_2: PSO rate constant. β: activity coefficient. ε: the Polanyi potential. kid: IPDT diffusion rate constant. C: diffusion layer thickness. Di: diffusion coefficient.

This finding is in accordance with Cu^{2+} ion uptake on modified nanocomposites [22]. The PSO kinetic model was used to derive the initial rate of sorption $h_0 = k_2 \cdot q_e^2$ and the half-sorption time $t_{1/2} = 1/(k_2 \cdot q_e)$ as the time needed for eliminating the half amount of the equilibrium value. Time is usually used as an indicator of the sorption rate. The high initial rate of adsorption h_0 (4.65 mg/(g·min)) and the briefer half-duration of adsorption (21.95 min) indicate that Cu^{2+} ions are adsorbed at a high rate [23]. In addition, the experimental data were modeled by the Elovich equation (Equation (6)) (Figure 3b), and the initial rate a and β values were 0.0863 and 57.37, respectively. Moreover, the correlation between the observed data and the Elovich equation is also confirmed by the R^2 value (i.e., 0.9687) [24]. The good agreement with the Elovich model supports that the adsorption mechanism is governed by chemisorption, supporting the PSO kinetic model [24,25]. Danesh et al. studied the elimination of Cu^{2+} ions by modified nanocomposites (GFLE) and proved that experimental data are well described by the second-order-kinetics model.

3.2.3. Intraparticle Diffusion/Transport Model (IPDT)

The IPDT (Equation (8)) may move the adsorbed Cu^{2+} metal ions from the preponderance of the solution to nanoparticles' interface. This process presents a step in the adsorption mechanism that acts as a restraint. Weber and Morris' diffusion model verifies the potential of intraparticle dissemination (Table 1) [24,25]. The C and k_{dif} model constants' values can be derived from the intercept and slope of q_t vs. $(t)^{1/2}$ linear plot, respectively (Figure 3d). Since q_t changes linearly with $t^{1/2}$, the uptake of Cu^{2+} metal ions at the Zr-G-C_3N_4 nanomaterial surface verifies the plausibility of the IPDT kinetic model. In addition, the regression coefficient (R^2) close to the unit) identifies the IPDT mode of diffusion. Moreover, C is a parameter measuring the thickness of the border layer. The substantial effect of the solution border layer on the uptake process is validated by the higher constant values given in Table 1 [23–25].

Moreover, the primary adsorption step has a greater rate than the final step, as affirmed by the values of k_{dif} shown in Table 1. The rapid rate of the primary step may be attributed to the transfer of Cu^{2+} metal ions through the solution up to the nanomaterials' external surface via the border layer. Simultaneously, the subsequent stage manifests the final equilibrium step as the IPDT drops due to the less solute concentration gradient resulting from the dwindling number of mesopores available for diffusion. In addition, the increased value of the C parameter in the final step suggests the existence of a border coating impact [23], confirming the role of intraparticle diffusion in the removal of Cu^{2+} metal ions by Zr-G-C_3N_4 [40].

The Boyd model is assessed for chemisorption kinetics to confirm the film diffusion influence. It is expressed by the formula given in Equation (8) [41]. The term F represents the ratio of metal ions adsorbed at any time t (min), and Bt is a mathematical function of F. In case the plot of Bt versus t is linear and passes through the origin, the rate of mass transfer is controlled by pore diffusion. Else a nonlinear plot—or linear without passing through the origin—the film diffusion or chemical reaction controls the adsorption rate [42]. In accordance, Bt vs. t graph that does not pass through the origin (Figure 3c) denotes a film diffusion or chemical reaction control on the adsorption rate [43].

The diffusion coefficient D_i can be calculated by the following formula [44]:

$$D_i = \frac{B}{\pi A}$$

where B is the slope of Bt vs. t in min^{-1}, and A is the adsorbent's surface area in $cm^2 \cdot g^{-1}$. The calculated D_i value is 9.166×10^{-9} $cm^{-2} \cdot g^{-1} \cdot min^{-1}$ which is outside the 10 to 11×10^{-11} $cm^{-2} \cdot g^{-1} \cdot min^{-1}$ range quantified for the intraparticle phenomenon. This corroborates the film diffusion involvement in the adsorption process [44] and that intraparticle diffusion is not the sole rate-controlling step of the process [45].

3.2.4. Uptake Isotherms of Copper Ions

The uptake isotherms reveal the affinity of the nanomaterial and its surface characteristics. Consequently, they are commonly employed to compare the ultimate adsorbed amount (q_m) of the adsorbent to the pollutants in wastewater. Various empirical and semiempirical equations are used to describe the behavior of the adsorption process until equilibrium is reached. Among these equations, the Langmuir model (Equation (9)) presumes that all active sites of the homogeneous surfaces possess equivalent adsorption energies, where the interaction between the sorbed species is insignificant. Table 2 presents the nonlinear form of this isotherm, where q_e represents the number of Cu^{2+} metal ions adsorbed and q_m and b characterize the complete monolayer packing and the Langmuir constant, respectively [46].

Furthermore, Freundlich's model (Equation (10)) is a semiempirical equation applicable to multilayer filling on heterogeneous surfaces characterizing a nonideal uptake [11].

These equations [14] rely upon the amount of adsorbate per gram of adsorbent (q_e) to the solute equilibrium concentration (C_e) [30]. The n and k_f Freundlich model's constants describe the adsorption intensity and relative uptake capability. Moreover, they establish the capacity of adsorption and the nonlinear behavior representing the concentration change and solution. The adsorption intensity depends on the values of n; i.e., $n < 1$ for low; $1 < n < 2$ for moderate, and $2 < n < 10$ for high adsorption capacities, respectively [47].

Table 2. Different equilibrium isotherms' constants for Cu^{2+} ion adsorption by Zr-G-C_3N_4 nanomaterial.

Equilibrium Model	Linear and Nonlinear Equilibrium Equations	Equation No.	Refs.	Parameters	Cu^{2+}
Langmuir	$q_e = \frac{q_m b C_e}{1 + b C_e}$, $R_L = \frac{1}{1 + a_L C_0}$	(9)	[26]	q_m (mol·kg^{-1})	2.262
				b (L·mol^{-1})	5.5×10^{-6}
				R_L	0.9514
				R^2	0.9907
Freundlich	$q_e = K_F (C_e)^{1/n}$	(10)	[48]	n	1.73
				k_F mmol·g^{-1}(mmol·L^{-1})$^{-0.578}$	9.87
				R^2	0.9634
Temkin	$q_e = \frac{RT}{\beta_T} Ln(K_T C_e)$	(11)	[49]	β_T (J·mol^{-1})	563.2
				k_T (L·mmol^{-1})	5.85
				R^2	0.9612
Dubinin–Radushkevich	$q_e = q_m \exp(-\beta \varepsilon^2)$, $\varepsilon = (RT ln(1 + \frac{1}{C_e}))^2$ $E = \frac{1}{\sqrt{2\beta}}$	(12)	[50]	β (mol^2·J^{-2})	1.95×10^{-8}
				q (mol·kg^{-1})	18.6
				E (J·mol^{-1})	5064
				R^2	0.9864

The Temkin isotherm (Equation (11)) compensates for solute–solute indirect interaction in the adsorption process, where the molecules' adsorption heat linearly decreases inversely proportional to the layers' coverage [31]. Table 2 provides the linear and nonlinear equations for the Temkin model. The coefficients β_T and K_T are, sequentially, the Temkin isotherm energy and constant.

The same table illustrates the Dubinin–Radushkevich model (D-R)'s linear and nonlinear expressions (Equation (12)) [51] as a final stage, where β is related to the adsorption energy in mol^2/kJ^2, and ε ($RTln(1 + 1/Ce)$) is the Polanyi potential in kJ/mol. The coefficient β value allows for the determination of the mean value of adsorption activation energy, $E = \frac{1}{\sqrt{2\beta}}$, and pinpoints whether adsorption nature is a physical or chemical process [32]. The nonlinear uptake isotherms of Cu^{2+} metal ions on the Zr-G-C_3N_4 nonmaterial surface are displayed in Figure 4, respectively.

The best-described isotherm model for Cu^{2+} metal ions' equilibrium data is mainly evaluated based on the value of R^2 shown in Table 2. This table shows that the Langmuir isotherm has the highest regression factor value (0.9907) and the maximum q_m, indicating the compliance of the adsorption mechanism. The Langmuir model's key characteristics are evaluated through the separation coefficient, R_L, to examine the isotherm's applicability, as expressed in Table 2.

The R_L values prescribe the adsorption isotherm behavior as linear, favorable, unfavorable or irreversible if ($R_L = 1$), ($0 < R_L < 1$), ($R_L > 1$), or ($R_L = 0$) respectively [17,29,32]. The results given in Table 2 reveal a favorable isotherm (i.e., $R_L = 0.9514$), manifesting that the Cu ions' adsorption follows the Langmuir model. The obtained values demonstrate that the D-R model concurs with the elimination procedure, as confirmed by the higher value of r² (0.9864) and the significant adsorption capacity of Cu^{2+} metal ions (118.3 mg/g), and the energy amplitude is about 5.063 kJ·mol^{-1} [22,32–34]. Similarly, the coefficient (R^2) corresponds to the Freundlich equilibrium model, which is more significant than 0.96, agreeing well with the obtained experimental results. This finding can be sustained by the n-value amplitude (i.e., 1.73), which suggests a moderately good adsorption capacity. Moreover, the negative free energy value suggests a spontaneous physisorption uptake process (Table 2).

In this study, the Elovich model advocated that the adsorption of copper ions on the surface of the nanomaterial accrues under the chemisorption process. This result, together with the experimental data fitting to the Langmuir isotherm and the PSO kinetics model, indicates that the adsorption process of Cu^{2+} ions onto Zr-G-C_3N_4 nanomaterial involves a chemisorption mechanism. This is in agreement with the chemisorption of Cu(II) ions on modified chitosan [52], magnetite (Fe_3O_4) [53], activated carbon [54], and fly ash-derived zeolites [55] as examples.

3.2.5. The Impact of pH on Cu^{2+} (II) Ion Uptake

The impact of pH on Cu^{2+} metal ion removal efficiency is a critical aspect of the uptake process. It modifies the active sites on the Zr-G-C_3N_4 sorbent surface, which is competent for Cu coordination and the solubility of Cu ions in the aqueous solution. Depending on the solution's pH, the species of copper present include: Cu^{2+}, $Cu(OH)^+$, $Cu(OH)^0_2$, $Cu(OH)_3^-$, and $Cu(OH)_4^{2-}$ [56]. To determine the optimal pH for eliminating Cu^{2+} metal ions, the pH value varies from 1 to 8. Figure 5 illustrates the effect of initial pH on the uptake of the tested metal ions, and the results indicate that the best uptake capability is attained at pH 5 (43.37 mg g^{-1}). The amount of Cu^{2+} ions removed increases significantly at pH = 5. This is due to the copper ions precipitation as $Cu(OH)_2$(s) at pH > 5 their existence in the hydrolyzed forms $Cu(OH)^0_2$, $Cu(OH)_3^-$, and $Cu(OH)_4^{2-}$ at higher pH values.

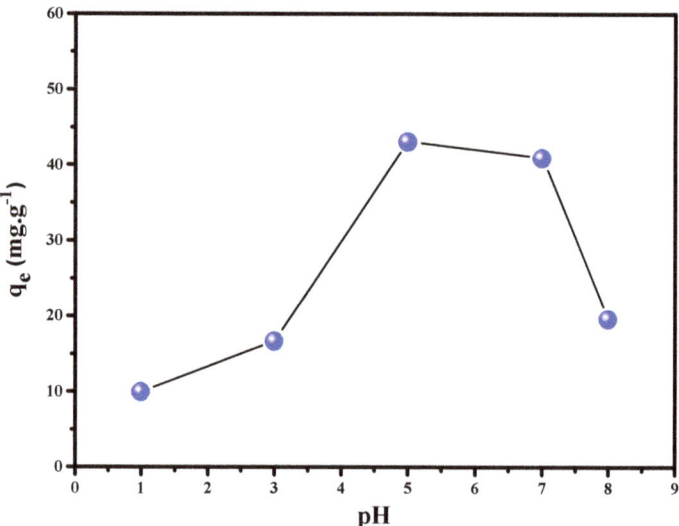

Figure 5. The pH impact on the uptake rate of Cu^{2+} metal ions by Zr-G-C_3N_4 nanomaterial.

At lower pH values, more hydrogen protons subsist to protonate active species on Zr-G-C_3N_4 nanomaterial surfaces and compete with Cu^{2+} ions in the suspension, which corroborates previous studies in the literature [57–60]. In addition, according to earlier studies, Cu^{2+} ions are only accessible in the divalent state at pH values below 5.0. Furthermore, it is important to highlight that there are numerous neutral hydrolysis species at pH 7.0 and above.

3.2.6. Uptake Mechanism of Copper Ions

To suggest a plausible uptake mechanism of Cu^{2+} ions on Zr-G-C_3N_4, Fourier transmission infrared spectra and EDS elemental mapping were recorded before and after the adsorption process. The FTIR spectrum of Zr-G-C_3N_4 before adsorption (Figure 6) reveals a broad band around 3266 cm^{-1}, which was ascribed to the stretching vibration of –OH groups of physically sorbed H_2O and the terminal amino groups on the Zr-G-C_3N_4 nanomaterial surface [40]. The band located at 887 cm^{-1} is assigned to the triazine ring mode. Further, the bands located at 1256, 1329, and 1429 cm^{-1} correspond to the bridged aromatic C–N stretching modes [41]. After Cu ion uptake, important modifications were noticed in the FTIR spectrum, as displayed in Figure 7. The band centered around 3266 cm^{-1} was less broadened, while the triazine ring mode band at 887 cm^{-1} shifted slightly to 881 cm^{-1}.

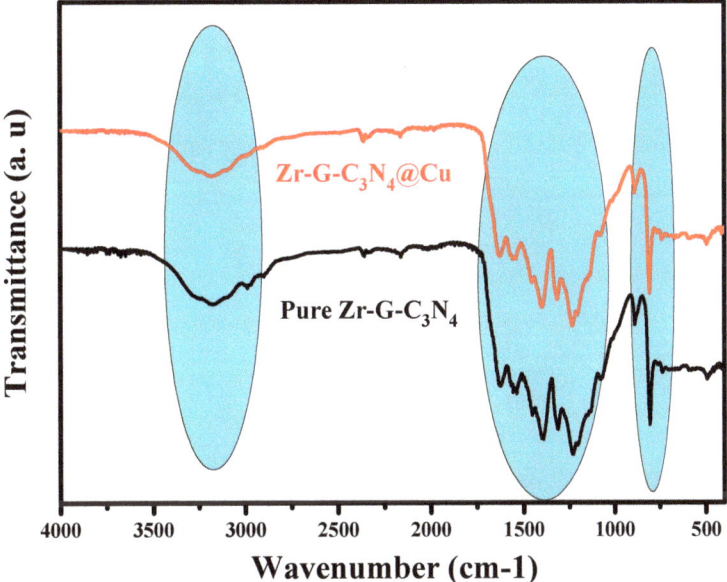

Figure 6. FTIR spectra before and after Cu^{2+} adsorption onto the surface of Zr-G-C_3N_4 nanomaterial.

Figure 7 illustrates SEM morphological observations with EDX chemical analysis and elemental mapping of the nanocomposite after the adsorption process. The homogeneous distribution of the Zr-G-C_3N_4 nanocomposite constituents (C, O, N, and Zr) shown in Figure 7d–g, while the adsorbed Cu, is clearly observed, as and Figure 7c. This strongly confirms the adsorption of Cu^{2+} onto the composite nanoparticles' surfaces.

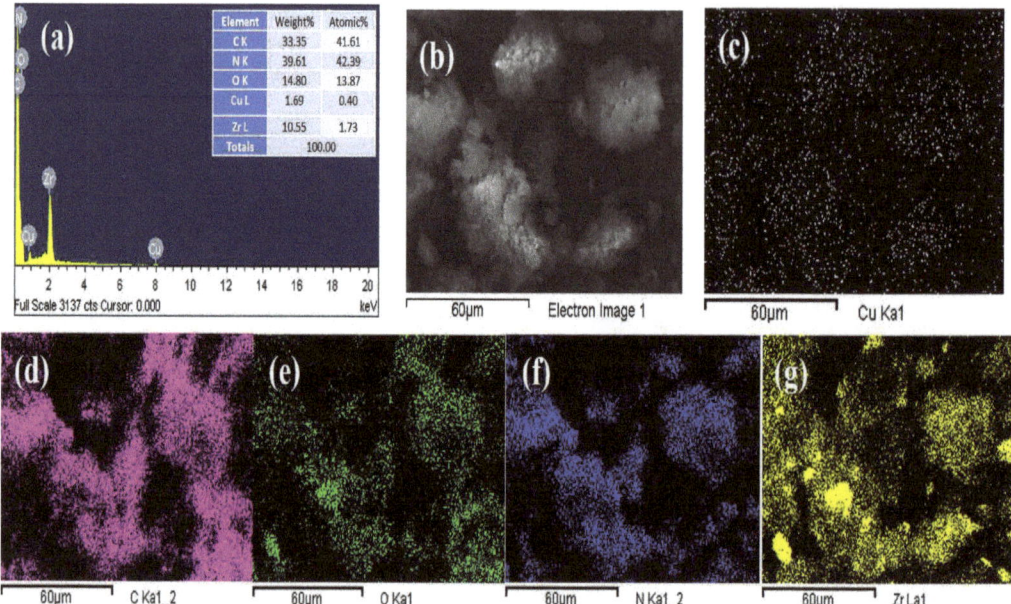

Figure 7. (a) EDX spectrum for elements composition; (b) SEM image; (c–g) elemental mapping of C, N, O, and Zr after Cu^{2+} adsorption onto $Zr\text{-}G\text{-}C_3N_4$ nanocomposite.

XPS analysis was utilized to investigate the surface chemical composition, the oxidation state of elements on the surface, and the interaction between ZrO_2 and $g\text{-}C_3N_4$. As displayed in Figure 8, the survey scan spectrum contains only the components of N, C, O, and Zr, hence confirming once again the purity of the as-fabricated composite material composed of graphitic carbon nitride and zirconium oxide. The XPS chemical profile spectra of individual fitting of peaks related to Zr, O, C, and N elements are shown in Figure 8. The peaks positioned at 285.7, 288.0, and 289.5 eV for C 1s (Figure 8) correspond to the covalent link of sp^2 hybridized graphic carbon, sp^2 carbon chemically connected to N (N–C=N), and dependence features due to $\pi - \pi *$ excitation, successively [61,62]. Figure 8 displays the distinctive high-resolution peaks for N 1s into the $Zr\text{-}G\text{-}C_3N_4$ sorbent at 389.9, and 404.4 eV is related to sp^2 hybridized aromatic N integrated into the triazine structure (C=N–C) and π excitation [63]. The high-resolution peaks for Zr 3d at 182.7 and 185.1 eV may be independently associated with Zr $3d_{5/2}$ and Zr $3d_{3/2}$ [64]. The O 1s spectrum of $ZrO_2@g\text{-}C_3N_4$ displays two peaks around 530.8 eV ascribed to the oxygen (O) crystal structure in zirconium oxide and 532.5 eV corresponding to the materially adsorbed oxygen, as shown in Figure 8 [65,66]. According to the XPS analysis, the $Zr\text{-}G\text{-}C_3N_4$ sorbent consists of Zr, O, N, and C, which demonstrates its purity and corroborates the literature [67,68]. The XPS after adsorption revealed a shift in the binding energies for N1s, C1s, and O1s to higher values with a reduction in peak intensities. This may indicate changes in electron clouds around these elements due to coordination with the copper ions. This is in agreement with previous reports of binding Tb to the nitride moiety, as justified by the reduction in the number of =N−C=N−, as a result of the breakage of a C=N π bond and the establishment of a new bond with the metal [69]. Similar results were also reported for Ni and Cr ion complexation interactions with the nitride species [70]. In addition, the adsorption of the Cu ions is validated by the presence of $Cu_{2p3/2}$ and $Cu_{2p1/2}$ peaks at 936.5 and 956.4 eV and an area under the curve ratio of 20,851:111,22 ≈ 2:1. Interestingly, the binding energies $Cu_{2p3/2}$ and $Cu_{2p1/2}$ are higher than the respective 932.8 and 952.6 eV values reported for the free ions [71] which may further confirm a successful adsorption process. From the XPS analysis, it can be inferred that the increase in the binding energy of O1s and N1s may be attributable to the coordination of the Cu ions to the N and O on the composite surface that share electrons with the copper ions. The increase in the C1s binding energy is a result of a reduction in the electron densities of neighboring carbons during the complexation of O and N to the metal ions [72]. Thus, strong surface complexation between the electrons present in the functional groups of the composite acting as Lewis bases that interact with the metal ions (Lewis acid) may come about [73].

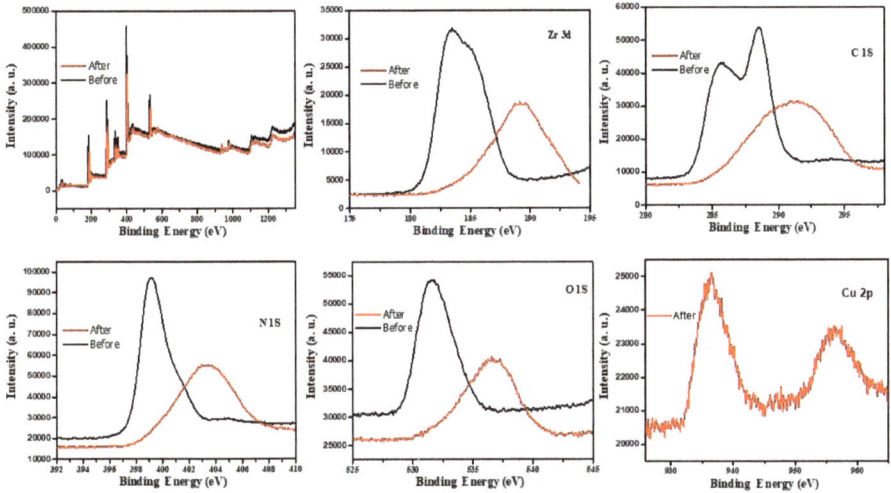

Figure 8. The XPS of the Zr-G-C_3N_4 before and after Cu ions adsorption.

The results obtained by FTIR, XPS, and SEM/EDS elemental mapping demonstrate that functional groups of Zr-G-C_3N_4 (OH, and -NH_2) and p-delocalized electrons of the triazine ring (C_3N_3) are most probably involved in the elimination of Cu (II). The plausible adsorption mechanism of Cu^{2+} ions by Zr-G-C_3N_4 nanomaterial is illustrated in Figure 9.

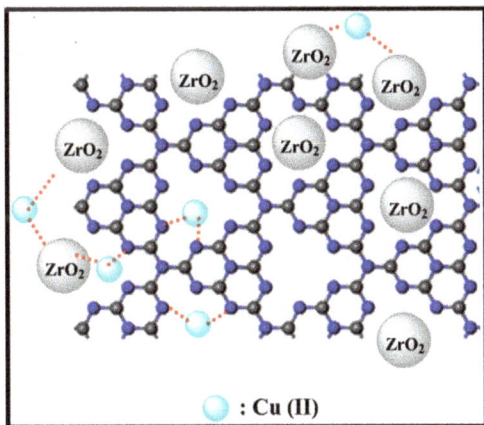

Figure 9. Plausible mechanism for Cu ions adsorption onto Zr-g-C_3N_4 nanoparticles' surface.

3.2.7. Regeneration Tests for Zr-g-C$_3$N$_4$ Adsorbent

Desorption analysis was performed to calculate the regeneration capacity of the adsorbent to establish its environmental applicability from an experimental and economic perspective. The reusability and the renewal of an adsorbent such as the Zr-G-C$_3$N$_4$ nanomaterial are essential when assessing scale-up and industrial applications. Furthermore, the cyclic reconditioning of this nanosorbent from the physi- or chemisorption reaction medium is vital. After the adsorption investigation, the synthesized nanoparticles were recovered by centrifugation and filtration, thoroughly rinsed with ultrapure hot water, and oven-dried at 105 °C. Afterward, the nanoparticles were soaked with 0.1 M NaOH [74] solution as a desorbing agent in a 50 mL agitated flask (i.e., 500 rpm). This desorbing agent easily releases Cu^{2+} ions from the surface of the Zr-G-C$_3$N$_4$ nanocomposite. Accordingly, the collected nanocomposite particles were reutilized for a new removal cycle. Figure 10 illustrates the performance experiment of the adsorption–desorption process for four consecutive cycles. For large-scale applications, the nanoadsorbents can be secured in a meticulous form through lodging in an unyielding support such as a synthetic or natural polymer to guarantee the nanomaterial trapping and nonrelease in the water system. Such arrangement can be adopted to relocate to the fixed-bed column for continuous flow treatment pilots [75–82].

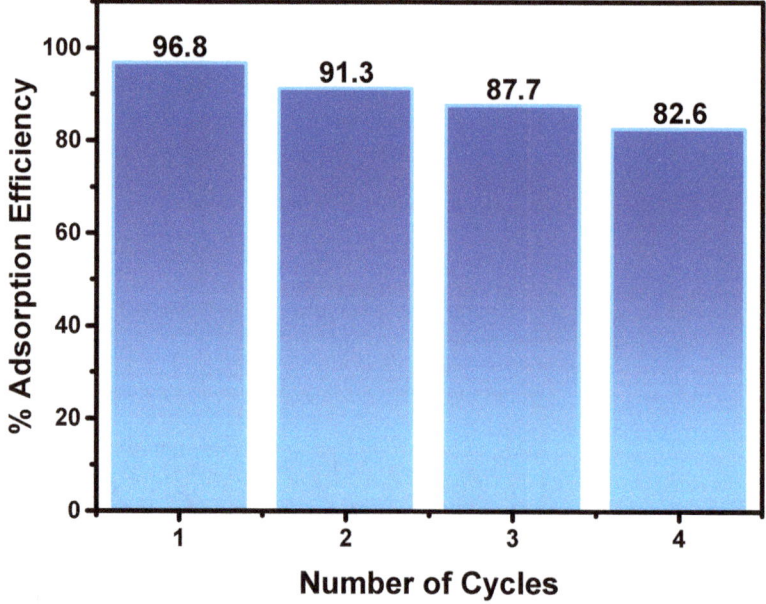

Figure 10. The adsorption–desorption process for four successive cycles.

The Zr-G-C$_3$N$_4$ nanocomposite was proven effective for removing Cu^{2+} ions, with an average value of around 88%. This deficiency may be associated with the disrobing agent's despoiler impact and the discharge procedure's weight waste of the nanocomposite [22]. Moreover, this phenomenon can be explained by the decrement in the number of available sites on the surface of the nanomaterial occupied by the adsorbed metal ions [42]. Accordingly, chemisorption was the main mode for the uptake of copper ions onto the nano-adsorbents. Moreover, the stability of the Zr-G-C$_3$N$_4$ nano-sorbent was checked by FTIR (Figure 7) after the reusability experiments. The results demonstrate that the FTIR spectrum of nanosorbent continues unmoved following the recycling tests. Thus, the mesoporous Zr-G-C$_3$N$_4$ nanocomposite synthesized by a simple-step ultrasonication technique can efficiently and continually eliminate hazardous metal ions from wastewater. Thus, adsorption employing such a composite composed of a metallic center and attached organic functional via robust bonding stands out as a viable, economic, applicable, and reproducible approach [76] for efficient Cu ion removal.

To emphasize the remarkable efficiency of Zr-G-C$_3$N$_4$ nanomaterial for the removal of Cu^{2+} ions, a comprehensive comparison with the literature is illustrated in Table 3.

Table 3. Comparison of Cu ion uptake capability by Zr-G-C$_3$N$_4$ with other sorbents.

Adsorbents	Cu^{2+} Uptake (mg/g)	t (min)	pH	References
GFLE	193.40	105	1 and 40 °C	[22]
ɣ Fe$_2$O$_3$ nanoparticles	26.00	240	6	[77]
Graphene oxide	75.00	1440	<5.95	[59]
MgO-CaO-Al$_2$O$_3$-SiO$_2$-CO$_2$ system	16.70	168	-	[78]
MnO$_2$ nanowires	75.48	30	1.9 to 5.3	[79]
CO$_3$·Mg-Al LDH	70.7	-	4	[80]
COSAC	17.67	60	5	[45]
Zr-G-C$_3$N$_4$	144.1	48	5	This work

3.2.8. Comparative Study

The as-fabricated Zr-G-C$_3$N$_4$ mesoporous nanomaterial demonstrates excellent efficiency for the removal of Cu^{2+} metal ions in aqueous solutions (see Table 3).

The as-fabricated nanocomposite exhibits a very high uptake capability of 144.1 mg/g in a relatively shorter time (48 min) under the optimal experimental conditions. This achievement can be explicitly attributed to the material's high surface area (96 m^2/g), nanostructure (50 nm), and mesoporous nature. These findings affirm that the as-fabricated Zr-G-C$_3$N$_4$ by a simple and low-cost route can be an effective material for eliminating other potentially organic pollutants and toxic metals [74,75].

4. Conclusions

The present study emphasizes the capability of the mesoporous Zr-G-C$_3$N$_4$ sorbent fabricated by a simple ultrasonication route to effectively eliminate the Cu^{2+} metal ions. Adsorption equilibrium isotherms were expressed by Langmuir, Freundlich, and Dubinin–Radushkevich (D-R) adsorption models. The elimination of Cu ions was well fitted by the Langmuir model at a pH of 5.0. The adsorptive capacity of Zr-G-C$_3$N$_4$ for copper ions is comparable to—or even better than—many other nanoadsorbents, and surpasses 144 mg/g. The Cu^{2+} ion adsorption was significantly rapid, and the pseudo-second-order rate isotherm better describes the experimental data. The intraparticle diffusion model's results suggest a border coating impact, confirming intraparticle diffusion's role in Cu ions elimination by Zr-G-C$_3$N$_4$. In addition, the proven reusability and high regeneration capabilities, even after four adsorption–desorption cycles indicate the feasibility for the Cu ions' adsorption onto the Zr-G-C$_3$N$_4$ nanosorbent. Based on the obtained results, the produced nanoadsorbents have the potential to be used as suitable adsorbents for eliminating Cu^{2+} ions.

Author Contributions: L.K. and A.M.: conceptualization and methodology; L.K., A.M. and N.B.H.: writing—original draft preparation; L.K., A.M., K.K.T., N.B.H. and M.B.: writing—review and editing; L.K., A.A.A. and M.B.: supervision and editing. All authors have read and agreed to the published version of the manuscript.

Funding: This research received no external funding.

Data Availability Statement: Not applicable.

Acknowledgments: The authors extend their appreciation to the Deanship of Scientific Research at Imam Mohammad Ibn Saud Islamic University (IMSIU) for funding and supporting this work through Research Partnership Program no RP-21-09-66.

Conflicts of Interest: The authors declare no conflict of interest.

References

1. Anbia, M.; Ghassemian, Z. Removal of Cd(II) and Cu(II) from aqueous solutions using mesoporous silicate containing zirconium and iron. *Chem. Eng. Res. Des.* **2011**, *89*, 2770–2775. [CrossRef]
2. Anbia, M.; Haqshenas, M. Adsorption studies of Pb(II) and Cu(II) ions on mesoporous carbon nitride functionalized with melamine-based dendrimer amine. *Int. J. Environ. Sci. Technol.* **2015**, *12*, 2649–2664. [CrossRef]
3. Teodoro, F.S.; Soares, L.C.; Filgueiras, J.G.; de Azevedo, E.R.; Patiño-Agudelo, J.; Adarme, O.F.H.; da Silva, L.H.M.; Gurgel, L.V.A. Batch and continuous adsorption of Cu(II) and Zn(II) ions from aqueous solution on bi-functionalized sugarcane-based biosorbent. *Environ. Sci. Pollut. Res.* **2021**, *29*, 26425–26448. [CrossRef] [PubMed]
4. Wang, R.-H.; Zhu, X.-F.; Qian, W.; Yu, Y.-C.; Xu, R.-K. Effect of pectin on adsorption of Cu(II) by two variable-charge soils from southern China. *Environ. Sci. Pollut. Res.* **2015**, *22*, 19687–19694. [CrossRef]
5. Meng, J.; Feng, X.; Dai, Z.; Liu, X.; Wu, J.; Xu, J. Adsorption characteristics of Cu(II) from aqueous solution onto biochar derived from swine manure. *Environ. Sci. Pollut. Res.* **2014**, *21*, 7035–7046. [CrossRef]
6. Dong, J.; Du, Y.; Duyu, R.; Shang, Y.; Zhang, S.; Han, R. Adsorption of copper ion from solution by polyethylenimine modified wheat straw. *Bioresour. Technol. Rep.* **2019**, *6*, 96–102. [CrossRef]
7. Trikkaliotis, D.G.; Christoforidis, A.K.; Mitropoulos, A.C.; Kyzas, G.Z. Adsorption of copper ions onto chitosan/poly(vinyl alcohol) beads functionalized with poly(ethylene glycol). *Carbohydr. Polym.* **2020**, *234*, 115890. [CrossRef]
8. Khan, J.; Lin, S.; Nizeyimana, J.C.; Wu, Y.; Wang, Q.; Liu, X. Removal of copper ions from wastewater via adsorption on modified hematite (α-Fe_2O_3) iron oxide coated sand. *J. Clean. Prod.* **2021**, *319*, 128687. [CrossRef]
9. Ye, X.; Shang, S.; Zhao, Y.; Cui, S.; Zhong, Y.; Huang, L. Ultra-efficient adsorption of copper ions in chitosan–montmorillonite composite aerogel at wastewater treatment. *Cellulose* **2021**, *28*, 7201–7212. [CrossRef]
10. Su, C.; Berekute, A.K.; Yu, K.-P. Chitosan@ TiO_2 composites for the adsorption of copper (II) and antibacterial applications. *Sustain. Environ. Res.* **2022**, *32*, 1–15. [CrossRef]
11. Reddad, Z.; Gerente, C.; Andres, Y.; Le Cloirec, P. Adsorption of Several Metal Ions onto a Low-Cost Biosorbent: Kinetic and Equilibrium Studies. *Environ. Sci. Technol.* **2002**, *36*, 2067–2073. [CrossRef] [PubMed]
12. Gavrilescu, M. Removal of Heavy Metals from the Environment by Biosorption. *Eng. Life Sci.* **2004**, *4*, 219–232. [CrossRef]
13. Wan, J.; Chen, L.; Li, Q.; Ye, Y.; Feng, X.; Zhou, A.; Long, X.; Xia, D.; Zhang, T.C. A novel hydrogel for highly efficient adsorption of Cu (II): Synthesis, characterization, and mechanisms. *Environ. Sci. Pollut. Res.* **2020**, *27*, 26621–26630. [CrossRef] [PubMed]
14. Al-Saydeh, S.A.; El-Naas, M.H.; Zaidi, S.J. Copper removal from industrial wastewater: A comprehensive review. *J. Ind. Eng. Chem.* **2017**, *56*, 35–44. [CrossRef]
15. Liu, Y.-L.; Yang, L.; Li, P.; Li, S.-J.; Li, L.; Pang, X.-X.; Ye, F.; Fu, Y. A novel colorimetric and "turn-off" fluorescent probe based on catalyzed hydrolysis reaction for detection of Cu2+ in real water and in living cells. *Spectrochim. Acta Part A Mol. Biomol. Spectrosc.* **2020**, *227*, 117540. [CrossRef]
16. Ismael, M. A review on graphitic carbon nitride (g-C3N4) based nanocomposites: Synthesis, categories, and their application in photocatalysis. *J. Alloys Compd.* **2020**, *846*, 156446. [CrossRef]
17. Obregón, S. Exploring nanoengineering strategies for the preparation of graphitic carbon nitride nanostructures. *FlatChem* **2023**, *38*, 100473. [CrossRef]
18. Zhang, M.; Yang, Y.; An, X.; Hou, L.-A. A critical review of g-C3N4-based photocatalytic membrane for water purification. *Chem. Eng. J.* **2021**, *412*, 128663. [CrossRef]
19. Shahbazi, A.; Younesi, H.; Badiei, A. Functionalized SBA-15 mesoporous silica by melamine-based dendrimer amines for adsorptive characteristics of Pb (II), Cu (II) and Cd (II) heavy metal ions in batch and fixed bed column. *Chem. Eng. J.* **2011**, *168*, 505–518. [CrossRef]
20. Khezami, L.; Elamin, N.; Modwi, A.; Taha, K.K.; Amer, M.S.; Bououdina, M. Mesoporous Sn@ TiO_2 nanostructures as excellent adsorbent for Ba ions in aqueous solution. *Ceram. Int.* **2022**, *48*, 5805–5813. [CrossRef]
21. Mustafa, B.; Modwi, A.; Ismail, M.; Makawi, S.; Hussein, T.; Abaker, Z.; Khezami, L. Adsorption performance and Kinetics study of Pb(II) by RuO_2–ZnO nanocomposite: Construction and Recyclability. *Int. J. Environ. Sci. Technol.* **2022**, *19*, 327–340. [CrossRef]
22. Abdulkhair, B.; Salih, M.; Modwi, A.; Adam, F.; Elamin, N.; Seydou, M.; Rahali, S. Adsorption behavior of barium ions onto ZnO surfaces: Experiments associated with DFT calculations. *J. Mol. Struct.* **2021**, *1223*, 128991. [CrossRef]
23. Ghiloufi, I.; El Ghoul, J.; Modwi, A.; El Mir, L. Ga-doped ZnO for adsorption of heavy metals from aqueous solution. *Mater. Sci. Semicond. Process.* **2016**, *42*, 102–106. [CrossRef]
24. Modwi, A.; Khezami, L.; Taha, K.; Al-Duaij, O.; Houas, A. Fast and high efficiency adsorption of Pb(II) ions by Cu/ZnO composite. *Mater. Lett.* **2017**, *195*, 41–44. [CrossRef]

25. Khezami, L.; Taha, K.K.; Modwi, A. Efficient Removal of Cobalt from Aqueous Solution by Zinc Oxide Nanoparticles: Kinetic and Thermodynamic Studies. *Z. Nat. A* **2017**, *72*, 409–418. [CrossRef]
26. Khezami, L.; Modwi, A.; Ghiloufi, I.; Taha, K.K.; Bououdina, M.; ElJery, A.; El Mir, L. Effect of aluminum loading on structural and morphological characteristics of ZnO nanoparticles for heavy metal ion elimination. *Environ. Sci. Pollut. Res.* **2020**, *27*, 3086–3099. [CrossRef] [PubMed]
27. Singh, J.; Singh, S.; Gill, R. Applications of biopolymer coatings in biomedical engineering. *J. Electrochem. Sci. Eng.* **2022**, *13*, 63–81. [CrossRef]
28. Nevarez-Rascon, A.; González-Lopez, S.; Acosta-Torres, L.S.; Nevarez-Rascon, M.M.; Borunda, E.O. Synthesis, biocompatibility and mechanical properties of ZrO2-Al2O3 ceramics composites. *Dent. Mater. J.* **2016**, *35*, 392–398. [CrossRef]
29. Sayed, E.T.; Abdelkareem, M.A.; Alawadhi, H.; Elsaid, K.; Wilberforce, T.; Olabi, A. Graphitic carbon nitride/carbon brush composite as a novel anode for yeast-based microbial fuel cells. *Energy* **2021**, *221*, 119849. [CrossRef]
30. Svoboda, L.; Bednář, J.; Dvorský, R.; Panáček, A.; Hochvaldová, L.; Kvítek, L.; Malina, T.; Konvičková, Z.; Henych, J.; Němečková, Z.; et al. Crucial cytotoxic and antimicrobial activity changes driven by amount of doped silver in biocompatible carbon nitride nanosheets. *Colloids Surf. B Biointerfaces* **2021**, *202*, 111680. [CrossRef]
31. Xavier, J.R. Electrochemical and Mechanical Investigation of Newly Synthesized NiO-ZrO 2 Nanoparticle–Grafted Polyure-thane Nanocomposite Coating on Mild Steel in Chloride Media. *J. Mater. Eng. Perform.* **2021**, *30*, 1554–1566. [CrossRef]
32. Channu, V.R.; Kalluru, R.R.; Schlesinger, M.; Mehring, M.; Holze, R. Synthesis and characterization of ZrO2 nanoparticles for optical and electrochemical applications. *Colloids Surf. A Physicochem. Eng. Asp.* **2011**, *386*, 151–157. [CrossRef]
33. Yu, Y.; Yan, W.; Wang, X.; Li, P.; Gao, W.; Zou, H.; Wu, S.; Ding, K. Surface engineering for extremely enhanced charge separation and photocatalytic hydrogen evolution on g-C3N4. *Adv. Mater.* **2018**, *30*, 1705060. [CrossRef] [PubMed]
34. Gonçalves, N.; Carvalho, J.; Lima, Z.; Sasaki, J. Size–strain study of NiO nanoparticles by X-ray powder diffraction line broadening. *Mater. Lett.* **2012**, *72*, 36–38. [CrossRef]
35. Wang, X.; Wang, S.; Hu, W.; Cai, J.; Zhang, L.; Dong, L.; Zhao, L.; He, Y. Synthesis and photocatalytic activity of SiO_2/g-C_3N_4 composite photocatalyst. *Mater. Lett.* **2014**, *115*, 53–56. [CrossRef]
36. Modwi, A.; Abbo, M.A.; Hassan, E.A.; Houas, A. Effect of annealing on physicochemical and photocatalytic activity of Cu5% loading on ZnO synthesized by sol–gel method. *J. Mater. Sci. Mater. Electron.* **2016**, *27*, 12974–12984. [CrossRef]
37. Lagergren, S.K. About the theory of so-called adsorption of soluble substances. *Sven. Vetenskapsakad. Handingarl.* **1898**, *24*, 1–39.
38. Abdelrahman, E.A.; El-Reash, Y.A.; Youssef, H.M.; Kotp, Y.H.; Hegazey, R. Utilization of rice husk and waste aluminum cans for the synthesis of some nanosized zeolite, zeolite/zeolite, and geopolymer/zeolite products for the efficient removal of Co(II), Cu(II), and Zn(II) ions from aqueous media. *J. Hazard. Mater.* **2021**, *401*, 123813. [CrossRef]
39. Chien, S.H.; Clayton, W.R. Application of Elovich Equation to the Kinetics of Phosphate Release and Sorption in Soils. *Soil Sci. Soc. Am. J.* **1980**, *44*, 265–268. [CrossRef]
40. Ali, I.; Peng, C.; Ye, T.; Naz, I. Sorption of cationic malachite green dye on phytogenic magnetic nanoparticles functionalized by 3-marcaptopropanic acid. *RSC Adv.* **2018**, *8*, 8878–8897. [CrossRef]
41. Wang, X.; Cai, W.; Lin, Y.; Wang, G.; Liang, C. Mass production of micro/nanostructured porous ZnO plates and their strong structurally enhanced and selective adsorption performance for environmental remediation. *J. Mater. Chem.* **2010**, *20*, 8582. [CrossRef]
42. Acharya, J.; Sahu, J.; Mohanty, C.; Meikap, B. Removal of lead(II) from wastewater by activated carbon developed from Tamarind wood by zinc chloride activation. *Chem. Eng. J.* **2009**, *149*, 249–262. [CrossRef]
43. Ahmad, M.A.; Rahman, N.K. Equilibrium, kinetics and thermodynamic of Remazol Brilliant Orange 3R dye adsorption on coffee husk-based activated carbon. *Chem. Eng. J.* **2011**, *170*, 154–161. [CrossRef]
44. Kumar, K.V.; Ramamurthi, V.; Sivanesan, S. Biosorption of malachite green, a cationic dye onto Pithophora sp., a fresh water algae. *Dye. Pigment.* **2006**, *69*, 102–107. [CrossRef]
45. Podder, M.; Majumder, C. Biosorption of As (III) and As (V) on the surface of TW/$MnFe_2O_4$ composite from wastewater: Kinetics, mechanistic and thermodynamics. *Appl. Water Sci.* **2017**, *7*, 2689–2715. [CrossRef]
46. Ismail, M.; Jobara, A.; Bekouche, H.; Allateef, M.A.; Ben Aissa, M.A.; Modwi, A. Impact of Cu Ions removal onto MgO nanostructures: Adsorption capacity and mechanism. *J. Mater. Sci. Mater. Electron.* **2022**, *33*, 12500–12512. [CrossRef]
47. Brdar, M.; Šćiban, M.; Takači, A.; Došenović, T. Comparison of two and three parameters adsorption isotherm for Cr(VI) onto Kraft lignin. *Chem. Eng. J.* **2012**, *183*, 108–111. [CrossRef]
48. Freundlich, H. Over the adsorption in solution. *J. Phys. Chem.* **1906**, *57*, 1100–1107.
49. Temkin, M.I. Adsorption equilibrium and the kinetics of processes on nonhomogeneous surfaces and in the interaction between adsorbed molecules. *Zh. Fiz. Chim.* **1941**, *15*, 296–332.
50. Dubinin, M. The equation of the characteristic curve of activated charcoal. *Dokl. Akad. Nauk. SSSR.* **1947**, *55*, 327–329.
51. Hu, Q.; Zhang, Z. Application of Dubinin–Radushkevich isotherm model at the solid/solution interface: A theoretical analysis. *J. Mol. Liq.* **2019**, *277*, 646–648. [CrossRef]
52. Nikiforova, T.E.; Kozlov, V.A.; Telegin, F.Y. Chemisorption of copper ions in aqueous acidic solutions by modified chitosan. *Mater. Sci. Eng. B* **2021**, *263*, 114778. [CrossRef]
53. Bae, M.; Lee, H.; Yoo, K.; Kim, S. Copper(I) selective chemisorption on magnetite (Fe_3O_4) over gold(I) ions in chloride solution with cyanide. *Hydrometallurgy* **2021**, *201*, 105560. [CrossRef]

54. Gao, X.; Wu, L.; Xu, Q.; Tian, W.; Li, Z.; Kobayashi, N. Adsorption kinetics and mechanisms of copper ions on activated carbons derived from pinewood sawdust by fast H$_3$PO$_4$ activation. *Environ. Sci. Pollut. Res.* **2018**, *25*, 7907–7915. [CrossRef] [PubMed]
55. Buema, G.; Trifas, L.-M.; Harja, M. Removal of Toxic Copper Ion from Aqueous Media by Adsorption on Fly Ash-Derived Zeolites: Kinetic and Equilibrium Studies. *Polymers* **2021**, *13*, 3468. [CrossRef] [PubMed]
56. Kumar, G.P.; Kumar, P.A.; Chakraborty, S.; Ray, M. Uptake and desorption of copper ion using functionalized polymer coated silica gel in aqueous environment. *Sep. Purif. Technol.* **2007**, *57*, 47–56. [CrossRef]
57. Salam, M.A.; Al-Zhrani, G.; Kosa, S.A. Simultaneous removal of copper (II), lead (II), zinc (II) and cadmium (II) from aqueous solutions by multi-walled carbon nanotubes. *Comptes Rendus Chim.* **2012**, *15*, 398–408. [CrossRef]
58. Toscano, G.; Caristi, C.; Cimino, G. Sorption of heavy metal from aqueous solution by volcanic ash. *Comptes Rendus Chim.* **2008**, *11*, 765–771. [CrossRef]
59. Ren, X.; Li, J.; Tan, X.; Wang, X. Comparative study of graphene oxide, activated carbon and carbon nanotubes as adsorbents for copper de-contamination. *Dalton Trans.* **2013**, *42*, 5266–5274. [CrossRef]
60. Bohli, T.; Ouederni, A.; Fiol, N.; Villaescusa, I. Evaluation of an activated carbon from olive stones used as an adsorbent for heavy metal removal from aqueous phases. *Comptes Rendus Chim.* **2015**, *18*, 88–99. [CrossRef]
61. Mao, N.; Jiang, J.-X. MgO/g-C$_3$N$_4$ nanocomposites as efficient water splitting photocatalysts under visible light irradiation. *Appl. Surf. Sci.* **2019**, *476*, 144–150. [CrossRef]
62. Ge, L.; Han, C. Synthesis of MWNTs/g-C$_3$N$_4$ composite photocatalysts with efficient visible light photocatalytic hydrogen evolution activity. *Appl. Catal. B Environ.* **2012**, *117-118*, 268–274. [CrossRef]
63. Ji, H.; Chang, F.; Hu, X.; Qin, W.; Shen, J. Photocatalytic degradation of 2,4,6-trichlorophenol over g-C$_3$N$_4$ under visible light irradiation. *Chem. Eng. J.* **2013**, *218*, 183–190. [CrossRef]
64. Velu, S.; Suzuki, K.; Gopinath, C.S.; Yoshida, H.; Hattori, T. XPS, XANES and EXAFS investigations of CuO/ZnO/Al$_2$O$_3$/ZrO$_2$ mixed oxide catalysts. *Phys. Chem. Chem. Phys.* **2002**, *4*, 1990–1999. [CrossRef]
65. Ragupathy, P.; Park, D.H.; Campet, G.; Vasan, H.N.; Hwang, S.-J.; Choy, J.-H.; Munichandraiah, N. Remarkable Capacity Retention of Nanostructured Manganese Oxide upon Cycling as an Electrode Material for Supercapacitor. *J. Phys. Chem. C* **2009**, *113*, 6303–6309. [CrossRef]
66. Toghan, A.; Modwi, A. Boosting unprecedented indigo carmine dye photodegradation via mesoporous MgO@ g-C$_3$N$_4$ nanocomposite. *J. Photochem. Photobiol. A Chem.* **2021**, *419*, 113467. [CrossRef]
67. Muhmood, T.; Xia, M.; Lei, W.; Wang, F.; Khan, M.A. Efficient and stable ZrO$_2$/Fe modified hollow-C 3 N 4 for photodegradation of the herbicide MTSM. *RSC Adv.* **2017**, *7*, 3966–3974. [CrossRef]
68. Ismael, M.; Wu, Y.; Wark, M. Photocatalytic activity of ZrO$_2$ composites with graphitic carbon nitride for hydrogen production under visible light. *New J. Chem.* **2019**, *43*, 4455–4462. [CrossRef]
69. Liu, X.; Zhang, S.; Liu, J.; Wei, X.; Yang, T.; Chen, M.; Wang, J. Rare-Earth Doping Graphitic Carbon Nitride Endows Distinctive Multiple Emissions with Large Stokes Shifts. *CCS Chem.* **2022**, *4*, 1990–1999. [CrossRef]
70. Guo, J.; Chen, T.; Zhou, X.; Xia, W.; Zheng, T.; Zhong, C.; Liu, Y. Synthesis, Cr(VI) removal performance and mechanism of nanoscale zero-valent iron modified potassium-doped graphitic carbon nitride. *Water Sci. Technol.* **2020**, *81*, 1840–1851. [CrossRef]
71. Zou, X.; Silva, R.; Goswami, A.; Asefa, T. Cu-doped carbon nitride: Bio-inspired synthesis of H2-evolving electrocatalysts using graphitic carbon nitride (g-C$_3$N$_4$) as a host material. *Appl. Surf. Sci.* **2015**, *357*, 221–228. [CrossRef]
72. Zhang, L.; Guo, J.; Huang, X.; Wang, W.; Sun, P.; Li, Y.; Han, J. Functionalized biochar-supported magnetic MnFe$_2$O$_4$ nanocomposite for the removal of Pb (ii) and Cd (ii). *RSC Adv.* **2019**, *9*, 365–376. [CrossRef] [PubMed]
73. Shen, C.; Chen, C.; Wen, T.; Zhao, Z.; Wang, X.; Xu, A. Superior adsorption capacity of g-C$_3$N$_4$ for heavy metal ions from aqueous solutions. *J. Colloid Interface Sci.* **2015**, *456*, 7–14. [CrossRef]
74. Awual, M.R.; Yaita, T.; Suzuki, S.; Shiwaku, H. Ultimate selenium (IV) monitoring and removal from water using a new class of organic ligand based composite adsorbent. *J. Hazard. Mater.* **2015**, *291*, 111–119. [CrossRef] [PubMed]
75. Aziz, F.; El Achaby, M.; Lissaneddine, A.; Aziz, K.; Ouazzani, N.; Mamouni, R.; Mandi, L. Composites with alginate beads: A novel design of nano-adsorbents impregnation for large-scale continuous flow wastewater treatment pilots. *Saudi J. Biol. Sci.* **2020**, *27*, 2499–2508. [CrossRef]
76. Şimşek, S.; Derin, Y.; Kaya, S.; Şenol, Z.M.; Katin, K.P.; Özer, A.; Tutar, A. High-performance material for the effective removal of uranyl ion from solution: Computationally sup-ported experimental studies. *Langmuir* **2022**, *38*, 10098–10113. [CrossRef] [PubMed]
77. Ozdemir, S.; Turkan, Z.; Kilinc, E.; Bayat, R.; Soylak, M.; Sen, F. Preconcentrations of Cu (II) and Mn (II) by magnetic solid-phase extraction on Bacillus cereus loaded γ-Fe2O3 nanomaterials. *Environ. Res.* **2022**, *209*, 112766. [CrossRef]
78. Morozova, A.G.; Lonzinger, T.M.; Skotnikov, V.A.; Mikhailov, G.G.; Kapelyushin, Y.; Khandaker, M.U.; Alqahtani, A.; Bradley, D.A.; Sayyed, M.I.; Tishkevich, D.I.; et al. Insights into sorption–mineralization mechanism for sustainable granular composite of MgO-CaO-Al$_2$O$_3$-SiO$_2$-CO$_2$ based on nanosized adsorption centers and its effect on aqueous Cu (II) removal. *Nanomaterials* **2021**, *12*, 116. [CrossRef]
79. Claros, M.; Kuta, J.; El-Dahshan, O.; Michalička, J.; Jimenez, Y.P.; Vallejos, S. Hydrothermally synthesized MnO$_2$ nanowires and their application in Lead (II) and Copper (II) batch ad-sorption. *J. Mol. Liq.* **2021**, *325*, 115203. [CrossRef]

80. Baaloudj, O.; Nasrallah, N.; Kebir, M.; Guedioura, B.; Amrane, A.; Nguyen-Tri, P.; Nanda, S.; Assadi, A.A. Artificial neural network modeling of cefixime photodegradation by synthesized CoBi$_2$O$_4$ na-noparticles. *Environ. Sci. Pollut. Res.* **2021**, *28*, 15436–15452. [CrossRef]
81. Azzaz, A.A.; Jellali, S.; Akrout, H.; Assadi, A.A.; Bousselmi, L. Dynamic investigations on cationic dye desorption from chemically modified lignocellulosic material using a low-cost eluent: Dye recovery and anodic oxidation efficiencies of the desorbed solutions. *J. Clean. Prod.* **2018**, *201*, 28–38. [CrossRef]
82. Yang, X.; Kameda, T.; Saito, Y.; Kumagai, S.; Yoshioka, T. Investigation of the mechanism of Cu(II) removal using Mg-Al layered double hydroxide intercalated with carbonate: Equilibrium and pH studies and solid-state analyses. *Inorg. Chem. Commun.* **2021**, *132*, 108839. [CrossRef]

Disclaimer/Publisher's Note: The statements, opinions and data contained in all publications are solely those of the individual author(s) and contributor(s) and not of MDPI and/or the editor(s). MDPI and/or the editor(s) disclaim responsibility for any injury to people or property resulting from any ideas, methods, instructions or products referred to in the content.

Article

Applying Linear Forms of Pseudo-Second-Order Kinetic Model for Feasibly Identifying Errors in the Initial Periods of Time-Dependent Adsorption Datasets

Hai Nguyen Tran [1,2]

1 Center for Energy and Environmental Materials, Institute of Fundamental and Applied Sciences, Duy Tan University, Ho Chi Minh 700000, Vietnam; trannguyenhai@duytan.edu.vn or trannguyenhai2512@gmail.com
2 Faculty of Environmental and Chemical Engineering, Duy Tan University, Da Nang 550000, Vietnam

Abstract: Initial periods of adsorption kinetics play an important role in estimating the initial adsorption rate and rate constant of an adsorption process. Several adsorption processes rapidly occur, and the experimental data of adsorption kinetics under the initial periods can contain potential errors. The pseudo-second-order (PSO) kinetic model has been popularly applied in the field of adsorption. The use of the nonlinear optimization method to obtain the parameters of the PSO model can minimize error functions during modelling compared to the linear method. However, the nonlinear method has limitations in that it cannot directly recognize potential errors in the experimental points of time-dependent adsorption, especially under the initial periods. In this study, for the first time, the different linear types (Types 1–6) of the PSO model are applied to discover the error points under the initial periods. Results indicated that the fitting method using its linear equations (Types 2–5) is really helpful for identifying the error (doubtful) experimental points from the initial periods of adsorption kinetics. The imprecise points lead to low adjusted R^2 (adj-R^2), high reduced χ^2 (red-χ^2), and high Bayesian information criterion (BIC) values. After removing these points, the experimental data were adequately fitted with the PSO model. Statistical analyses demonstrated that the nonlinear method must be used for modelling the PSO model because its red-χ^2 and BIC were lower than the linear method. Type 1 has been extensively applied in the literature because of its very high adj-R^2 value (0.9999) and its excellent fitting to experimental points. However, its application should be limited because the potential errors from experimental points are not identified by this type. For comparison, the other kinetic models (i.e., pseudo-first-order, pseudo-nth-order, Avrami, and Elovich) are applied. The modelling result using the nonlinear forms of these models indicated that the fault experimental points from the initial periods were not detected in this study.

Keywords: adsorption kinetics; pseudo-second-order model; nonlinear method; linear method

Citation: Tran, H.N. Applying Linear Forms of Pseudo-Second-Order Kinetic Model for Feasibly Identifying Errors in the Initial Periods of Time-Dependent Adsorption Datasets. *Water* **2023**, *15*, 1231. https://doi.org/10.3390/w15061231

Academic Editor: Cidália Botelho

Received: 9 February 2023
Revised: 9 March 2023
Accepted: 16 March 2023
Published: 21 March 2023

Copyright: © 2023 by the author. Licensee MDPI, Basel, Switzerland. This article is an open access article distributed under the terms and conditions of the Creative Commons Attribution (CC BY) license (https://creativecommons.org/licenses/by/4.0/).

1. Introduction

Adsorption is a combination process (adsorption and desorption occur simultaneously) [1]. Adsorption kinetics is commonly conducted under batch experiments [2]. Some adsorption processes occur very fast. Guo et al. [3] found that the adsorption process of Cd^{2+} ions onto biosorbent (derived from maize straw modified with succinic anhydride) rapidly occurred under the first period of adsorption kinetics, with approximately 70.0–96.6% of total Cd^{2+} ion solutions (C_o = 100, 300, and 500 mg/L at pH = 5.8) being removed within 1 min. Zbair et al. [4] reported that the adsorption process of bisphenol A or diuron by Argan-nutshell-based hydrochar reached a fast equilibrium within 4 min of contact. An analogous result has been reported by many scholars who invested in the adsorption of Cd^{2+} ions onto orange-peel-derived biochar [5] and rhodamine B dye onto white-sugar-based activated carbon [6]. Therefore, some authors [3,5,6] took a very short time (1 min, 2 min, etc.) for adsorption kinetics instead of a longer one (i.e., starting from 10 min) [7].

Clearly, the initial periods of time-dependent adsorption datasets play an important role in establishing the adsorption rate constant and initial rate of adsorption kinetics. Hubbe et al. [8] called them "early" data points. However, some potential errors or technical mistakes can occur if samples are taken uncarefully and analysed for a very short time (i.e., 1 min, 2 min, etc.). The adsorption process often includes two phases: solid (material or adsorbent) and liquid. The liquid phase often contains solute (or adsorbate) and solvent (commonly water). In general, after withdrawing small fractions from the mixture of the solid and liquid at an interval of time, the liquid phase is separated from the mixture by the filtration or centrifugal method. A dilution step is additionally required if necessary. The available sites in a material can continue to adsorb adsorbate molecules in the withdrawn fractions if the separation process is delayed or not performed instantly. As a result, the concentrations of adsorbate at given times (C_t) decrease, and the amount of adsorbate adsorbed by an adsorbent at times (q_t) increases. This means that the values q_t at the initial periods might be higher than those expected if researchers do not conduct experiments carefully. They can be defined as the error points in datasets and affect the results of modelling adsorption on kinetics. It might be hard to verify those doubtful results by existing techniques, and this is a current limitation in this field.

In the study of kinetic adsorption, two common (reaction-called) models that are used to model time-dependent experiment data of the adsorption process are the pseudo-first-order (PFO) model and pseudo-second-order (PSO) model [9–11]. Lima et al. [2] concluded in their review that the PSO model describes most experimental data better than the PFO model. Some authors [12,13] indicated that the PSO model was initially established by Blanchard and co-workers [14] in 1984 to describe the time-dependent process of heavy metal adsorption by clinoptilolite. However, other authors [8] found that Coleman and co-workers [15] proposed a similar model earlier in 1956. Its differential equation is commonly expressed as Equation (1) [12]. After taking integration by applying the boundary conditions (q_t = 0 at t = 0, and q_t = q_t at t = t), the nonlinear form of the PSO model is obtained as Equation (2) [16].

$$\frac{dq_t}{dt} = k_2(q_{e(2)} - q_t)^2 \tag{1}$$

$$q_t = \frac{q_{e(2)}^2 k_2 t}{1 + q_{e(2)} k_2 t} \tag{2}$$

$$q_t = \frac{C_o - C_t}{m} \times V \tag{3}$$

where k_2 (kg/(mol × min)) is the rate constant of this model; $q_{e(2)}$ and q_t (mol/kg) are the amounts of pollutant in the solution adsorbed by the material at equilibrium and any time t (min) is obtained from Equation (3); C_o and C_t (mol/L) are the concentrations of the pollutant at the beginning and any time t (min); m (kg) is the dry mass of the material; and V (L) is the volume of the pollutant used in the study of the adsorption kinetics.

Six types of linear forms of the PSO model are found in the literature [12,17–19]. Although Type 5 and Type 6 have been reported elsewhere [17,19], they are not commonly applied in the literature compared to Type 1. Type 1, which might have been introduced first by Ho and McKay [20], has been intensively applied in the literature because the R^2 or r^2 value obtained from this type is very close to 1 [2,12].

$$\frac{t}{q_t} = \left(\frac{1}{q_{e(2)}}\right)t + \frac{1}{k_2 q_{e(2)}^2} \quad (\text{Type1; the plot of } t/q_t vs. t) \tag{4}$$

$$\frac{1}{q_t} = \left(\frac{1}{k_2 q_{e(2)}^2}\right)\frac{1}{t} + \frac{1}{q_{e(2)}} \quad (\text{Type2; the plot of } 1/q_t vs. 1/t) \tag{5}$$

$$q_t = \left(\frac{-1}{k_2 q_{e(2)}}\right)\frac{q_t}{t} + q_{e(2)} \quad \text{(Type3; the plot of } q_t vs. q_t/t) \tag{6}$$

$$\frac{q_t}{t} = -(k_2 q_{e(2)})q_t + k_2 q_{e(2)}^2 \quad \text{(Type4; the plot of } q_t/t vs. q_t) \tag{7}$$

$$\frac{1}{t} = k_2 q_{e(2)}^2 \frac{1}{q_t} - k_2 q_{e(2)} \quad \text{(Type5; the plot of } 1/t vs. 1/q_t) \tag{8}$$

$$\frac{1}{(q_{e(2)} - q_t)} = k_2 t + \frac{1}{q_{e(2)}} \quad \text{(Type6; the plot of } 1/(q_e - q_t) vs. t) \tag{9}$$

For comparison, several statistical analyses were applied along with the determination coefficient (R^2; Equation (10)) and adjusted determination coefficient (adj-R^2; Equation (11)). They include chi-squared (χ^2; Equation (12)), reduced chi-square (red-χ^2; Equation (13)), and Bayesian information criterion (BIC; Equation (14)) [2,12,16] A model with higher adj-R^2 value and lower red-χ^2 indicates a better fitting than the others. Because of the difference among the parameters of the models used, the magnitude (or the absolute value) of ΔBIC (BIC of model 1–BIC of model 2) is used for selecting the best fitting model [2,16].

$$R^2 = \frac{\sum(y_{exp} - y_{exp-mean})^2 - \sum(y_{exp} - y_{model})^2}{\sum(y_{exp} - y_{exp-mean})^2} = 1 - \frac{\sum(y_{exp} - y_{model})^2}{\sum(y_{exp} - y_{exp-mean})^2} \tag{10}$$

$$\text{adj} - R^2 = 1 - (1 - R^2) \times \left(\frac{N-1}{\text{DOF}}\right) \tag{11}$$

$$\text{red} - \chi^2 = \frac{\sum(y_{exp} - y_{model})^2}{\text{DOF}} \tag{12}$$

$$\text{BIC} = N \times \ln\left(\frac{\sum(y_{exp} - y_{model})^2}{N}\right) + P\ln(N) \tag{13}$$

where y_{exp} is q_t (mol/kg) obtained from experiments; $y_{exp-mean}$ is an average of y_{exp} values used for modelling; y_{model} is q_t calculated based on the PSO model; N is the number of experimental points; P is the number of parameters of the model (I.e., $P = 2$ for the PSO model); and DOF is the degrees of freedom.

The nonlinear optimization method is often suggested to model the adsorption kinetic model because it can minimize error functions [1,2,12,16]. However, this method cannot provide information on potential errors in adsorption kinetic datasets. This study aimed to evaluate whether it is feasible to apply the linear forms of the pseudo-second-order kinetic model for feasibly identifying errors in the initial periods of kinetic adsorption. This hypothesis is introduced for the first time. The kinetic adsorption data were also fitted to some common models: the pseudo-first-order, pseudo-n^{th}-order, Avrami, and Elovich ones. Commercial activated carbon and paracetamol (PRC) were used as the target adsorbent and adsorbate.

2. Experimental Conditions and Procedures

The adsorption process of paracetamol (PRC, purchased from Sigma-Aldrich, St. Louis, MO, USA) using commercial activated carbon (CAC) was carried out. The surface area and total pore volume of CAC were 1275 m^2/g and 0.670 cm^3/g, respectively. The experiment of kinetics adsorption was conducted in triplicate, and the result of q_t was averaged. The adsorption condition was maintained at 25 °C, pH = 7.0, and initial PRC (3.45 mmol/L). The solutions (1.0 M NaOH and 1.0 M HCl) were used to adjust the pH of PRC solution before and during the adsorption process.

Approximately 1.0 g of CAC was added to an Erlenmeyer flask containing 1 L of PRC solution (3.45 mmol/L and pH = 7.0). The mixture of the solid/liquid was shaken at 150 rpm and 30 °C. The contact time range for studying adsorption kinetic was 1, 5, 15, 30, 45, 60, 90, 120, 180, 240, 300, 360, 720, 1440, 2880, and 4320 min (n = 16). After the desirable time, the solid/liquid mixture was separated using 0.45 μm filter paper. The concentration of PRC in solution was detected by a high-performance liquid chromatography (HPLC; Dionex 3000 Thermo, Sunnyvale, CA, USA) coupled with a photo-diode array. The q_t value was calculated based on the mass balance equation (Equation (3)). The blank samples (without the presence of CAC in the PRC solution) were conducted simultaneously.

3. Pseudo-Second-Order Model

Two regions (kinetic and equilibrium) are observed in Figure 1a. The first region plays an important part in adsorption kinetics because information on the adsorption rate and relevant rate constant is often identified at this region. The second region is plateau points. The PSO and PFO models often tend to fit the adsorption kinetic dataset when the second region (considering specific cases: q_t values are nearly identical at this region) presents in the kinetic curve (the plot of q_t vs. t) [2].

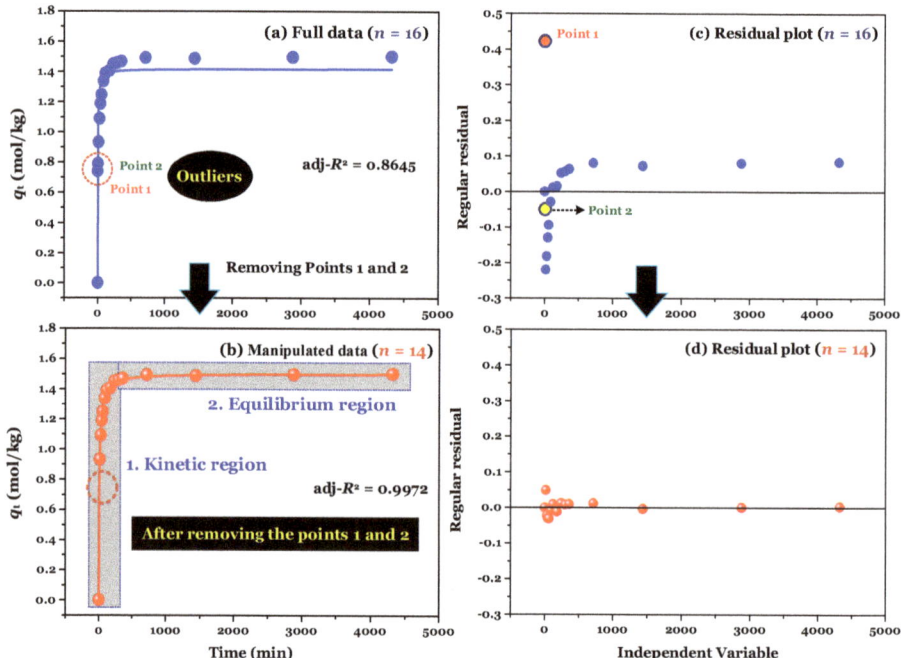

Figure 1. Adsorption kinetics of paracetamol by commercial activated carbon fitted by the PSO model.

The result of modelling (using the nonlinear method) is shown in Figure 1a. Its adj-R^2 value was only 0.8645, suggesting that the adsorption process was not well-described by the PSO model. Clearly, the application of the nonlinear method did not provide information as to where the experimental points of time-dependent adsorption are errors.

The results (n = 16; full raw experimental data) of modelling (using the linear method) are provided in Figure 2a,c (for Type 1 and 2), Figure 3a,c (for Type 3 and 4), and Figure 4a,c (for Type 5 and 6). Clearly, Type 1 was not helpful for recognizing errors in the experimental points. This is reflected through a very high adj-R^2 value of 0.9999 (Table 1). A similar conclusion was obtained for Type 6. In contrast, the fault experimental points (denoted as Point 1 and Point 2 in the corresponding figures) were well identified by the others (Types 2, 3, 4, and 5). This is because their R^2 values only ranged from 0.4571 to 0.5627 (Table 1).

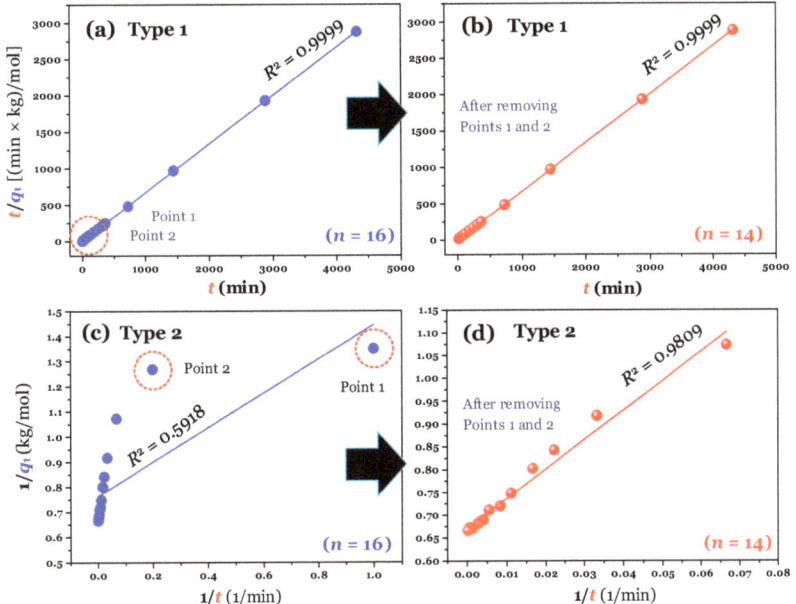

Figure 2. The linear forms (Type 1 and Type 2) of the pseudo-second-order model (**a**,**c**) before ($n = 16$) and (**b**,**d**) after ($n = 14$; Points 1 and 2 removed) data manipulated.

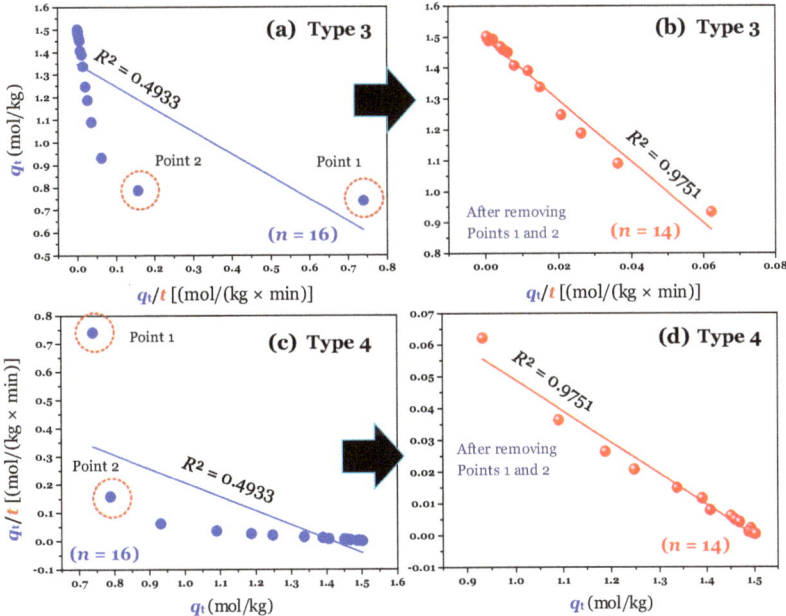

Figure 3. The linear forms (Type 3 and Type 4) of the pseudo-second-order model (**a**,**c**) before ($n = 16$) and (**b**,**d**) after ($n = 14$; Points 1 and 2 removed) data manipulated.

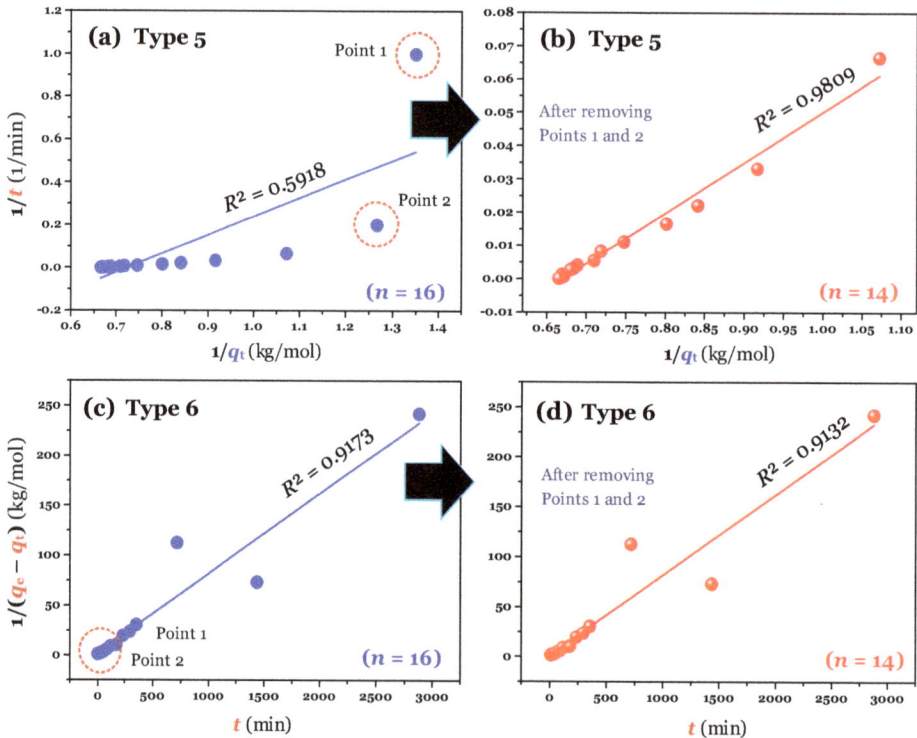

Figure 4. The linear forms (Type 5 and Type 6) of the pseudo-second-order model (**a**,**c**) before (n = 16) and (**b**,**d**) after (n = 14; Points 1 and 2 removed) data manipulated.

Considering Point 1 and Point 2 in all figures as potential errors, a manipulation was performed by removing Points 1 and 2 from the experimental data. As expected, after manipulating (n = 14), the experimental data of the time-dependent adsorption was adequately described by the PSO model, with the adj-R^2 values being 0.9972 (for the nonlinear method), and 0.9731−0.9793 (for the linear form of Type 2–5; Table 1). Figure 1b shows that the experimental data (after removing Points 1 and 2 in Figure 1a) were well described by the nonlinear form of the PSO model (adj-R^2 = 0.9972). Although the evaluation of regular residuals can be used to identify the potential errors within a dataset [11], it is not helpful for many cases. For example, Figure 1c shows that the outlier identification is only visible for Point 1.

Notably, the application of the linear form of Type 1 can lead to misconclusions. This is because its adj-R^2 values did not change before (adj-R^2 = 0.9999) and after manipulation (0.9999; Table 1). The difference between the k_2 (0.0711 kg/(mol × min)) value of full data and the k_2 value (0.0624 kg/(mol × min)) of manipulated data obtained from Type 1 was only 13% (Table 1). However, the difference was remarkable for Type 2 (170%), Type 3 (167%), Type 4 (137%), and Type 5 (148%). A similar percentage of the high difference was reported for the case of using the nonlinear method (105%). The result suggested that the use of the linear form (Type 1) of the PSO model can achieve a very high adj-R^2 value and well describe the experimental data (Figure S1). However, it is very hard to verify whether the parameters of the PSO model are highly believable and accurate. In the literature, some authors have analysed the problems of the linear form of Type 1 [2,21]. They found that although the R^2 value of Type 1 is very high (even reaching 1), it is a spurious correlation [2,21].

Table 1. Parameters of the pseudo-second-order model obtained from the nonlinear form and its six linear forms.

	Result of Modelling										Result of Re-Modelling				
	Full Raw Data ($n = 16$) *					Manipulated Data ($n = 14$) **					Revisited Data ($n = 16$) ***				
	k_2	$q_{e(2)}$	adj-R^2	χ^2	BIC	k_2	$q_{e(2)}$	adj-R^2	χ^2	BIC	k_2	$q_{e(2)}$	adj-R^2	χ^2	BIC
1. Nonlinear form	0.2034	1.419	0.8645	2.2×10^{-2}	−61.6	0.0635	1.501	0.9972	4.2×10^{-4}	−113.3	0.0635	1.501	0.9985	3.65×10^{-4}	−131.0
2. Linear forms															
Type 1	0.0711	1.504	0.9999	3.1×10^{-2}	−52.1	0.0624	1.504	0.9999	4.6×10^{-4}	−104.4	0.0625	1.504	0.9999	4.0×10^{-4}	−121.9
Type 2	0.856	1.309	0.5627	3.4×10^{-2}	−50.8	0.0701	1.489	0.9793	6.0×10^{-4}	−100.8	0.0633	1.504	0.9999	4.0×10^{-4}	−121.9
Type 3	0.746	1.348	0.4571	3.1×10^{-2}	−52.3	0.0673	1.495	0.9731	6.0×10^{-4}	−100.8	0.0641	1.500	0.9953	3.9×10^{-4}	−122.1
Type 4	0.3498	1.418	0.4571	2.6×10^{-2}	−55.2	0.0656	1.498	0.9731	6.0×10^{-4}	−100.8	0.0638	1.501	0.9953	3.9×10^{-4}	−122.1
Type 5	0.4543	1.382	0.5627	2.6×10^{-2}	−54.8	0.0684	1.492	0.9793	6.0×10^{-4}	−100.8	0.0632	1.505	0.9999	4.0×10^{-4}	−121.7
Type 6	0.0804	0.703	0.9109	6.3×10^{-1}	−3.60	0.0804	0.672	0.9053	6.3×10^{-4}	−92.9	0.0805	0.750	0.9173	5.1×10^{-1}	−6.938

Notes: * Case 1: results obtained from the raw datasets. ** Case 2: results obtained after removing the outlets (Points 1 and 2; $q_t = 0.741$ and 0.789 mol/kg); and *** Case 3: results obtained after using the new Points 1 ($q_t = 0.131$ mol/kg; Table 3) and 2 (0.484 mol/kg; Table 3) estimated from the PSO model ($k_2 = 0.0635$ and $q_{e(2)} = 1.501$); the unit of q_e (mol/kg) and k_2 (kg/(mol × min)); $q_{e,exp}$ (1.495 ± 0.0061 mol/kg) obtained from the experiment.

Table 2. Parameters of some models obtained based on the three cases: full raw data, manipulated data, and revisited data.

	Unit	Result of Modeling (Using the Nonlinear Method)		
		Full Raw Data (n = 16) *	Manipulated Data (n = 14) **	Revisited Data (n = 16) ***
$q_{e,exp}$	mol/kg	~1.495	~1.495	~1.495
1. Pseudo-second-order (PSO) model				
$q_{e(2)}$	mol/kg	1.419	1.501	1.501
k_2	kg/(mol × min)	0.2034	0.0635	0.0635
adj-R^2	-	0.8645	0.9972	0.9985
red-χ^2	-	2.16×10^{-2}	4.20×10^{-4}	3.65×10^{-4}
BIC	-	−61.6	−113.3	−131.1
2. Pseudo-first-order (PFO) model				
$q_{e(1)}$	mol/kg	1.368	1.436	1.428
k_1	1/min	0.1792	0.0512	0.0576
adj-R^2	-	0.7506	0.9597	0.9710
red-χ^2	-	3.98×10^{-2}	6.09×10^{-3}	6.86×10^{-3}
BIC	-	−51.3	−85.6	−81.16
3. Elovich model				
α	mol/(kg × min)	126.1	1364	2.913
α	mg/(g × min)	19,073	206,235	440.4
β	kg/mol	9.299	11.11	6.335
adj-R^2	-	0.9435	0.9480	0.8700
red-χ^2	-	9.03×10^{-3}	7.75×10^{-3}	3.08×10^{-2}
BIC	-	−76.5	−81.5	−55.6
4. Avrami model				
$q_{e(AV)}$	mol/kg	1.536	1.496	1.472
k_{AV}	1/min	0.0973	0.0618	0.0510
n_{AV}	-	0.2913	0.4571	0.6312
adj-R^2	-	0.9801	0.9995	0.9935
red-χ^2	-	3.17×10^{-3}	7.35×10^{-5}	1.54×10^{-3}
BIC	-	−95.4	−162.1	−107.7
5. Pseudo-n^{th}-order (PNO) model				
$q_{e(PNO)}$	mol/kg	1.933	1.522	1.515
k_{PNO}	kg^{m-1}/(min × mol^{m-1})	0.0268	0.0659	0.0631
m	-	7.299	2.247	2.134
adj-R^2	-	0.9605	0.9979	0.9986
red-χ^2	-	6.31×10^{-3}	3.21×10^{-4}	3.25×10^{-4}
BIC	-	−83.8	−137.0	−134.1

Notes: * Case 1: results obtained from the raw datasets; ** Case 2: results obtained after removing the outlets (Points 1 and 2; q_t = 0.741 and 0.789 mol/kg) from the raw datasets; and *** Case 3: results obtained after removing Points 1–2 and submitting the new Point 1 (q_t = 0.131 mol/kg; Table 3) and Point 2 (0.484 mol/kg; Table 3) estimated from the PSO model (k_2 = 0.0635 and $q_{e(2)}$ = 1.501) into the raw datasets.

Table 3. Comparison of the result estimation of some q_t values at the initial periods obtained by different kinetic models and $q_{t(exp)}$ from the experiment.

Time (min)	$q_{t(exp)}$	q_t Values Estimated Based on the Models					Difference (%) between $q_{t(exp)}$ and $q_{t(model)}$				
		PFO	PSO	PNO	Elovich	Avrami	PFO	PSO	PNO	Elovich	Avrami
1 min (Point 1)	0.741	0.072	0.131	0.151	0.866	0.042	164.7	140.0	132.4	−15.7	178.7
5 min (Point 2)	0.789	0.324	0.484	0.525	1.011	0.197	83.4	47.8	40.2	−24.7	120.0
15	0.933	0.770	0.883	0.905	1.110	0.517	19.1	5.43	3.03	−17.40	57.4
30	1.090	1.127	1.112	1.114	1.173	0.855	−3.290	−1.965	−2.129	−7.258	24.2
45	1.188	1.293	1.217	1.211	1.209	1.076	−8.446	−2.441	−1.928	−1.770	9.846
60	1.247	1.369	1.278	1.268	1.235	1.221	−9.327	−2.389	−1.635	1.005	2.116

To compare the fitting between the nonlinear and linear methods, the statistical analyses (red-χ^2 and BIC) were applied. The results (Table 1) demonstrated that the values of red-χ^2 and BIC obtained from the nonlinear method were lower than those from the linear method. Therefore, the utilization of the nonlinear method can minimize some error functions during the modelling process. A similar conclusion has been reported by some scholars [2,18,19].

It can be concluded that the application of the linear method for calculating the parameters of the PSO model is not highly recommended compared to the nonlinear method. However, the linear method (Type 2, Type 3, Type 4, and Type 5) is very helpful for identifying errors in the experimental points (outliers). The use of Type 1 of the linear form of the PSO model should be undertaken cautiously. This is because it is very hard to identify experimental points and verify the feasibility of the parameters obtained (especially the k_2 value).

4. Other Common Adsorption Kinetic Models

A question is whether other adsorption kinetic models are helpful for identifying errors in the experimental points of time-dependent adsorption. Therefore, it is necessary to use other models for modelling the experimental data. Apart from the PSO model, several models have been used for modelling the time-dependent adsorption data. They include the two- and three-parameter models [8,22,23]. The pseudo-first-order model [24] and Elovich kinetic model [25] are expressed as Equation (14) and Equation (15), respectively [12]. The three-parameter ones are the pseudo-n^{th}-order model (herein called the PNO model) (Equation (16)) [22,23] (also known as the general-order kinetic model (Equation (17)) [16] and the Avrami-fractional-order model (Equation (18)) [26]. More detailed information on these models has been reported in some documents [16].

$$q_t = q_{e(1)}[1 - \exp(-k_1 t)] \tag{14}$$

$$q_t = \frac{1}{\beta}\ln(1 + \alpha\beta t) \tag{15}$$

$$q_t = q_{e(PNO)} - \frac{q_{e(PNO)}}{\left[1 + (m-1)q_{e(PNO)}^{(m-1)} k_{PNO} t\right]^{\frac{1}{m-1}}} \tag{16}$$

$$q_t = q_{e(PNO)}\left\{1 - \left[\frac{1}{1 + (m-1)q_{e(PNO)}^{(m-1)} k_{PNO} t}\right]^{\frac{1}{m-1}}\right\} \tag{17}$$

$$q_t = q_{e(AV)}\left[1 - \exp(-k_{AV} t)^{n_{AV}}\right] \tag{18}$$

where q_t (mol/kg) is defined in Equation (2); k_1 (1/min), k_{PNO} (kg^{m-1}/(min × mol^{m-1})), and k_{AV} (1/min) are the adsorption rate constants of the PFO, PNO, and Avrami models, respectively; $q_{e(1)}$, $q_{e(PNO)}$, and $q_{e(AV)}$ are the adsorption capacities of CAC towards paracetamol at equilibrium (mol/kg) estimated by the PFO, PNO, and Avrami models, respectively; m and n_{AV} are the exponents of the PNO and Avrami models; α (mol/(kg × min)) and β (kg/mol) are the initial rate and the desorption constant, respectively, of the adsorption kinetics of the Elovich model.

In general, the PFO and PSO models are frequently used in the literature. Unlike the PSO model, the linear form of the PFO model (Equation (19) or Equation (20)) indicates two unknown parameters that are $q_{e(1)}$ and k_1. The parameter $q_{e(1)}$ must be selected before applying the linear method. However, it is not easy to select $q_{e(1)}$ appropriately. If $q_{e(1)}$ is lower than (or equal to) q_t, the result of calculating $\ln(q_{e(1)} - q_t)$ is an error (#NUM!). By selecting $q_{e(1)}$ slightly higher than the highest value of q_t, the result of fitting the linear method is provided in Figure S2. Clearly, the errors in the experimental datasets of time-dependent adsorption (i.e., Points 1 and 2) were not detected through the linear fitting of the PFO model (Figure S2).

$$\ln(q_{e(1)} - q_t) = -k_1 t + \ln(q_{e(1)}) \tag{19}$$

$$\log(q_{e(1)} - q_t) = -\frac{k_1}{2.303} t + \log(q_{e(1)}) \tag{20}$$

For the three-parameter models, the application of the linear method for computing the parameters is not suitable. Therefore, the nonlinear optimization method is applied for calculating the parameters of those kinetic models using full data (n = 16; Figure 1a) and manipulated data (n = 14; Figure 1b). Furthermore, to obtain a fair comparison between the two- and three-parameter models, the BIC statistics were used to assess the best-fitting model [2].

The result of modelling is provided in Figure S3 and Table 2. For the full data (n = 16), the fitting model followed the order of the Avrami model (BIC = −95.4) > PNO model (−83.8) > Elovich model (−76.5) > PSO (−61.6) > PFO (−51.3). Notably, some models indicated a doubtful result. For example, the m of the PNO was 7.299; this means that the overall order of the adsorption process was up to seven (an impossible result for the adsorption kinetics). However, the results (Figure S3 and Table 2) do not indicate which of the experimental points of the time-dependent adsorption are errors. The conclusion is consistent with the previous finding in Section 2.

Furthermore, Points 1 and 2 in Figure 1a are assumed to be outliers (errors). After removing them from the raw data, the result of modelling (based on Figure 1b; n = 14) indicates that the best-fitting models were in the following order: the Avrami model (BIC = −162.1) > PNO model (−137.0) > PSO model (−113.3) > PFO model (−85.6) > Elovich model (−81.5). The three-parameter models (i.e., the Avrami and PNO models) indicated the best description for the current experimental data compared to the two-parameter ones. The overall order of the adsorption kinetics (obtained based on the PNO model; m = 2.247) was close to two (feasibly considering it as the PSO model).

5. Re-Modelling Using Points 1 and 2 Estimated from the PSO Model

After removing the two outliers, the results of modelling (n = 14) are listed in Table 2. Using the parameters of the models, the values q_t at 1 min (Point 1) and 5 min (Point 2) are estimated. The results (Table 3) indicated a remarkable difference between the q_t value obtained from the experiment ($q_{t(exp)}$) and the q_t value estimated from the model ($q_{t(model)}$). Taking the case of the PSO model as a typical example, the values $q_{t(exp)}$ at 1 min and 5 min (0.741 and 0.789 mol/kg) are overwhelming higher than those obtained from the PSO model (0.131 and 0.484 mol/kg, respectively). The differences between $q_{t(exp)}$ and $q_{t(model)}$ are 140% for Point 1 and 47.8% for Point 2 (Table 3). The result confirms again that they (Points 1 and 2) were errors and outliers.

The q_t values for Point 1 (0.131 mol/kg) and Point 2 (0.484 mol/kg) are estimated by the PSO model. After submitting these estimated points into the datasets, the results of re-modelling are provided in Figure S4 and Table 2 (revisited data). The fitting model followed the decreasing order: the PNO model (BIC = −134.1) > PSO model (−131.1) > Avrami model (−107.7) > PFO model (−81.16) > Elovich model (−55.6). The results indicate that the Elovich and Avrami models can be considered empirical equations used for physically fitting the experiential data of time-dependent adsorption. In particular, the initial rate α of the Elovich model (Table 2) indicated a remarkable change of 19,073 mg/(g × min) for the full raw data (n = 16), 206,235 mg/(g × min) for the manipulated data (n = 14), and 440.4 mg/(g × min) for the manipulated data (n = 16). The exponent n_{AV} of the Avrami model that does not provide information on the overall order of the adsorption process only serves an empirically adjusted parameter. Unlike those models, the PNO model (especially its exponent m) is helpful for initially checking whether the datasets are feasible or not. The overall order of the adsorption process (m) was 2.134, which is close to the second order.

Because the linear forms of the PSO model are helpful for identifying the outlets from the initial period of the databases, it is necessary to re-verify this conclusion. After submitting the two estimated points into (Point 1 (q_t = 0.131 mol/kg) and Point 2 (0.484 mol/kg)) the datasets, the results of re-modelling using six linear forms of the PSO model are provided in Figure S5 and Table 1. As expected, the errors (Points 1 and 2) in the experimental datasets are not observed in Types 1–5, suggesting that the application of the linear form (i.e., Types 1–5) of the PSO model can help to identify these potential errors in the initial adsorption period.

In contrast, Type 1 of the PSO model fitted the experimental datasets well (adj-R^2 = 0.9999) when considering three cases: the full raw data (n = 16), manipulated data (n = 14), and manipulated data (n = 16) in Table 1; therefore, it is impossible to verify whether there are errors in the datasets or not. Type 6 should not be used because it must first estimate the unknown q_e value as in the case of the linear form of the PFO model. Figure 5 provides a brief summary of the validation and re-verification processes of the time-dependent adsorption data. Adsorption kinetic curves (q_t vs. t) must indicate an equilibrium region (as showed in Figure 5).

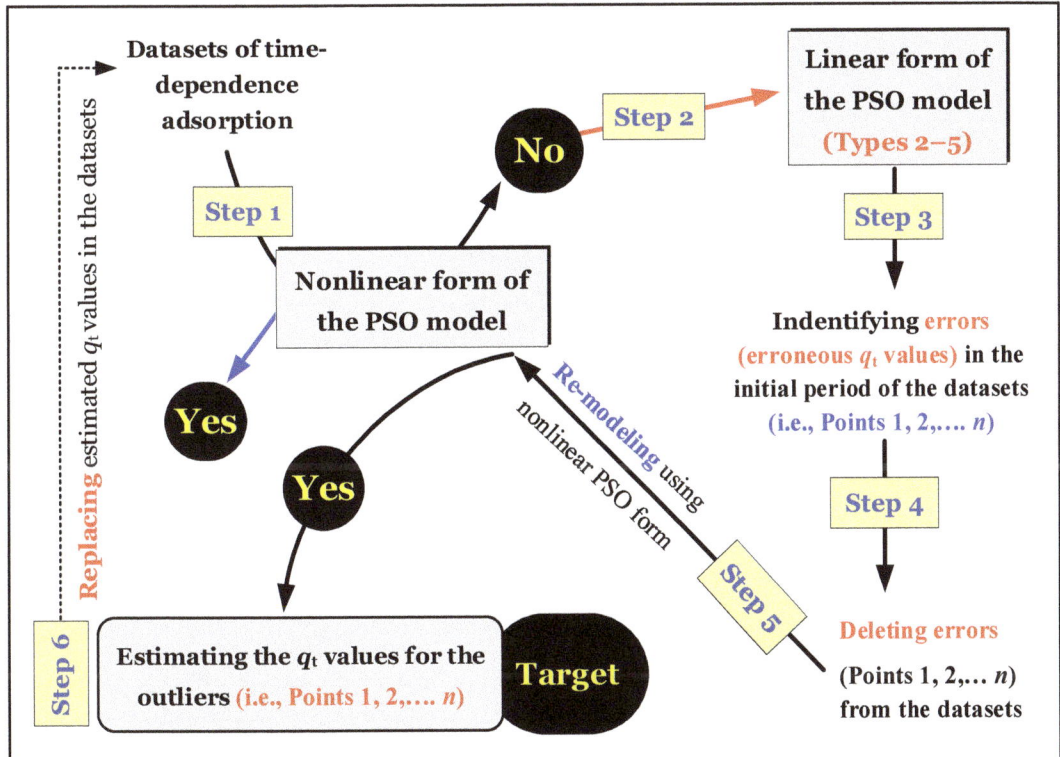

Figure 5. Processes for validation and re-verification of time-dependent adsorption data (Note: adsorption kinetic curves (q_t vs. t) must indicate two regions: kinetic and equilibrium; equilibrium is a plateau region).

Notably, the conclusions of this work might be feasible for some normal adsorption processes. Although other contributions ($n-\pi$ interaction, hydrogen bonding formation, and $\pi-\pi$ interaction) existed in the adsorption process of organic pollutants (i.e., PRC), pore filling has been identified as a primary mechanism of PRC into CAC [27] and other porous carbonaceous materials (such as spherical and nonspherical biochars [28]). Those mechanisms were acknowledged as physical adsorption because of low magnitudes of standard adsorption enthalpy change ($\Delta H° = 6.36$–24.1 kJ/mol) [27,28]. Therefore, the process for the validation and re-verification of the time-dependent adsorption data (Figure 5) needs to be applied.

In contrast, the adsorption process of heavy metals (i.e., Cd) onto some $CaCO_3$-enriched materials occurred very fast at the first period of adsorption. The primary adsorption was surface precipitation such as $Cd(CO_3)$ and/or $(Cd,Ca)CO_3$ [5,29]. This mechanism was known as nonactivated chemisorption that often occurs very rapidly and has low activation energy [29,30]. It might not be necessary to recheck the initial periods of the time-dependent adsorption datasets. This is because it is the nature of nonactivated chemisorption [3,30].

6. Conclusions

Applying the nonlinear optimization method for modelling the time-dependent experimental data is recommended. However, this nonlinear method does not provide relevant information on errors in the experimental points (potential outliers). This method should only be applied after the outliers are identified and removed. Results showed that the utilization of the linear forms (Types 2–5) of the PSO model was appropriate for identifying errors in the experimental points (at the initial adsorption period) of time-dependent adsorption. This conclusion is valid when the plot of q_t vs. t indicates two regions: kinetics and equilibrium (equilibrium is a plateau region).

Type 1 was physically well-fitted to the full raw experimental data, manipulated data (adj-$R^2 = 0.9999$; $n = 14$), and revisited data (adj-$R^2 = 0.9999$; $n = 16$). However, it (adj-R^2 or r^2) has been denoted as a spurious correlation in some cases. The application of Type 1 for calculating the parameters ($q_{e(2)}$ and k_2) should be avoided because it is hard to validate the accuracy of these parameters. For Type 6, the q_e value must be firstly estimated (a similar case of the linear form of the PFO model); therefore, its parameters obtained did not bring a high accuracy.

The PNO model can give some general information on doubts in the datasets. For example, its exponent m (7.299) is very high when considering the full raw data but suitable when considering the manipulated data ($m = 2.247$) and the revisited data ($m = 2.134$ that can be called the overall order of the adsorption process). However, this model did not provide where the outlets are located in the datasets.

Supplementary Materials: The following supporting information can be downloaded at: https://www.mdpi.com/article/10.3390/w15061231/s1, Figure S1. Physical fitting lines of the experimental data of kinetic adsorption by the linear form (Type 1) of the PSO model; Figure S2. The linear plot of PFO model obtained by considering the full data ($n = 16$) and the manipulated data ($n = 14$); Figure S3. Adsorption kinetics of paracetamol by commercial activated carbon fitted by various adsorption kinetic models; Figure S4. Re-modelling adsorption kinetics after submitting Point 1 ($q_t = 0.131$ mol/kg) and Point 2 (0.484 mol/kg) estimated from the PSO model to the datasets (Note: the modelling results for the case of the revisited data are provided in Table 2); Figure S5. Re-modelling results using the six linear forms of the PSO model after submitting Point 1 (0.131 mol/kg) and Point 2 (0.484 mol/kg) estimated from the PSO model to the six datasets (Note: the modelling results for the case of the revisited data are provided in Table 1).

Funding: This research received no external funding.

Data Availability Statement: Not applicable.

Conflicts of Interest: The author declares no conflict of interest.

References

1. Azizian, S. Kinetic models of sorption: A theoretical analysis. *J. Colloid Interface Sci.* **2004**, *276*, 47–52. [CrossRef] [PubMed]
2. Lima, E.C.; Sher, F.; Guleria, A.; Saeb, M.R.; Anastopoulos, I.; Tran, H.N.; Hosseini-Bandegharaei, A. Is one performing the treatment data of adsorption kinetics correctly? *J. Environ. Chem. Eng.* **2020**, *9*, 104813. [CrossRef]
3. Guo, H.; Zhang, S.; Kou, Z.; Zhai, S.; Ma, W.; Yang, Y. Removal of cadmium(II) from aqueous solutions by chemically modified maize straw. *Carbohydr. Polym.* **2015**, *115*, 177–185. [CrossRef] [PubMed]
4. Zbair, M.; Bottlinger, M.; Ainassaari, K.; Ojala, S.; Stein, O.; Keiski, R.L.; Bensitel, M.; Brahmi, R. Hydrothermal Carbonization of Argan Nut Shell: Functional Mesoporous Carbon with Excellent Performance in the Adsorption of Bisphenol A and Diuron. *Waste Biomass Valorization* **2020**, *11*, 1565–1584. [CrossRef]
5. Tran, H.N.; You, S.-J.; Chao, H.-P. Effect of pyrolysis temperatures and times on the adsorption of cadmium onto orange peel derived biochar. *Waste Manag. Res.* **2016**, *34*, 129–138. [CrossRef] [PubMed]
6. Xiao, W.; Garba, Z.N.; Sun, S.; Lawan, I.; Wang, L.; Lin, M.; Yuan, Z. Preparation and evaluation of an effective activated carbon from white sugar for the adsorption of rhodamine B dye. *J. Clean. Prod.* **2020**, *253*, 119989. [CrossRef]
7. Georgin, J.; da Boit Martinello, K.; Franco, D.S.P.; Netto, M.S.; Piccilli, D.G.A.; Foletto, E.L.; Silva, L.F.O.; Dotto, G.L. Efficient removal of naproxen from aqueous solution by highly porous activated carbon produced from Grapetree (*Plinia cauliflora*) fruit peels. *J. Environ. Chem. Eng.* **2021**, *9*, 106820. [CrossRef]
8. Hubbe, M.A.; Azizian, S.; Douven, S. Implications of apparent pseudo-second-order adsorption kinetics onto cellulosic materials: A review. *BioResources* **2019**, *14*, 7582–7626. [CrossRef]
9. Guo, X.; Wang, J. A general kinetic model for adsorption: Theoretical analysis and modeling. *J. Mol. Liq.* **2019**, *288*, 111100. [CrossRef]
10. Wang, J.; Guo, X. Adsorption kinetic models: Physical meanings, applications, and solving methods. *J. Hazard. Mater.* **2020**, *390*, 122156. [CrossRef]
11. Revellame, E.D.; Fortela, D.L.; Sharp, W.; Hernandez, R.; Zappi, M.E. Adsorption kinetic modeling using pseudo-first order and pseudo-second order rate laws: A review. *Clean. Eng. Technol.* **2020**, *1*, 100032. [CrossRef]
12. Tran, H.N.; You, S.-J.; Hosseini-Bandegharaei, A.; Chao, H.-P. Mistakes and inconsistencies regarding adsorption of contaminants from aqueous solutions: A critical review. *Water Res.* **2017**, *120*, 88–116. [CrossRef] [PubMed]
13. Kumar, K.V. A note on the comments by Dr. Y.S. Ho on "Remediation of soil contaminated with the heavy metal (Cd^{2+})". *J. Hazard. Mater.* **2006**, *136*, 993–994. [CrossRef] [PubMed]
14. Blanchard, G.; Maunaye, M.; Martin, G. Removal of heavy metals from waters by means of natural zeolites. *Water Res.* **1984**, *18*, 1501–1507. [CrossRef]
15. Coleman, N.T.; McClung, A.C.; Moore, D.P. Formation Constants for Cu(II)-Peat Complexes. *Science* **1956**, *123*, 330–331. [CrossRef] [PubMed]
16. Lima, É.C.; Dehghani, M.H.; Guleria, A.; Sher, F.; Karri, R.R.; Dotto, G.L.; Tran, H.N. Chapter 3—Adsorption: Fundamental aspects and applications of adsorption for effluent treatment. In *Green Technologies for the Defluoridation of Water*; Dehghani, M.H., Karri, R., Lima, E., Eds.; Elsevier: Amsterdam, The Netherlands, 2021; pp. 41–88.
17. Hamdaoui, O.; Saoudi, F.; Chiha, M.; Naffrechoux, E. Sorption of malachite green by a novel sorbent, dead leaves of plane tree: Equilibrium and kinetic modeling. *Chem. Eng. J.* **2008**, *143*, 73–84. [CrossRef]
18. Kumar, K.V. Linear and non-linear regression analysis for the sorption kinetics of methylene blue onto activated carbon. *J. Hazard. Mater.* **2006**, *137*, 1538–1544. [CrossRef]
19. Lin, J.; Wang, L. Comparison between linear and non-linear forms of pseudo-first-order and pseudo-second-order adsorption kinetic models for the removal of methylene blue by activated carbon. *Front. Environ. Sci. Eng.* **2009**, *3*, 320–324. [CrossRef]
20. Ho, Y.S.; McKay, G. Sorption of dye from aqueous solution by peat. *Chem. Eng. J.* **1998**, *70*, 115–124. [CrossRef]
21. Xiao, Y.; Azaiez, J.; Hill, J.M. Erroneous Application of Pseudo-Second-Order Adsorption Kinetics Model: Ignored Assumptions and Spurious Correlations. *Ind. Eng. Chem. Res.* **2018**, *57*, 2705–2709. [CrossRef]
22. Tseng, R.-L.; Wu, P.-H.; Wu, F.-C.; Juang, R.-S. A convenient method to determine kinetic parameters of adsorption processes by nonlinear regression of pseudo-n^{th}-order equation. *Chem. Eng. J.* **2014**, *237*, 153–161. [CrossRef]
23. Hu, Q.; Zhang, Z. Prediction of half-life for adsorption kinetics in a batch reactor. *Environ. Sci. Pollut. Res.* **2020**, *27*, 43865–43869. [CrossRef] [PubMed]
24. Lagergren, S.K. About the theory of so-called adsorption of soluble substances. *Sven. Vetenskapsakad. Handingarl* **1898**, *24*, 1–39.
25. Roginsky, S.; Zeldovich, Y.B. The catalytic oxidation of carbon monoxide on manganese dioxide. *Acta Phys. Chem.* **1934**, *1*, 554.
26. Avrami, M. Kinetics of Phase Change. I General Theory. *J. Chem. Phys.* **1939**, *7*, 1103–1112. [CrossRef]
27. Nguyen, D.T.; Tran, H.N.; Juang, R.-S.; Dat, N.D.; Tomul, F.; Ivanets, A.; Woo, S.H.; Hosseini-Bandegharaei, A.; Nguyen, V.P.; Chao, H.-P. Adsorption process and mechanism of acetaminophen onto commercial activated carbon. *J. Envi. Chem. Eng.* **2020**, *8*, 104408. [CrossRef]
28. Tran, H.N.; Tomul, F.; Ha, N.T.H.; Nguyen, D.T.; Lima, E.C.; Le, G.T.; Chang, C.-T.; Masindi, V.; Woo, S.H. Innovative spherical biochar for pharmaceutical removal from water: Insight into adsorption mechanism. *J. Hazard. Mater.* **2020**, *394*, 122255. [CrossRef]

29. Van, H.T.; Nguyen, L.H.; Nguyen, V.D.; Nguyen, X.H.; Nguyen, T.H.; Nguyen, T.V.; Vigneswaran, S.; Rinklebe, J.; Tran, H.N. Characteristics and mechanisms of cadmium adsorption onto biogenic aragonite shells-derived biosorbent: Batch and column studies. *J. Environ. Manag.* **2019**, *241*, 535–548. [CrossRef]
30. Bai, J.; Fan, F.; Wu, X.; Tian, W.; Zhao, L.; Yin, X.; Fan, F.; Li, Z.; Tian, L.; Wang, Y.; et al. Equilibrium, kinetic and thermodynamic studies of uranium biosorption by calcium alginate beads. *J. Environ. Radioact.* **2013**, *126*, 226–231. [CrossRef]

Disclaimer/Publisher's Note: The statements, opinions and data contained in all publications are solely those of the individual author(s) and contributor(s) and not of MDPI and/or the editor(s). MDPI and/or the editor(s) disclaim responsibility for any injury to people or property resulting from any ideas, methods, instructions or products referred to in the content.

Article

Factorial Design Statistical Analysis and Optimization of the Adsorptive Removal of COD from Olive Mill Wastewater Using Sugarcane Bagasse as a Low-Cost Adsorbent

Fatima Elayadi [1,2], Mounia Achak [2,3,*], Wafaa Boumya [4], Sabah Elamraoui [2], Noureddine Barka [4], Edvina Lamy [5], Nadia Beniich [6] and Chakib El Adlouni [1]

1. Laboratory of Marine Biotechnologies and Environment, Sciences Faculty, Chouaïb Doukkali University, El Jadida 24000, Morocco
2. Science Engineer Laboratory for Energy, National School of Applied Sciences, Chouaïb Doukkali University, El Jadida 24000, Morocco; sabahamraoui0@gmail.com
3. Chemical & Biochemical Sciences, Green Process Engineering, CBS, Mohammed VI Polytechnic University, Ben Guerir 43150, Morocco
4. Multidisciplinary Research and 12 Innovation Laboratory, FP Khouribga, Sultan Moulay Slimane University of Beni Mellal, Beni-Mellal 23000, Morocco
5. EA 4297 TIMR Laboratory, Department of Chemical Engineering, Centre de Recherche de Royallieu, University of Technology of Compiegne—Sorbonne University, Rue du Dr Schweitzer, CS 60319, 60200 Compiegne, France
6. Dynamical Systems Laboratory, Sciences Faculty, Chouaïb Doukkali University, El Jadida 24000, Morocco
* Correspondence: achak_mounia@yahoo.fr; Tel.: +212-6614-74231

Abstract: This work highlights the elimination of chemical oxygen demand (COD) from olive mill wastewater using sugarcane bagasse. A 2^{5-1} fractional factorial design of experiments was used to obtain the optimum conditions for each parameter that influence the adsorption process. The influence of the concentration of sugarcane bagasse, solution pH, reaction time, temperature, and agitation speed on the percent of COD removal were considered. The design experiment describes a highly significant second-order quadratic model that provided a high removal rate of 55.07% by employing optimized factors, i.e., a temperature of 60 °C, an adsorbent dose of 10 g/L, a pH of 12, a contact time of 1 h, and a stirring speed of 80 rpm. The experimental data acquired at optimal conditions were confirmed using several isotherms and kinetic models to assess the solute interaction behavior and kind of adsorption. The results indicated that the experimental data were properly fitted with the pseudo-first-order kinetic model, whereas the Langmuir model was the best model for explaining the adsorption equilibrium.

Keywords: adsorption; sugarcane bagasse; olive mills wastewater; factorial design

1. Introduction

Olive mill wastewater (OMW) is an important environmental problem due to the strong color, low pH, and high concentrations of chemical oxygen demand (COD), biochemical oxygen demand (BOD), and phenolic compounds [1,2]. The discharge of this effluent into soil or rivers without treatment leads to damages to the environment that are attributed to their potential risk to flora or depletion of clean water reservoirs [3]. Hence, the treatment of OMW is a great challenge in a problematic environment. There are numerous approaches that can be used to remove organic pollutants and reduce COD levels from OMW such as coagulation–flocculation, aerobic and anaerobic biodegradation, solar distillation, adsorption, and infiltration percolation [4–10]. Among these techniques, adsorption is a well-established technique that benefits from advantages such as a simple design, low-cost, eco-friendly, high ability, non-generation of secondary pollutants, and reusability [11]. Activated carbon is one of the most efficient adsorbents used to treat many kinds of effluents. Nevertheless, it is relatively expensive, and its regeneration is difficult, which has necessitated the search for alternative adsorbents [12].

Therefore, several industrial and agricultural wastes exhibit considerable potential for the removal of pollutants. Several agricultural by-products are investigated for COD removal from wastewater, including bamboo [13], date pit [14], raw bagasse [15], areca catechu fronds [16], and date palm waste [17]. Among the efficient agro-industrial residues is sugarcane bagasse, which has a high cellulose (45%), hemicellulose (28%), and lignin (18%) content [18]. These materials contain reactive functional groups such as carboxylic and hydroxyl groups with the potential to adsorb organic loads [19]. Several studies have been conducted to investigate the sugarcane bagasse as an adsorbent for different pollutants, including Congo red, Pb(II), phenol from water, and phenol from OMW [20–24].

Although several studies have been conducted using sugarcane bagasse to remove several organic dye contaminants, no investigation has been carried out on the mathematical modeling and statistical optimization of the COD removal from OMW using sugarcane bagasse. Therefore, the present work aimed at the statistical optimization and modeling of key experimental parameters using fractional factorial design and response surface methodology (RSM) to attain optimum COD removal. Moreover, it recognizes the effects of five parameters, such as adsorbent dosage, pH, agitation time, stirring speed, and temperature as well as their interactions on the adsorption process of COD by sugarcane bagasse. Additionally, the adsorption kinetics, equilibrium, and thermodynamics were also studied.

2. Materials and Methods

2.1. OMW Sample

An OMW solution was obtained from an olive oil mill in Marrakech City, Morocco, during the campaign 2018/2019. This factory operates with the modern two- and three-phase olive oil extraction technology. The OMW used is characterized by a COD of 347.8 g(O_2)/L, a conductivity of 15.3 ms/cm, a phenolic content of 15.29 g/L and a pH of 4.27. The supplied effluent was immediately kept refrigerated at -4 °C to prevent any alteration in its physicochemical characteristics.

2.2. Sugarcane Bagasse Material

Sugarcane bagasse was collected from a local plant in the region of El Jadida, Morocco. The collected adsorbent was oven-dried for 24 h at 70 °C before being ground with a domestic electrical miller and sieved through a 50 μm screen. The obtained powder was repeatedly washed with distilled water to remove all impurities, and it was then oven dried at 100 °C for 48 h (Figure 1). The dried powder was used in experiments without any further treatment.

Figure 1. Sugarcane bagasse material steps.

The sugarcane bagasse functional groups were characterized by FTIR using Perkin-Elmer 1720-x spectrometer. The sample was mixed with KBr at a mass ratio of 1:100 and finely powdered and pressed to pellets. The infrared spectra were recorded in the range of 4000–500 cm^{-1} with 2 cm^{-1} resolution. Therefore, X-ray fluorescence analysis was employed to investigate the chemical composition of the adsorbent using the "Axion" spectrometer.

2.3. Batch Adsorption Experiments

All reagents used in the preparation and the adsorption studies were of analytical grade. The elimination of COD from OMW using sugarcane bagasse was studied in batch mode using shut-top pyrex bottles comprising 100 mL of OMW and an appropriate mass of adsorbent, which were stirred during the desired time in an incubator (Stuart SI50). The remaining COD content was determined from the UV-Vis absorbance characteristic using a Jenway 6320D UV/Vis spectrophotometer. The wavelength of maximum absorption (λ_{max}) was 620 nm for this measurement. The quantity adsorbed and the removal efficiency of COD were calculated by measuring the solution's concentration before and after adsorption using the following equations:

$$q_e = \frac{(C_o - C_e) \cdot V}{m} \quad (1)$$

$$\%Rem = \frac{(C_o - C_e)}{m} \times 100 \quad (2)$$

where q_e is the amount of COD removed by the adsorbent (mg/g); C_o and C_e are, respectively, the concentrations of COD before and after the batch adsorption study (mg/L); m is the mass of sugarcane bagasse (g); and V is the volume of the solution (L).

The nonlinear optimization method was used to fit equilibrium and kinetics data to the corresponding models. For this study, non-linear regression was applied using 8.0 software for fitting the curve. The best fit of the chosen non-linear models was determined by the use of four well-known error functions, namely the coefficient of determination (R^2), adjusted determination coefficient (adj-R^2), reduced chi-square (red-χ^2), and Bayesian information criterion (BIC). A model with a higher adj-R^2 value, lower red-χ^2 and BIC value indicates a better fitting than the others [25].

The equations of all error functions used are expressed as follows:

$$R^2 = 1 - \frac{\Sigma(q_{i,exp} - q_{i,model})^2}{\Sigma(q_{i,exp} - q_{i,exp-mean})^2} \quad (3)$$

$$adj - R^2 = 1 - (1 - R^2) \times \left(\frac{N-1}{DOF}\right) \quad (4)$$

$$red - \chi^2 = \frac{\Sigma(q_{i,exp} - q_{i,model})^2}{DOF} \quad (5)$$

$$BIC = N \times \ln\left(\frac{\Sigma(q_{i,exp} - q_{i,model})^2}{N}\right) + P\ln(N) \quad (6)$$

where $q_{i,exp}$ is the experimental adsorbed capacity value; $q_{i,model}$ is the modeled value; $q_{i,exp-mean}$ is an average of $q_{i,exp}$ values used for modelling; N is the number of experimental points; P is the number of model parameters; and DOF is the degrees of freedom.

2.4. Experimental Design

Batch experiments based on a 2^{5-1} fractional factorial design were conducted randomly to study the influence of the experimental variables on the percentage of COD removal (% Rem). The studied factors which are focused on this research; sugarcane bagasse dose (X_1), solution pH (X_2), contact time (X_3), stirring speed (X_4) and temperature (X_5). Table 1 shows the five experimental variables and their chosen levels. After performing batch experiments for the optimization process, the regression analysis was conducted to attain the study's statistical parameters with 95% confidence intervals using the Minitab 18 statistical software. The variance of the regression equation (mathematical relation between dependent and independent variables) is the most important analysis by the ANOVA method to find the desired function of COD adsorption. RSM is used as a sequential process to show the relation between the studied independent factors and the response to determine the set of optimal experimental parameters.

Table 1. Process factors and their levels.

Factors	Variable	Unit	Levels	
			Low (−)	High (+)
X_1	Adsorbent dose	g/L	10	60
X_2	pH	-	2	12
X_3	Contact time	h	1	24
X_4	Stirring speed	rpm	80	300
X_5	Temperature	°C	25	60

3. Results and Discussion

3.1. Analysis of Factorial Design

2^{5-1} Fractional factorial design was adopted using Minitab 18 to optimize the influence of the investigated parameters; sugarcane bagasse dose (X_1), solution pH (X_2), contact time (X_3), stirring speed (X_4) and temperature (X_5) on the elimination of COD from OMW. This design yields in 16 experiments with all possible combinations of X_1, X_2, X_3, X_4 and X_5. COD removal efficiency (Y) was measured for each of these experiments as shown in Table 2. The response obtained was correlated using the second-order polynomial model, expressed by Equation (7):

$$Y = b_0 + b_1X_1 + b_2X_2 + b_3X_3 + b_4X_4 + b_5X_5 + b_{12}X_1X_2 + b_{13}X_1X_3 + b_{14}X_1X_4 + b_{15}X_1X_5 + b_{23}X_2X + b_{24}X_2X_4 \\ + b_{34}X_3X_4 + b_{25}X_2X_5 + b_{35}X_3X_5 + b_{45}X_4X_5 \quad (7)$$

where Y is the COD removal efficiency response, b_0 is a constant, b_i correspond to linear coefficient of X_i, and b_{ij} is the interaction coefficient.

Table 2. Matrix design with experimental and predicted COD removal values.

N°	X_1	X_2	X_3	X_4	X_5	Y (Experimental)	Y (Predicted)
1	−1	−1	−1	−1	+1	36.39	37.46
2	1	−1	−1	−1	−1	41.26	43.47
3	−1	1	−1	−1	−1	54.27	51.69
4	1	1	−1	−1	+1	44.69	43.75
5	−1	−1	1	−1	−1	35.24	35.54
6	1	−1	1	−1	+1	42.40	43.24
7	−1	1	1	−1	+1	35.53	30.66
8	1	1	1	−1	−1	41.23	40.85
9	−1	−1	−1	1	−1	40.71	40.53
10	1	−1	−1	1	+1	42.12	43.90
11	−1	1	−1	1	+1	33.73	32.73
12	1	1	−1	1	−1	39.82	39.12
13	−1	−1	1	1	+1	33.23	29.86
14	1	−1	1	1	−1	31.12	34.75
15	−1	1	1	1	−1	39.25	38.14
16	1	1	1	1	+1	42.12	41.79

The statistical calculations and regression analysis were conducted to fit the response function with the experimental data. The regression coefficient values obtained are given in the final regression equation, after putting the values of all coefficients, as follows (Equation (8)):

$$Y (\%) = 37.85 - 2.183X_1 + 0.39X_2 - 0.562X_3 + 0.05958X_4 + 0.2182X_5 - 0.01712X_1X_3 - 0.0529X_1X_3 + 0.102X_1X_4 - 0.0528X_1X_5 - 0.05223X_2X_3 - 0.01976X_2X_4 - 0.01153X_2X_5 - 0.003092X_3X_4 - 0.00436X_3X_5 - 0.000623X_4X_5 \quad (8)$$

This equation expresses the COD removal efficiency as a function of investigated experimental factors and enables fixing experimental conditions for each targeted COD removal efficiency. From Table 2, the values predicted obtained by the relation (8) are compared with those of experimental results, which show a good agreement between the two sets of values. In Equation (8), the positive signs of the coefficients for X_2 and X_4 factors and the X_1X_4 interaction imply their positive effect on the response, while the negative signs of the coefficients for X_1, X_3, and X_5 factors as well as the X_1X_2, X_1X_3, X_1X_5, X_2X_3, X_2X_5, X_3X_4, X_3X_5, and X_4X_5 interactions represent the negative effect on the response. Based on the equation, the most influential factor for response was the sugarcane bagasse dose with a coefficient value of 2.183. The sign (-) means that each one-point decrease will have an effect of 2.183 on the COD removal efficiency value.

To check the quality of the model fitting, an ANOVA analysis was performed using the F-value and p-value (Table 3). Significant effects of the model terms were identified based on the p-values less than 0.05. Since the F-value for the 95% confidence interval, one degree of freedom, and 16 factorial runs, was equal to 4.54, all terms of the model with a F-value greater than 4.54 are considered statistically significant. The COD removal model was determined to be greatly significant from Fisher's test (F-value of 22.24) and a smaller p-value (<0.0001). Besides that, the significance of the model terms can be determined through the Pareto chart (Figure 2). The Student's t-test was performed to evaluate the significance of the regression coefficients. For the 95% confidence interval and one degree of freedom, it was observed that the t-value was equal to 2.12. Student's t-test values for model terms are displayed in horizontal columns. All term effects are significant if their absolute values exceed the vertical line (2.12). According to the obtained F value, p-value and Pareto chart, it seems that the main effects X_1, X_2, X_3, X_4 and X_5 factors as well as the X_2X_3, X_1X_4, X_3X_5, X_1X_2, X_3X_4, X_4X_5 and X_1X_3 interactions are statistically significant. In this way, the COD adsorption by sugarcane bagasse could be expressed using the following equation (Equation (9)):

$$Y (\%) = 21.5 - 0.2439 X_1 + 1.494 X_2 + 1.387 X_3 - 0.0451 X_4 + 0.4404 X_5 - 0.01712 X_1X_2 - 0.00497 X_1X_3 + 0.001892 X_1X_4 - 0.09223 X_2X_3 - 0.001131 X_3X_4 - 0.01268 X_3X_5 - 0.000743 X_4X_5 \quad (9)$$

To graphically verify the validity of the regression model obtained for COD removal, four validation indicator plots were created by plotting the differences between the predicted (model) and the observed (experimental) values (Figure 3).

The normal probability plot (Figure 3a) reveals that all points are reasonably near a straight line, suggesting that the predicted values of COD removal and the actual experimental data were in agreement, evidencing the normal distribution of the data and the validity of the regression model. The graphic plot of the residuals (Figure 3b) displays the predicted and observed values. The residual (vertical axis) is the difference between the experimental and the fitted values [26]. As shown in Figure 3b, the experimental points are reasonably aligned around zero, suggesting normal distribution. The histogram (Figure 3c) shows a random distribution of values with no noticeable shape or trend across all 16 runs conducted. The residuals versus the observation orders (Figure 3d) show that the residuals seem to be randomly scattered around zero.

Table 3. Analysis of ANOVA for COD removal. The underline shows the p-values < 0.05.

Source	DF	Adj SS	Adj MS	F-Value	p-Value
Model	16	3214.58	200.912	22.24	0.000
Linear	5	881.04	176.207	19.51	0.000
X_1	1	224.88	224.876	24.90	0.000
X_2	1	192.57	192.567	21.32	0.000
X_3	1	146.82	146.823	16.25	0.001
X_4	1	260.52	260.525	28.84	0.000
X_5	1	56.25	56.245	6.23	0.025
Interactions	10	2327.91	232.791	25.77	0.000
$X_1 \times X_2$	1	146.54	146.538	16.22	0.001
$X_1 \times X_3$	1	65.35	65.350	7.23	0.017
$X_1 \times X_4$	1	865.84	865.839	95.86	0.000
$X_1 \times X_5$	1	5.31	5.314	0.59	0.455
$X_2 \times X_3$	1	900.00	899.998	99.64	0.000
$X_2 \times X_4$	1	0.34	0.343	0.04	0.848
$X_2 \times X_5$	1	5.26	5.260	0.58	0.457
$X_3 \times X_4$	1	65.50	65.502	7.25	0.017
$X_3 \times X_5$	1	208.31	208.306	23.06	0.000
$X_4 \times X_5$	1	65.46	65.464	7.25	0.017
Error	15	135.49	9.033		
Total sum of squares	31	3350.07			
S square	R^2	R^2 (ajust)		R^2 (prev)	
6.2538	99.98%	87.54%		62.23%	

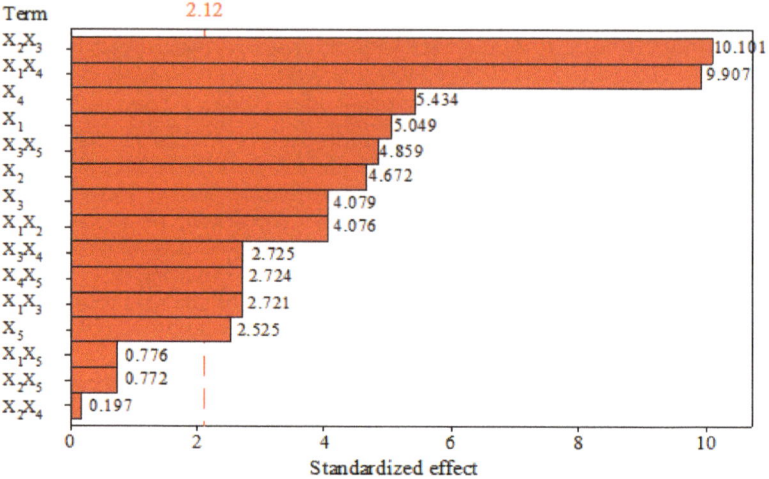

Figure 2. Pareto chart for standardized effects.

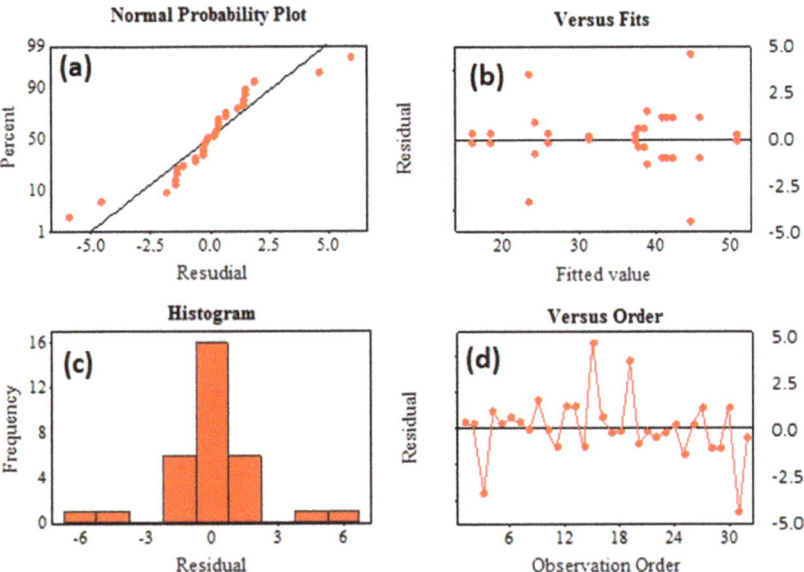

Figure 3. Validation indicator plots of the model for the experiments (**a**) Normal probability plot, (**b**) Versus fits (**c**) Histogram and (**d**) Versus order.

3.2. Response Surface Analysis

The significant interaction effect between each two input parameters on the % removal was presented by the 3D response surface and the contour of the plots (Figure 4A–G). The interaction between the solution pH (X_2) and contact time (X_3) is shown in Figure 4A. The graph demonstrated that better COD removal is achieved at acidic pH and long contact time. Further, the % removal decreased from 39 to 30% with increasing pH. This can be explained by the impact of pH on the ionic configuration of the functional groups presented in the sorbent. Previous pH drift tests indicated that the pH_{pzc} (point of zero charge) of sugarcane bagasse is equal to 5.0, which expresses that the surface of the bagasse is positively charged at a pH below 5 and negatively charged at a pH above 5 [21,27]. As a consequence, the electrostatic interaction between the adsorbent surface and the organic matter could be enhanced, which resulted in high adsorption of COD [28]. The interaction effect between the adsorbent dose (X_1) and agitation speed (X_4) shown in Figure 4B reveals that there will be a good COD removal equal to 44% when the adsorbent dose is 10 g/L and also at the stirring speed of 80 rpm. Nevertheless, an adsorbent dose greater than 10 g/L and an agitation speed above 80 rpm result in a substantial decrease in % removal. This phenomenon is due to the aggregation and glomeration of the sorbent particles and the reduction in the total surface area of the adsorbent [29].

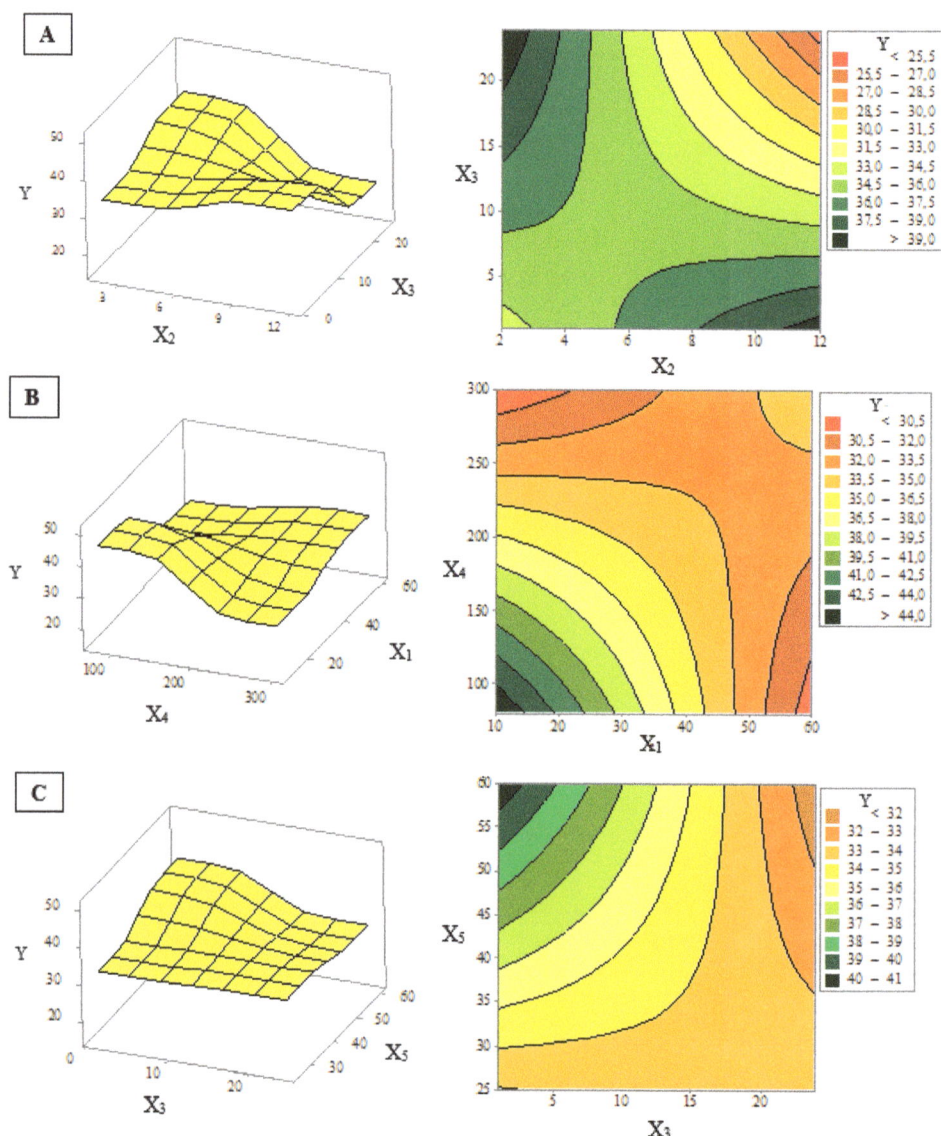

Figure 4. *Cont.*

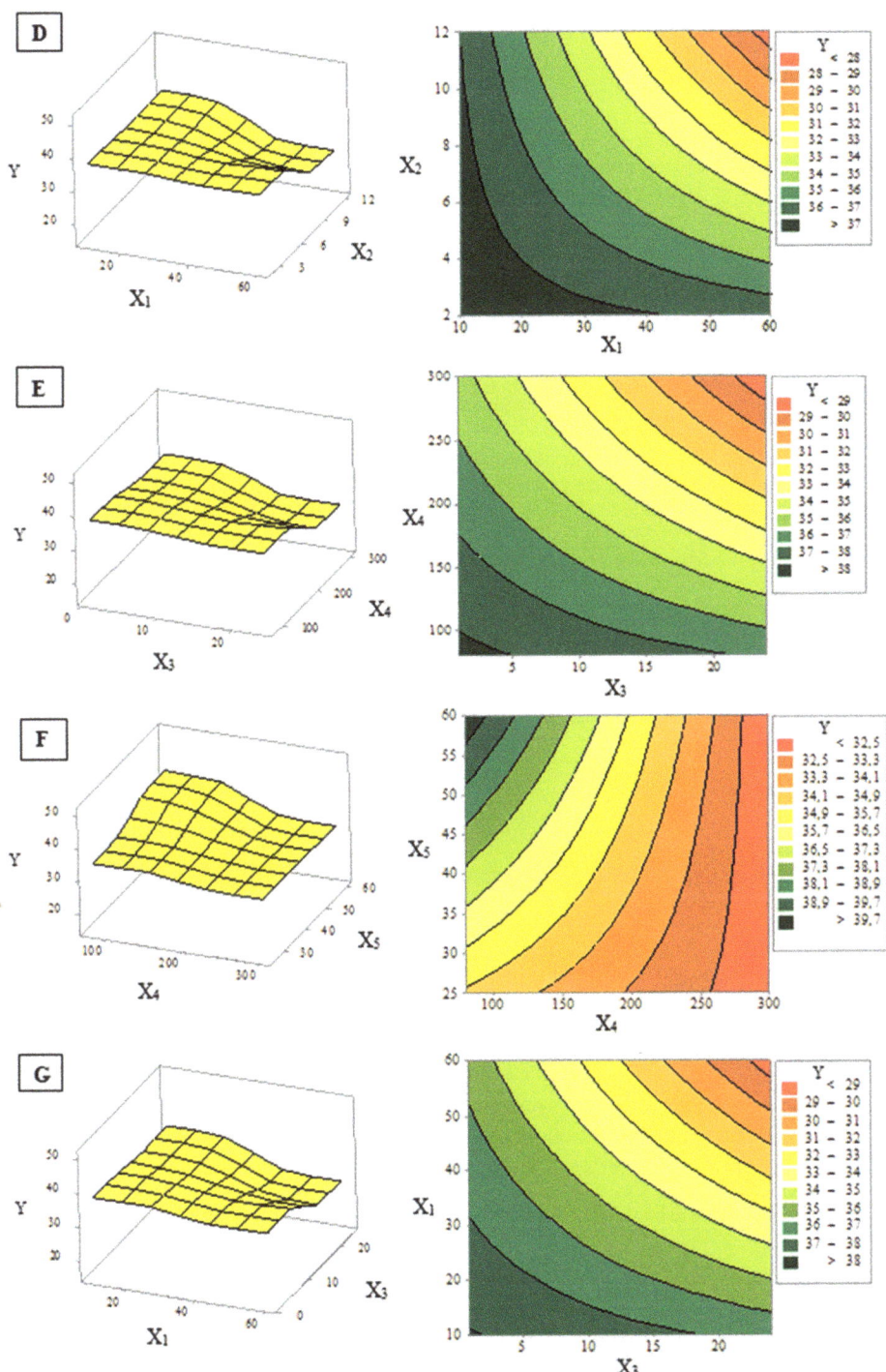

Figure 4. (**A–G**): Response surface and contour plots for the % Rem.

Figure 4C describes the interaction between the temperature (X_5) and reaction time (X_3) on the % removal. The % removal increases with an increase in temperature, whereas it decreases with increasing contact time. There are more active surface sites when the temperature increases because the adsorbent swells more as well. This result suggests that the COD adsorption is an endothermic process [30]. As shown in Figure 4D, the interaction between the pH (X_2) and adsorbent dose (X_1) has a negative effect on % removal. An increase in either of these two parameters reduces the COD removal efficiency. The possible reason for such observations is that the changes in pH could lead to changes in the properties of the surface of adsorbent [31], and the agglomeration of the adsorbate particles occurs with an increase in the adsorbent dose. In Figure 4E, the % removal decreased with increasing contact time (X_3) and stirring speed (X_4). The % removal was very rapid in the beginning stages of contact time, reaching about 38%. This result was related to the availability of more active sites on the adsorbent and the fact that the gradient of concentration between the adsorbate molecules in the solution and the adsorbate molecules on the adsorbent is high, which improves COD diffusion to the adsorbent surface [32,33]. After 60 min, the removal decreases over time due to adsorption site saturation, and the adsorbate molecules may bind poorly to the active receptors on the adsorbent [34,35].

Furthermore, the obtained results represented in Figure 4F show that the optimum of % removal was found at a temperature of 60 °C and agitation speed of 80 rpm, giving 39% of the recovery. This tendency may be due to the breakdown of the expanding chain and flocs. However, the higher stirring speed can encourage the process of agglomeration [36]. Additionally, the % removal was decreased by increasing the adsorbent dosage (X_1) and contact time (X_3) (Figure 4G). The % removal is initially higher and more rapid, reaching up to more than 38% within the first 60 min and 10 g/L, which could be attributed to the greater availability of adsorption sites to bind COD and a greater adsorbent active sites/COD ratio [37]. Subsequently, it decreased slightly with the prolonging of the contact time and the increase in adsorbent dose, which could be caused by an aggregation of the adsorbent and the reduction of the surface area available to COD [20].

3.3. Optimization Process

The response optimization study used to determine the combination of optimal values of the factors for maximum COD removal is shown in Figure 5. The validation of the optimal conditions is regarded as the final step in the modeling approach to examine the precision and robustness of the investigated model. The optimization requests to determine the desired pH, contact time, temperature, stirring speed, and adsorbent concentration to attain significant desirability. The model optimization shows that the efficiency of COD removal increases with an increasing pH, contact time and temperature, while decreasing with an increasing adsorbent dose and stirring speed. The optimum conditions indicate that the experimental modeling yields a desirability close to 1 with 55.07% COD removal efficiency, in the subsequent conditions: 10 g/L doses of sugarcane bagasse, 1 h of contact time, 80 rpm of stirring speed, 60 °C of temperature, and pH 12 of the OMW solution.

3.4. Characterization of Sugarcane Bagasse

Sugarcane bagasse characterization is an important analysis for understanding the behavior or the mechanism of COD removal on its surface. X-ray fluorescence elemental analysis was used to identify and quantify the amount of the elements contained in sugarcane bagasse in order to ultimately determine its elemental composition. The mineral composition of sugarcane bagasse is presented in Table 4. The results demonstrate that sugarcane bagasse contains a high amount of silicon dioxide (SiO_2), as high as 62.23%, and low amounts of alkaline oxide, Al_2O_3 and P_2O_5.

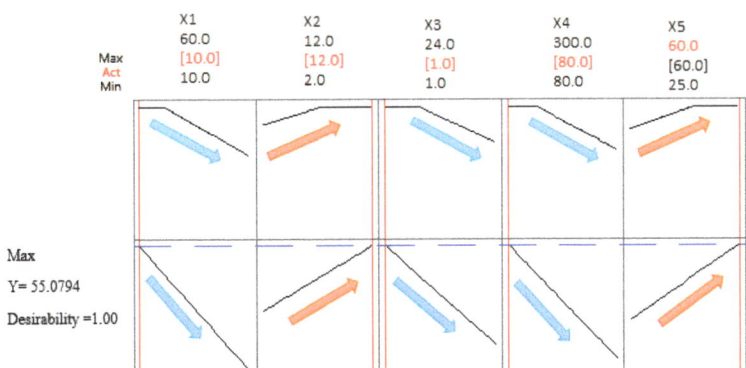

Figure 5. Optimization of the model obtained by desirability function.

Table 4. X-ray fluorescence analysis of sugarcane bagasse.

Composition	Concentration (%)
SiO_2	62.23
SO_3	24.00
CaO	11.50
MgO	0.68
Na_2O	0.60
K_2O	0.51
Al_2O_3	0.28
P_2O_5	0.1
Fe_2O_3	0.10

Therefore, the FTIR was utilized to investigate the functional groups of the sugarcane bagasse (Figure 6). The bands at 3330 and 2890 cm^{-1} indicate the presence of the O-H functional group and C-H stretching, respectively [20,38]. The peak at 1632 cm^{-1} is assigned to C=O vibrations in hemicellulose [39]. These groups are thought to play a very important role in the process of adsorption [40]. The bands that appeared at wave numbers between 1471 cm^{-1} and 1366 cm^{-1} are related to C=C-H indicating several bands in cellulose and xylose [41]. The bands appeared at 1241 cm^{-1} and 1029 cm^{-1} can be attributed to the CH=CH stretching of lignin [39] and C-O stretching in cellulose and hemicellulose [42], respectively. The bands appeared at 640 and 593 cm^{-1} in the FTIR spectra are attributed to the vibration of O-H groups out of the plane deformation [43].

3.5. Adsorption Isotherms

The specific relationship established between the COD remaining in solution (C_e) and the COD adsorbed by sugarcane bagasse (q_e) at equilibrium was analyzed by the models of Langmuir and Freundlich. The investigated adsorption isotherms for the COD adsorption on sugarcane bagasse are presented in Figure 7. Table 5 shows the calculated parameters for each of the sorption isotherm models obtained from non-linear regression forms. The best fit of the experimental data was analyzed based on R^2, adj-R^2, red-χ^2 and BIC. The obtained results displayed that the COD adsorption fitted well with the Langmuir isotherm model with the highest adj-R^2 of 0.994 and the lowest red-χ^2 and BIC values of 78.33 and 25.95, respectively. This result indicates that the adsorption process occurs on a homogeneous surface, and all adsorption sites are identical and energetically equivalent [44]. The maximum adsorption capacity (q_{max}) of COD onto sugarcane bagasse was determined to be 331.92 mg/g from the Langmuir model.

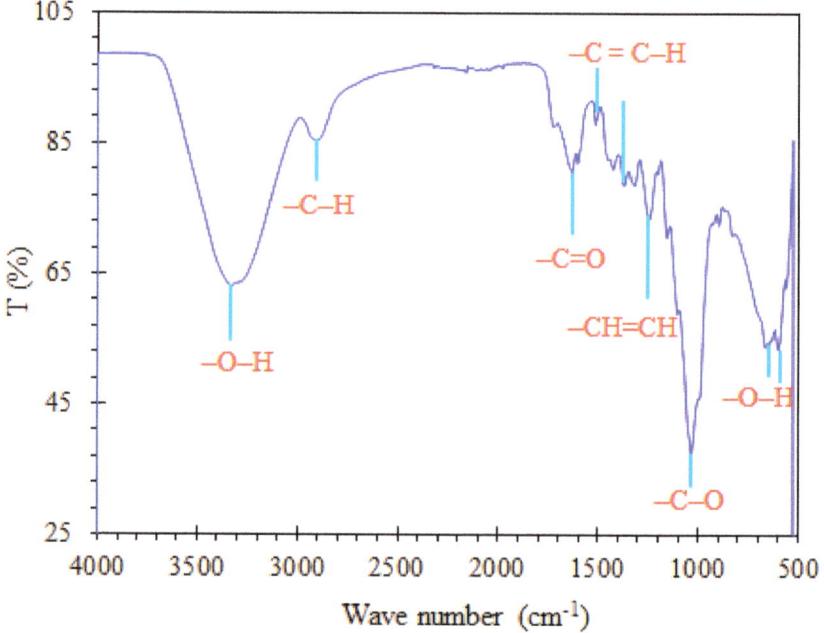

Figure 6. The FTIR spectra of sugarcane bagasse.

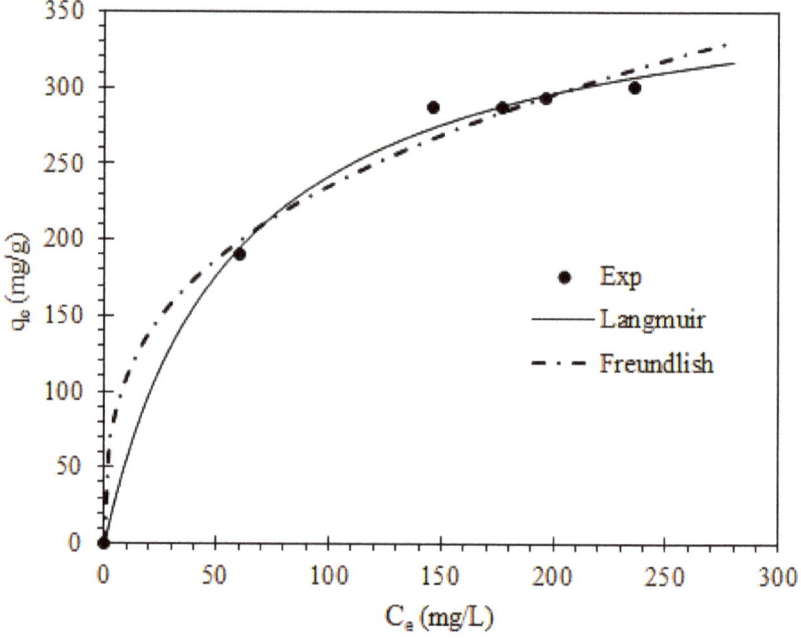

Figure 7. Isotherm plots for COD removal by sugarcane bagasse.

Table 5. Parameters and error function data for sorption isotherm models obtained from non-linear regression forms.

Models	Equation [a]	Parameters	Values
Langmuir	$q_e = \frac{q_{max} K_L C_e}{(1+K_L C_e)}$	qm (mg/g)	331.92
		K_L	0.019
		R^2	0.996
		Adj-R^2	0.994
		Red-χ^2	78.33
		BIC	25.59
Freundlich	$q_e = K_F C_e^{1/n}$	1/n	0.31
		K_F	51.13
		R^2	0.991
		Adj-R^2	0.985
		χ^2	208.11
		BIC	31.45

Note: [a] q_e (mg/g): mass of adsorbed molecule per unit mass of sugarcane bagasse, C_e (mg/L): concentration of no-adsorbed molecules, q_m and K_L are constants of the Langmuir model, and K_F and n are constants of Freundlich isotherm model.

The adsorption properties of sugarcane bagasse are compared with other adsorbents used for COD sorption reported in the literature, which are given in Table 6. The table indicates that the q_{max} value obtained in the present study is higher than that of other materials from previous studies. The q_{max} value found in this study reveals a very good adsorption capacity of the sugarcane bagasse, which falls as a promising adsorbent.

Table 6. Comparison of the q_{max} of COD removal by various adsorbents.

Adsorbent Types	q_{max} (mg/g)	Ref.
Sugarcane bagasse	331.92	This study
Date-pit activated carbon	252.81	[14]
Activated carbon	41.3	[17]
Raw bagasse	77.95	[15]
Bamboo-based activated carbon	24.39	[45]
Areca Nut Husk	64.94	[46]

3.6. Kinetic Studies

Kinetic modeling was undertaken to determine the rate of COD adsorption on the sugarcane bagasse and examine the controlling mechanisms of the adsorption process. The kinetic studies were carried out from 0 to 24 h, and the recorded data were studied with the pseudo-first-order and pseudo-second-order kinetic models. The rate constants values were valued from the non-linear plots and shown in Figure 8 and are also summarized in Table 7. The table indicated that the R^2 value obtained from the pseudo-first order kinetic equation was found to be higher than that of pseudo-second order. In addition, the calculated q_e value (337.45 mg/g) for the pseudo-first-order model is closer to the experimental value (326.29 mg/g) compared to the q_e value (405.9 mg/g) calculated for the pseudo-second-order model. The pseudo-first-order model also presented the highest adj-R^2 (0.967), lowest red-χ^2 (467.41) and BIC (43.01) values. This result advises that the experimental data demonstrated the best fit to the pseudo-first order model, and its applicability also indicates that a physical process might control the sorption process.

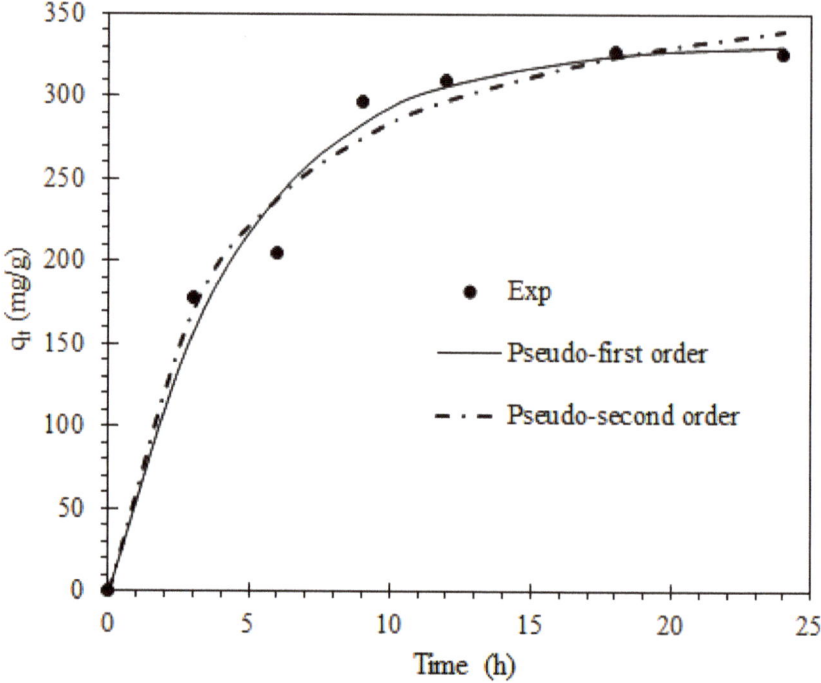

Figure 8. Kinetic plots for COD removal by bagasse.

Table 7. Parameters and error functions data for kinetic models studied obtained from non-linear regression forms.

Model	Equation [b]	Parameters	Value
Pseudo-first-order	$q_t = q_e\left(1 - e^{-k_1 t}\right)$	$q_{e.exp}$ (mg/g) $q_{e cal}$ (mg/g) k_1 R_1^2 Adj-R^2 χ^2 BIC	326.92 337.45 0.20 0.978 0.967 467.41 43.01
Pseudo-second-order	$q_t = \dfrac{K_2 q_e^2 t}{(1 + K_2 q_e t)}$	$q_{e cal}$ (mg/g) k^2 R_2^2 Adj-R^2 χ^2 BIC	405.9 0.0006 0.976 0.964 503.52 43.52

Note: [b] q_e and q_t (mg/g) are the amounts of COD removed at equilibrium and at time t, respectively; k_1 and k_2 are the adsorption rate constants.

4. Conclusions

In this work, the sugarcane bagasse prepared was utilized as an adsorbent for the elimination of COD from OMW by the adsorption. Sugarcane bagasse characterization was performed by FTIR and X-ray fluorescence. The 2^{1-5} fractional design associated to response surface methodology was successfully applied to optimize the effects of the operating variables of COD removal from OMW using the prepared adsorbent. The optimal conditions were found to be pH 12, the adsorbent dose of 10 g/L, a stirring speed of 80 rpm, a contact time of 1 h, and a temperature of 60 °C with a percentage of removal of 55.07%

and a desirability close to 1. The experimental data are well correlated by the Langmuir isotherm model with an R^2 of 0.996, and the maximum sorption capacity of 331.92 mg/g under optimal conditions. The kinetic data of the COD removal process were properly fitted with the pseudo-first-order model instead of pseudo-second-order model.

Author Contributions: F.E. carried out the experiments and prepared the draft manuscript. S.E.: and W.B. interpreted and discussed the results, and calculated the statistical parameters. N.B. (Nadia Beniich) analyzed and checked the statistical results. N.B. (Noureddine Barka) verified the analytical methods. C.E.A. and E.L. aided in interpreting the results and worked on the manuscript. M.A. corrected and wrote the final version of the manuscript. All authors have read and agreed to the published version of the manuscript.

Funding: This research received no external funding.

Data Availability Statement: The data presented in this study are available on request from the corresponding author.

Conflicts of Interest: The authors declare no conflict of interest.

References

1. Bayramoglu, M.; Kobya, M.; Can, O.T.; Sozbir, M. Operating cost analysis of electrocoagulation of textile dye wastewater. *Sep. Purif. Technol.* **2004**, *37*, 117–125. [CrossRef]
2. Sayan, E. Optimization and modeling of decolorization and COD reduction of reactive dye solutions by ultrasound-assisted adsorption. *Chem. Eng. J.* **2006**, *119*, 175–181. [CrossRef]
3. Rupani, P.F.; Singh, R.P.; Ibrahim, M.H.; Esa, N. Review of current palm oil mill effluent (POME) treatment methods: Vermicomposting as a sustainable practice. *World Appl. Sci. J.* **2010**, *11*, 70–81.
4. Oladipo, A.A.; Adeleye, O.J.; Oladipo, A.S.; Aleshinloye, A.O. Bio-derived MgO nanopowders for BOD and COD reduction from tannery wastewater. *J. Water Process Eng.* **2017**, *16*, 142–148. [CrossRef]
5. Elayadi, F.; El Adlouni, C.; Achak, M.; El Herradi, E.; El Krati, M.; Tahiri, S.; Naman, M.; Naman, F. Effects of raw and treated olive mill wastewater (OMW) by coagulation-flocculation, on the germination and the growth of three plant species (wheat, white beans, lettuce). *Mor. J. Chem.* **2019**, *7*, 111–122.
6. Khatamian, M.; Divband, B.; Shahi, R. Ultrasound assisted co-precipitation synthesis of Fe_3O_4/bentonite nanocomposite: Performance for nitrate, BOD and COD water treatment. *J. Water Process Eng.* **2019**, *3*, 100870. [CrossRef]
7. Aquilanti, L.; Taccari, M.; Bruglieri, D.; Osimani, A.; Clementi, F.; Comitini, F.; Ciani, M. Integrated biological approaches for olive mill wastewater treatment and agricultural exploitation. *Int. Biodeterior. Biodegrad.* **2014**, *88*, 162–168. [CrossRef]
8. Mastoras, P.; Vakalis, S.; Fountoulakis, M.S.; Gatidou, G.; Katsianou, P.; Koulis, G.; Thomaidis, N.S.; Haralambopoulos, D.; Stasinakis, A.S. Evaluation of the performance of a pilot-scale solar still for olive mill wastewater treatment. *J. Clean. Prod.* **2022**, *365*, 132695. [CrossRef]
9. Esteves, B.M.; Morales-Torres, S.; Madeira, L.M.; Maldonado-Hodar, F.J. Specific adsorbents for the treatment of OMW phenolic compounds by activation of bio-residues from the olive oil industry. *J. Environ. Manag.* **2022**, *306*, 114490. [CrossRef]
10. Alaoui, S.B.; Lamy, E.; Achak, M. Assessment of the impact of diluted and pretreated olive mill wastewater on the treatment efficiency by infiltration-percolation using natural bio-adsorbents. *Environ. Sci. Pollut. Res.* **2022**, *30*, 16305–16320. [CrossRef]
11. Jawad, A.H.; Abdulhameed, A.S.; Mastuli, M.S. Acid-factionalized biomass material for methylene blue dye removal: A comprehensive adsorption and mechanism study. *J. Taibah Univ. Sci.* **2020**, *14*, 305–313. [CrossRef]
12. Mohan, D.; Pittman, C.U., Jr. Activated carbons and low cost adsorbents for remediation of tri- and hexavalent chromium from water. *J. Hazard. Mater.* **2006**, *137*, 762–811. [CrossRef] [PubMed]
13. Hameed, B.H. Spent tea leaves: A new non-conventional and low-cost adsorbent for removal of basic dye from aqueous solutions. *J. Hazard. Mater.* **2009**, *161*, 753–759. [CrossRef]
14. El-Naas, M.H.; Al-Zuhair, S.; Abu Alhaija, M. Reduction of COD in refinery wastewater through adsorption on date-pit activated carbon. *J. Hazard. Mater.* **2010**, *173*, 750–757. [CrossRef]
15. Low, L.W.; Teng, T.T.; Alkarkhi, A.F.M.; Ahmad, A.; Morad, N. Optimization of the Adsorption Conditions for the Decolorization and COD Reduction of Methylene Blue Aqueous Solution using Low-Cost Adsorbent. *Water Air Soil Pollut.* **2011**, *214*, 185–195. [CrossRef]
16. Ismail, M.N.; Aziz, H.A.; Ahmad, M.A.; Yusoff, N.A. Optimization of Areca cathecu Fronds as Adsorbent for Decolorization and COD Removal of Wastewater through the Adsorption Process. *Sains Malaysiana* **2015**, *11*, 1609–1614.
17. Nayl, A.E.A.; Elkhashab, R.A.; El Malah, T.; Sobhy, M.; Yakout Mohamed, A.; El-Khateeb Mahmoud, M.S.; Ali Hazim, M.A. Adsorption studies on the removal of COD and BOD from treated sewage using activated carbon prepared from date palm waste. *Environ. Sci. Pollut. Res.* **2017**, *24*, 22284–22293. [CrossRef]

18. Mpatani, F.M.; Aryee, A.A.; Kani, A.N.; Wen, K.; Dovi, E.; Qu, L.; Li, Z.; Han, R. Removal of methylene blue from aqueous medium by citrate modified bagasse: Kinetic, Equilibrium and Thermodynamic study. *Bioresour. Technol. Rep.* **2020**, *11*, 100463. [CrossRef]
19. Ge, M.; Du, M.; Zheng, L.; Wang, B.; Zhou, X.; Jia, Z.; Hu, G.; Jahangir Alam, S.M. A maleic anhydride grafted sugarcane bagasse adsorbent and its performance on the removal of methylene blue from related wastewater. *Mater. Chem. Phys.* **2017**, *192*, 147–155. [CrossRef]
20. Surafel, M.B.; Venkatesa, P.S.; Tsegaye, S.T.; Abraham, G.A. Sugarcane bagasse based activated carbon preparation and its adsorption efficacy on removal of BOD and COD from textile effluents: RSM based modeling, optimization and kinetic aspects. *Bioresour. Technol. Rep.* **2021**, *14*, 100664. [CrossRef]
21. Zhang, Z.; Moghaddam, L.; O'Hara, I.M.; Doherty, W.O.S. Congo Red adsorption by ball-milled sugarcane bagasse. *Chem. Eng. J.* **2011**, *178*, 122–128. [CrossRef]
22. Akl, M.A.A.; Dawy, M.B.; Serage, A.A. Efficient removal of phenol from water samples using sugarcane bagasse based activated carbon. *J. Anal Bioanal Tech.* **2014**, *5*, 189. [CrossRef]
23. Abdelhafez, A.; Jianhua, L. Removal of Pb (II) from aqueous solution by using biochars derived from sugar cane bagasse and orange peel. *J. Taiwan Inst. Chem. Eng.* **2016**, *61*, 367–375. [CrossRef]
24. Elayadi, F.; Boumya, W.; Achak, M.; Chhiti, Y.; M'hamdi Alaoui, F.E.; Barka, N.; El Adlouni, C. Experimental and modelling studies of the removal of phenolic compounds from olive mill wastewater by adsorption on sugarcane bagasse. *Environ. Chall.* **2021**, *4*, 100184. [CrossRef]
25. Tran, H.N. Applying Linear Forms of Pseudo-Second-Order Kinetic Model for Feasibly Identifying Errors in the Initial Periods of Time-Dependent Adsorption Datasets. *Water* **2023**, *15*, 1231. [CrossRef]
26. Tawfik, A.; Saleh, S.; Omobayo, A.; Asif, M.; Dafalla, H. Response Surface Optimization and Statistical Analysis of Phenols Adsorption On Diethylenetriamine modified Activated Carbon. *J. Clean. Prod.* **2018**, *182*, 960–968.
27. Moubarik, A.; Grimi, N. Valorization of olive stone and sugar cane bagasse by-products as biosorbents for the removal of cadmium from aqueous solution. *Food Res. Int.* **2015**, *73*, 169–175. [CrossRef]
28. de Sales, P.F.; Magriotis, Z.M.; Rossi, M.A.L.S.; Resende, R.F.; Nunes, C.R. Optimization by Response Surface Methodology of the adsorption of Coomassie Blue dye on natural and acid-treated clays. *J. Environ. Manag.* **2013**, *130*, 417e428. [CrossRef]
29. Leili, M.; Faradmal, J.; Kosravian, F.; Heydari, M. A Comparison study on the removal of phenol from aqueous solution using Organomodified Bentonite and commercial activated carbon. *Avicenna J. Environ. Health Eng.* **2015**, *2*, 2698. [CrossRef]
30. Mondal, N.K.; Roy, S. Optimization study of adsorption parameters for removal of phenol on gastropod shell dust using response surface methodology. *Clean Technol. Environ. Policy* **2016**, *18*, 429–447. [CrossRef]
31. Maleki, S.; Karimi-Jashni, A. Optimization of Ni(II) adsorption onto Cloisite Na+ clay using response surface2 methodology. *Chemosphere* **2019**, *246*, 125710. [CrossRef] [PubMed]
32. Olusegun, S.J.; de Sousa Lima, L.F.; Santina Mohallem, N.D. Enhancement of adsorption capacity of clay through spray drying and surface modification process for wastewater treatment. *Chem. Eng. J.* **2018**, *334*, 1719–1728. [CrossRef]
33. Hafshejani, L.D.; Hooshmand, A.; Naseri, A.A.; Mohammadi, A.S.; Abbasi, F.; Bhatnagar, A. Removal of nitrate from aqueous solution by modified sugarcanebagasse biochar. *Ecol. Eng.* **2016**, *95*, 101–111. [CrossRef]
34. Yangui, A.; Abderrabba, M. Towards a high yield recovery of polyphenols from olive mill wastewater on activated carbon coated with milk proteins: Experimental design and antioxidant activity. *J. Food Chem.* **2018**, *262*, 102–109. [CrossRef] [PubMed]
35. Mandal, A.; Bar, N.; Kumar Das, S. Phenol removal from wastewater using low-cost natural bioadsorbent neem (Azadirachta indica) leaves: Adsorption study and MLR modeling. *Sustain. Chem. Pharm.* **2020**, *17*, 100–308. [CrossRef]
36. Geetha Devi, M.; Shinoon Al-Hashmi, Z.S.; Chandra Sekhar, G. Treatment of vegetable oil mill effluent using crab shell chitosan as adsorbent. *Int. J. Environ. Sci. Technol.* **2012**, *9*, 713–718. [CrossRef]
37. Pourmortazavi, S.M.; Sahebi, H.; Zandavar, H.; Mirsadeghi, S. Fabrication of Fe3O4 nanoparticles coated by extracted shrimp peels chitosan as sustainable adsorbents for removal of chromium contaminates from wastewater: The design of experiment. *Compos B. J. Eng.* **2019**, *175*, 107390. [CrossRef]
38. Palin, D., Jr.; Rufato, K.B.; Linde, G.A.; Colauto, N.B.; Caetano, J.; Alberton, O.; Jesus, D.A.; Dragunski, D.C. Evaluation of Pb (II) biosorption utilizing sugarcane bagasse colonized by Basidiomycetes. *Environ. Monit Assess.* **2016**, *188*, 188–279. [CrossRef]
39. Lei, S.; Dongmei, C.; Shungang, W.; Zebin, Y. Adsorption Studies of Dimetridazole and Metronidazole onto Biochar Derived from Sugarcane Bagasse: Kinetic, Equilibrium, and Mechanisms. *J. Polym. Environ.* **2017**, *26*, 765–777. [CrossRef]
40. Srinivasan, A.; Viraraghavan, T. Decolorization of dye wastewaters by biosor-bents: A review. *J. Environ. Manag.* **2010**, *91*, 1915–1929. [CrossRef]
41. Sharma, P.; Kaur, H. Sugarcane bagasse for the removal of erythrosin B and methylene blue from aqueous waste. *Appl. Water Sci.* **2011**, *1*, 135–145. [CrossRef]
42. Kumar, R.; Mago, G.; Balan, V.; Wyman, C.E. Physical and chemical character-izations of corn stover and poplar solids resulting from leading pretreatment technologies. *Bioresour. Technol.* **2009**, *100*, 3948–3962. [CrossRef] [PubMed]
43. Granados, D.A.; Ruiz, R.A.; Vega, L.Y.; Chejne, F. Study of reactivity reduction in sugarcane bagasse as consequence of a torrefaction process. *Energy* **2017**, *139*, 818–827. [CrossRef]
44. Vimonses, V.; Lei, S.M.; Jin, B.; Chowd, C.W.K.; Saint, C. Kinetic study and equilibrium isotherm analysis of Congo Red adsorption by clay materials. *J. Chem. Eng.* **2009**, *148*, 354–364. [CrossRef]

45. Ahmad, A.A.; Hameed, B.H. Reduction of COD and color of dyeing effluent from a cotton textile mill by adsorption onto bamboo-based activated carbon. *J. Hazard. Mater.* **2009**, *172*, 1538–1543. [CrossRef]
46. Akter, S.; Sultana, F.; Kabir, M.R.; Brahma, P.P.; Tasneem, A.; Sarker, N.; Mst, M.; Akter, M.; Kabir, M.; Uddin, M.K. Comparative Study of COD Removal Efficacy from Pharmaceutical Wastewater by Areca Nut Husk Produced and Commercially Available Activated Carbons. *Asian J. Environ. Ecol.* **2021**, *14*, 47–56. [CrossRef]

Disclaimer/Publisher's Note: The statements, opinions and data contained in all publications are solely those of the individual author(s) and contributor(s) and not of MDPI and/or the editor(s). MDPI and/or the editor(s) disclaim responsibility for any injury to people or property resulting from any ideas, methods, instructions or products referred to in the content.

Article

Comparison of Phenol Adsorption Property and Mechanism onto Different Moroccan Clays

Younes Dehmani [1,*], Dison S. P. Franco [2], Jordana Georgin [2], Taibi Lamhasni [3], Younes Brahmi [4], Rachid Oukhrib [5], Belfaquir Mustapha [6], Hamou Moussout [6], Hassan Ouallal [7] and Abouarnadasse Sadik [1]

1 Laboratory of Chemistry and Biology Applied to the Environment, Faculty of Sciences of Meknes, Moulay Ismail University, Meknes 50050, Morocco; abouarnadasse@yahoo.fr
2 Department of Civil and Environmental, Universidad de la Costa, CUC, Calle 58 #55-66, Barranquilla 50366, Colombia; francodison@gmail.com (D.S.P.F.); jordanageorgin89@gmail.com (J.G.)
3 Institut National des Sciences de l'Archéologie et du Patrimoine (INSAP), BP 6828, Madinat al Irfane, Avenue Allal El Fassi, Angle rues 5 et 7, Rabat-Instituts, Rabat 10000, Morocco; t.lamhasni@gmail.com
4 HTMR-Lab, Mohammed VI Polytechnic University (UM6P), Benguerir 43150, Morocco; younes.brahmi@um6p.ma
5 Team of Physical Chemistry and Environment, Faculty of Sciences, IBN ZOHR University, Agadir 80000, Morocco; rachid.oukhrib@edu.uiz.ac.ma
6 Laboratory of Advanced Materials and Process Engineering, Faculty of Sciences, University Ibn Tofail, PB: 133, Kenitra 14000, Morocco; moussouthammou@gmail.com (B.M.); mbelfaquir@gmail.com (H.M.)
7 Laboratory of Physic-Chemistry Materials, Faculty of Sciences and Techniques, Department of Chemistry Fundamental and Applied, Moulay Ismaïl University, PO Bosc 509, Errachidia 52000, Morocco; hassanouallalaghbalou@gmail.com
* Correspondence: dehmaniy@gmail.com; Tel.: +212-678937801

Citation: Dehmani, Y.; Franco, D.S.P.; Georgin, J.; Lamhasni, T.; Brahmi, Y.; Oukhrib, R.; Mustapha, B.; Moussout, H.; Ouallal, H.; Sadik, A. Comparison of Phenol Adsorption Property and Mechanism onto Different Moroccan Clays. Water 2023, 15, 1881. https://doi.org/10.3390/w15101881

Academic Editor: Hai Nguyen Tran

Received: 18 April 2023
Revised: 7 May 2023
Accepted: 12 May 2023
Published: 16 May 2023

Copyright: © 2023 by the authors. Licensee MDPI, Basel, Switzerland. This article is an open access article distributed under the terms and conditions of the Creative Commons Attribution (CC BY) license (https://creativecommons.org/licenses/by/4.0/).

Abstract: This study focuses on the removal of phenol from aqueous media using Agouraï clay (Fes-Meknes-Morocco region) and Geulmima clay (Draa Tafilalet region). The characterization of the clay by Fourier Transform Infrared (FTIR) Spectroscopy, X-ray diffraction (XRD), N2 adsorption (BET), Scanning Electron Microscopy (SEM), and Thermogravimetric and differential thermal analysis (DTA/GTA) indicates that it is mainly composed of quartz, kaolinite, and illite. The results showed that raw Clay Agourai (RCA) and raw Clay Geulmima (RCG) adsorbed phenol very quickly and reached equilibrium after 30 min. Thermodynamic parameters reveal the physical nature of the adsorption, the spontaneity, and the sequence of the process. However, the structure and structural characterization of the solid before and after phenol adsorption indicated that the mechanism of the reaction was electrostatic and that hydrogen bonding played an important role in RCG, while kinetic modeling showed the pseudo-second-order model dynamics. The physics-statistics modeling was employed for describing the isotherm adsorption for both systems. It was found that the monolayer model with two different energy sites best describes adsorption irrespective of the system. The model indicates that the receptor density of each clay direct influences the adsorption capacity, demonstrating that the composition of the clay is the main source of the difference. Thermodynamic simulations have shown that the adsorption of phenol is spontaneous and endothermic, irrespective of the system. In addition, thermodynamic simulations show that the RCG could be adsorbed even further since the equilibrium was not achieved for any thermodynamic variable. The strength of this study lies in the determination of the adsorption mechanism of phenol on clay materials and the optimum values of temperature and pH.

Keywords: phenol adsorption; different clays; adsorption mechanism; physics-statistics modeling

1. Introduction

Water is the essential element of life, and the quality and quantity of water on earth ask critical questions of all social actors [1]. Water pollution by various organic and inorganic pollutants is a global problem that requires a solution in the source of pollutants and a solution before releasing them into aquatic ecosystems [2]. The pollutants that alter water quality are various and diverse and especially include phenol and phenolic compounds widely used in several industrial and agricultural fields [3]. The remarkably high toxicity of phenol has prompted the services in charge of environmental protection to normalize its concentration in water to reduce its impacts on humans and the environment [4]. Despite the non-high concentrations of these compounds in the aquatic environment, their toxic effects and their impact on the environment are remarkable, and several studies in the literature show the effects of these pollutants on health and the environment [5]. There are numerous techniques for cleaning up water that has been contaminated by organic substances such as degradation, biological treatment, membrane treatment, and catalytic oxidation. However, adsorption is the technique of choice for the removal of phenol [6]. The major advantage of this technique is the possibility of using several types of adsorbent and the simplicity of the process [6]. Clays have proven to be very effective in removing phenol from wastewater [6]. Clay minerals are the most important materials because of their known properties of adsorption and retention of pollutants [7], as well as their ease of modification and/or functionalization [7].

Despite the huge potential of clays, Morocco is still behind in the use of these processes to produce new high-value clay materials that could lead to innovative applications [8–10]. Some studies have focused on adsorption on clays. Agnieszka Gładysz-Płaska used red clay to adsorb uranium [11], and the adsorption of cationic and anionic dyes on clays has been the subject of work by Islem Chaari and collaborators [12]. Meichen Wang was interested in the adsorption of benzo[a]pyrene on modified clays [13]. Several works on the adsorption of dyes on clays and modified clays are mentioned in the work of Abida Kausar [14]. The work of Ali Q. Selim and co-workers [15] aimed at adsorbing phosphates on chemically modified carbonaceous clays. Using a natural clay (kaolinite), Nouria Nabbou tried to remove fluoride from groundwater [16].

There is a special property that characterizes porous solids, especially clay. This characteristic is the ability to adsorb heavy metals and organic matter contained in aqueous solutions, as well as a cationic exchange capacity [15]. This arises mainly due to their natural acidity and their high specific surface. Its performance is affected by several parameters: temperature, pH, and the properties of the adsorbed elements. The Moroccan kaolin clays of the regions of Guelmima (Darra Tafilalet region) and Agourai (Fes Meknes region) are part of this class. They are currently used in the ceramic industry, but they are also potential candidates for use in pollution control.

It is in this general context that this study has as its main objective the valorization and comparison of two Moroccan clays, one from the region of Daraâ Tafilalet and one from the region of Fes Meknes, as natural adsorbents for the retention of phenol from an aqueous solution. This work is divided into two parts. The first part is devoted to the characterization of the raw material. In the second part, we present the adsorption experiments, the methodology adopted and the results obtained, and their interpretations. The general conclusion summarizes the main results of this research work.

2. Materials and Methods

2.1. Sampling Area

The clays used in this work come from the town of Agouraï, Fes-Meknes region, and the town of Geulmima, Draa Tafilalet region (Morocco), as shown on the map in Figure 1. The sampled clays were sieved, washed, and dried in the oven for one night at a temperature of 130 °C.

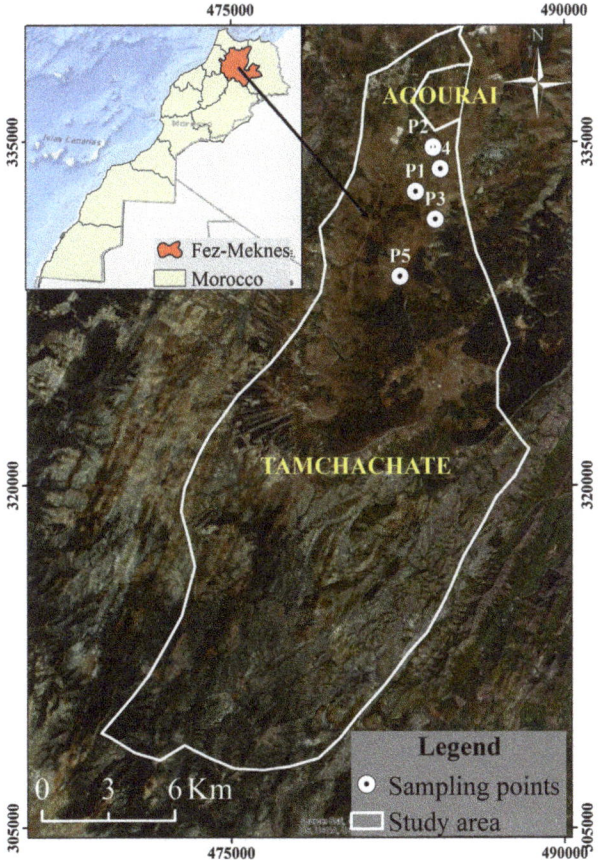

Figure 1. Geographical site of the clay used.

2.2. Adsorption Batch

The phenol adsorption experiments were conducted in accordance with the following protocol: a solution of phenol (20 mL) with a concentration that varied between 10 and 500 mg L^{-1} was brought into contact with a mass of 0.1 g of solid at a constant temperature (T = 30, 40, and 50 °C) for pH = 4 and with a stirring rate of 600 rpm during the adsorption period. At the end of the assay, the mixtures were filtered and analyzed by UV/Vis spectroscopy. The residual concentration was determined based on a UV spectrometer calibration curve/Visible at λ = 270 nm (Shimadzu UV-1240). Equation (1) was used for the determination of the amount adsorbed:

$$Q_{ads} = \frac{(C_0 - C_e)}{m_{ads}} \times V_{sol} \tag{1}$$

where Q_{ads} is the adsorption capacity (mg g), C_0 is the initial concentration of the phenol (mg L), C_e is the residual concentration of the phenol (mg L), m_{ads} is the mass of adsorbent used (g), and V_{sol} is the volume of the solution (L).

2.3. Kinetic Modelling

The experimental results were adjusted to nonlinear models in order to ascertain the adsorption mechanism of phenol on both solids (RCA and RCG). The modeling of the adsorption kinetics was done using the pseudo-first-order and pseudo-second-order models [17]:

$$q_t = q_q(1 - e^{-tk_1}) \quad (2)$$

$$q_t = q_e^2 k_2 \frac{t}{q_e k_2 t + 1} \quad (3)$$

where q_t is the adsorption capacity according to time (mg g^{-1}), q_e is the adsorption capacity at equilibrium (mg g^{-1}), k_1 is the pseudo-first-order kinetics constant (min^{-1}), k_2 is the pseudo-second-order kinetic constant (g mg^{-1} min^{-1}), and t is the time (min).

2.4. Isotherm Modeling through the Physic-Statistics Approach

In this study, the adsorption of phenol onto several adsorbents was compared using a physical-statistical (Phys-Stat) modeling approach (RCA and RCG). The use of the grand canonical ensemble serves as the foundation for the Phys-stat models. which defines systems that, at a constant temperature, can exchange particles with their environment.

2.4.1. Monolayer Model with Single Energy Site (MLO)

The most commonly employed model is the monolayer model with one energy site (MLO). This model takes into consideration that a variable number of phenol molecules (n, dimensionless) is adsorbed in one type of receptor site with energy ($-\varepsilon_1$), reflecting directly in the quantity of adsorbed molecules, represented by the receptor density (N_m, mg g^{-1}), according to Equation (4) [18]:

$$q_e = \frac{nN_m}{1+\left(\frac{C_{1/2}}{C_e}\right)^n} = \frac{q_m}{1+\left(\frac{C_{1/2}}{C_e}\right)^n} \quad (4)$$

where $C_{1/2}$ is the concentration at half-saturation (mg L^{-1}) and q_m is the adsorption capacity at saturation (mg g^{-1}).

2.4.2. Monolayer Model with Two Energy Sites (MLT)

When considering that a variable number of phenol molecules can be adsorbed onto two energetically different sites ($-\varepsilon_1$ and $-\varepsilon_2$) the monolayer model with two energy sites (MLT) is obtained. In this case, a different number of phenol molecules (n_1 and n_2, dimensionless) are adsorbed onto different receptors sites with different densities (N_{m1} and N_{m2}, mg g^{-1}), according to Equation (5) [19]:

$$q_e = \frac{n_1 N_{m1}}{1+\left(\frac{C_1}{C_e}\right)^{n_1}} + \frac{n_2 N_{m2}}{1+\left(\frac{C_2}{C_e}\right)^{n_2}} = \frac{q_{m1}}{1+\left(\frac{C_1}{C_e}\right)^{n_1}} + \frac{q_{m2}}{1+\left(\frac{C_2}{C_e}\right)^{n_2}} \quad (5)$$

where C_1 and C_2 are the concentrations at half-saturation of the first and second receptor sites (mg L^{-1}), respectively, and q_{m1} and q_{m2} are the adsorption capacity at saturation of the first and second receptor sites (mg g^{-1}), respectively.

2.4.3. Double-Layer Model with One Energy Site (DLO)

Another possibility is the formation of dual layers. This hypothesis emerges from the consideration that the phenol molecules (n, dimensionless) can form a dual layer of molecules with a single energy of adsorption ($-\varepsilon_1$) with a single density of receptor site (N_m, mg g^{-1}) according to Equation (6) [20]:

$$q_e = nN_m \frac{\left(\frac{C_e}{C_{1/2}}\right)^n + 2\left(\frac{C_e}{C_{1/2}}\right)^{2n}}{1+\left(\frac{C_e}{C_{1/2}}\right)^n + \left(\frac{C_e}{C_{1/2}}\right)^{2n}} = q_m \frac{\left(\frac{C_e}{C_{1/2}}\right)^n + 2\left(\frac{C_e}{C_{1/2}}\right)^{2n}}{1+\left(\frac{C_e}{C_{1/2}}\right)^n + \left(\frac{C_e}{C_{1/2}}\right)^{2n}} \quad (6)$$

where $C_{1/2}$ (mg L^{-1}) is the concentration at half-saturation and q_m is the adsorption capacity at saturation (mg g^{-1}).

2.4.4. Double-Layer Model with Two Energy Sites (DLT)

It is also possible that the phenol molecules (n, dimensionless) can form a double-layer, with each layer having different adsorption energy ($-\varepsilon_1$ and $-\varepsilon_2$), with the same receptor sites having the same density (N_m, mg g^{-1}), here presented by Equation (7) [21]:

$$q_e = nN_m \frac{\left(\frac{C_e}{C_1}\right)^n + 2\left(\frac{C_e}{C_2}\right)^{2n}}{1+\left(\frac{C_e}{C_1}\right)^n + \left(\frac{C_e}{C_2}\right)^{2n}} = q_m \frac{\left(\frac{C_e}{C_1}\right)^n + 2\left(\frac{C_e}{C_2}\right)^{2n}}{1+\left(\frac{C_e}{C_1}\right)^n + \left(\frac{C_e}{C_2}\right)^{2n}} \tag{7}$$

where C_1 and C_2 are the half-saturation concentration (mg L^{-1}), and q_m is the adsorption capacity at saturation (mg g^{-1}).

2.4.5. Multilayer Model (MM)

Last, there is the possibility of forming a multilayer of adsorption. This model considers that the anchorage of the phenol molecules (n, dimensionless) depends on two energies, one corresponding to the first layer ($-\varepsilon_1$) and the other to all layers beyond the first ($-\varepsilon_2$), where the energy of adsorption of the first layer is always higher than the other layers. Similar to the other described model, the multilayer model considers only one density receptor site and the same number of molecules per layer [22]. The model is described according to the following Equations:

$$q_e = nN_m\left(\frac{F_1+F_2+F_3+F_4}{G}\right) = q_m\left(\frac{F_1+F_2+F_3+F_4}{G}\right) \tag{8}$$

$$F_1 = -\frac{2\left(\frac{C_e}{C_1}\right)^{2n}}{\left(1-\left(\frac{C_e}{C_1}\right)^n\right)} + \frac{\left(\frac{C_e}{C_1}\right)^n\left(1-\left(\frac{C_e}{C_1}\right)^n\right)}{\left(1-\left(\frac{C_e}{C_1}\right)^n\right)^2} \tag{9}$$

$$F_2 = \frac{2\left(\frac{C_e}{C_1}\right)^n\left(\frac{C_e}{C_2}\right)^n\left(1-\left(\frac{C_e}{C_2}\right)^{nN_2}\right)}{\left(1-\left(\frac{C_e}{C_2}\right)^n\right)} \tag{10}$$

$$F_3 = \frac{\left(\frac{C_e}{C_1}\right)^n\left(\frac{C_e}{C_2}\right)^{2n}\left(\frac{C_e}{C_2}\right)^{nN_2}N_2}{\left(1-\left(\frac{C_e}{C_2}\right)^n\right)} \tag{11}$$

$$F_4 = \frac{\left(\frac{C_e}{C_1}\right)^n\left(\frac{C_e}{C_2}\right)^{2n}\left(1-\left(\frac{C_e}{C_2}\right)^{nN_2}\right)}{\left(1-\left(\frac{C_e}{C_1}\right)^n\right)^2} \tag{12}$$

$$G = \frac{\left(1-\left(\frac{C_e}{C_1}\right)^{2n}\right)}{\left(1-\left(\frac{C_e}{C_1}\right)^n\right)} + \frac{\left(\frac{C_e}{C_1}\right)^n\left(\frac{C_e}{C_2}\right)^n\left(1-\left(\frac{C_e}{C_2}\right)^{nN_2}\right)}{\left(1-\left(\frac{C_e}{C_2}\right)^n\right)} \tag{13}$$

where N_m is the density receptor of the receptor site (mg g^{-1}), N_2 is the number of formed layers, C_1 and C_2 are the half-saturation concentration (mg L^{-1}), and q_m is the adsorption capacity at saturation (mg g^{-1}).

2.5. Parameter Estimation and Model Evaluation

The steric parameters were estimated using Matlab scripting programming. The built-in Matlab functions were utilized for this: particleswarm for the initial guess estimation, and lsqnonlin or nlinfit for incorrect particleswarm estimation. The difference between *lsqnonlin* and *nlinfit* is in the restrictions, where the first is used for a limited boundary and the second performs optimization for an undefined boundary. For the model evaluation, the coefficient of correlation (R^2), adjusted coefficient of determination (R^2_{adj}), minimum squared error (MSR, (mg g^{-1})^2), average relative error (ARE, %), and Bayesian Criterion Indicator (BIC) were employed [22]:

$$R^2 = 1 - \frac{\sum_{i=1}^{n}\left(q_{exp} - q_{pred}\right)^2}{\sum_{i=1}^{n}\left(q_{exp} - \bar{q}_{exp}\right)^2} \quad (14)$$

$$ARE = \frac{100\%}{n}\sum_{i=1}^{n}\left|\frac{q_{exp} - q_{pred}}{q_{exp}}\right| \quad (15)$$

$$MSR = \frac{1}{n-p}\sum_{i=1}^{n}\left(q_{exp} - q_{pred}\right)^2 \quad (16)$$

$$BIC = n\,Ln\left(\frac{RSS}{n}\right) + p\,Ln(n) \quad (17)$$

where q_{exp} is the experimental adsorption capacity at the equilibrium (mg g^{-1}) and q_{pred} is the predicted adsorption capacity at the equilibrium (mg g^{-1}), n is the number of experimental data points, and p is the number of parameters used in the model.

2.6. Structure Characterization

First, an "Axion" type X-ray fluorescence spectrometer with a dispersion of 1 kW wavelength was used to measure the X-ray fluorescence. The UATRS laboratory and CNRST in Rabat, Morocco, performed this chemical analysis. Then, using an X'PERT MPD-PRO wide-angle X-ray powder diffractometer equipped with a diffracted beam monochromator and Ni-filtered CuK radiation (λ = 1.5406), X-ray diffraction (XRD) patterns were captured. With a counting period of 2.0 s and increments of 0.02°, the 2θ angle was scanned from 4° to 30°. Then, using a Fourier transform infrared spectrometer, the Fourier transforms infrared (FTIR) spectra of RCG and CCG were characterized (VERTEX70). The samples were made in a conventional manner on KBr discs from extremely well-dried mixes containing about 4% (w/w). FTIR spectra between 4000 and 400 cm^{-1} were captured. Using LABSYS/Evo thermal, Thermogravimetric Analysis/Differential Thermal Analysis (TGA/DTA) studies were performed in an air environment. The samples were heated linearly (T = T0 + t) at a rate of 20 °C/min from ambient to 600 °C. In order to determine the textural characteristics, BET Nitrogen adsorption measurements were acquired using Micromeritics ASAP 2010. SEM images were acquired using a Quanta 200 EIF microscope equipped with standard secondary electron (SE) and backscattered electron (BSE) detectors. The electron beam is produced by a conventional tungsten electron source, which can resolve features as small as 3 nm under optimal operating conditions. The microscope is equipped with an Oxford Inca energy dispersive X-ray (EDX) system for elemental chemical analysis.

3. Results

3.1. X-ray Fluorescence

Our findings showed that, of the clays under study, silica, alumina, and calcium oxide made up around 66% of the composition of the RCG clays and 50 percent of the RCA clays, respectively. The concentrations of alkali and alkaline earth metals were lower in both clay solids (Table 1). This chemical analysis showed a 10.5% lower alumina content compared to the alumina content of fire clay: 45% [15,23]. Moreover, the weight of calcite in the clay of the Fes Meknes region is higher than that of Geulmima. The presence of calcite reduces the shrinkage of the agglomerate.

Table 1. Chemical composition of the studied clays.

Adsorbent	Oxide (Weight %)								
	SiO$_2$	Al$_2$O$_3$	CaO	MgO	Fe$_2$O$_3$	S	BaO	P$_2$O$_5$	L.O.I
RCA	38.74	10.60	16.80	0.92	6.00	0.19	0.02	0.10	16.61
RCG	48.39	16.16	2.17	1.75	15.44	3.55	0.26	1.39	10.09

3.2. X-ray Patterns of RCA and RCG

The diffractograms of RCG and RCA are shown in Figure 2. The diffractogram of RCG shows that Quartz (Q), Kaolinite (K), and Illite make up the majority of the clay (I). Conversely, the XRD diffractogram of RCA shows that the composition consists of kaolinite (Al$_2$Si$_2$O$_5$(OH)$_4$), illite [(K, H$_3$O) Al$_2$Si$_3$AlO$_{10}$(OH)$_2$], quartz (SiO$_2$), and calcite (CaCO$_3$). However, a comparison of the diffraction patterns shows the existence of two strong peaks, the first of which is connected to quartz and the other to calcite, indicating that this clay is heterogeneous [15,24,25].

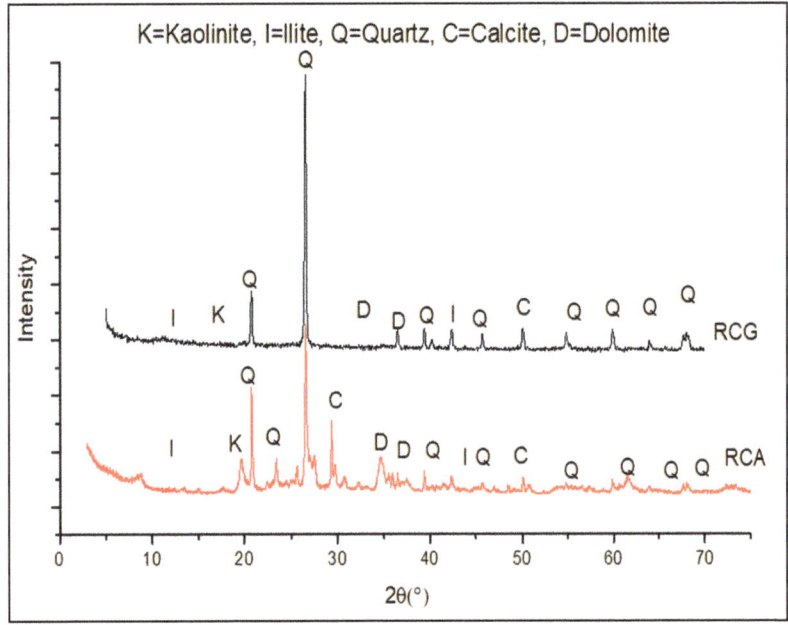

Figure 2. XRD patterns of RCA and RCG.

3.3. FTIR Spectra of RCA and RCG

Figure 3 shows the FTIR spectra of RCA and RCG. Both spectra show a broad absorption band around 3425 cm^{-1} due to the stretching and bending vibrations, respectively, of adsorbed H$_2$O [26]. It is possible to assign the bands at 3649, 3690, and 3650 cm^{-1} to the hydroxyl group's stretching vibration in various environments (Al, AlOH), (Al, MgOH), or (Al, FeOH) [27]. For the RCA clay, a band located at 1380 cm^{-1} indicates the presence of calcite, while the bands detected at 2870, 1800, 1435, 870, and 715 cm^{-1} are attributable to the deformation and elongation vibrations of calcite (CaCO$_3$) [28]. Vibrational distortion of the Si-O-Al bond is responsible for the band seen at 800 cm^{-1}. The bands at 470 and

520 cm^{-1} are related to the deformation vibrations of the Si-O bond in quartz, while the band at 1030 cm^{-1} is associated with the elongation vibration of the Si-O-Si bond in kaolinite or quartz [29]. The bands at 420 cm^{-1}, 470 cm^{-1}, and 525 cm^{-1} correspond to the deformations of Si-O-Fe, Si-O-Mg, and Si-O-Al. Due to the Si-O stretching, a band seen at 1020 cm^{-1} is appropriate [28]. The bands centered at 985, 836, 797, 674, and 508 cm^{-1} can be attributed to the vibrational deformation of Al-OH-Al, Si-O-Al / Al-Mg-OH, cristobalite, Si-O-Mg, and Mg-OH bonds, respectively [30]. In addition to the Al-OH-Al strain vibration bonds, the band centered at 915 cm^{-1} is also attributable to the presence of kaolinite. There are quartz-corresponding absorption bands at 797 and 779 cm^{-1} [31].

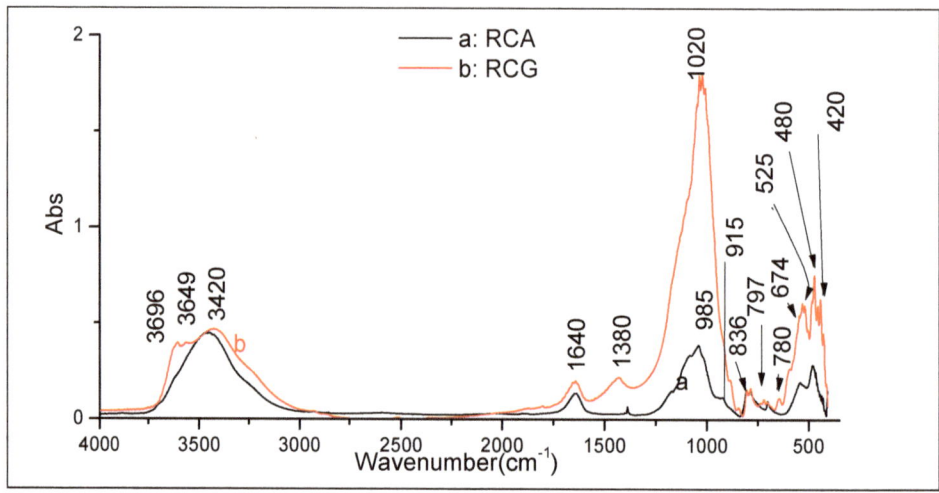

Figure 3. FTIR spectra of RCA and RCG.

3.4. TGA/TDA

Since both solids only experience a mass loss due to water adsorbed on their surfaces, the TGA curves (Figure 4) demonstrate the remarkable thermal stability of the two clays up to 600 °C. The DTG thermogram for the Geulmima clay (RCG) does, in fact, show the existence of an endothermic peak at 120 °C that corresponds to the dehydration and mass loss (1.29%) of the physisorbed water. The thermogram of Agourai clay (RCA) shows that the global mass measured is evaluated at 27% by an endothermic peak at 75 °C and 128 °C. This loss corresponds to the elimination of surface water. Thermal analysis reveals that the two clays under study vary in that there is an exothermic peak between 280 and 380 °C. Another endothermic peak between 450 and 550 °C, which is accompanied by a minor loss of mass, has been linked to the dehydroxylation of the raw clay for the RCA. This peak has been linked to the removal of organic materials [32,33].

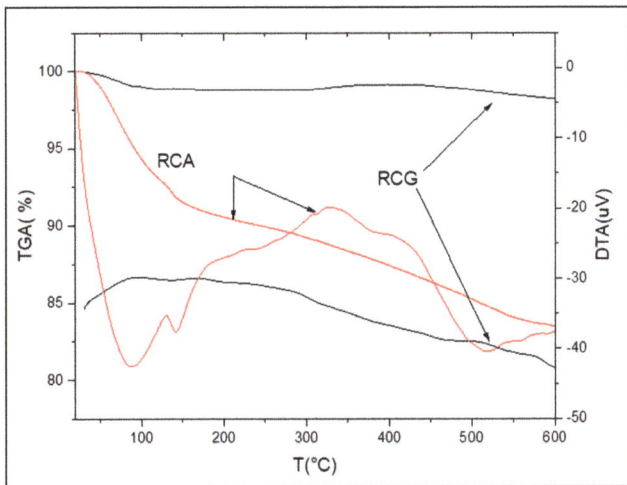

Figure 4. TGA/DTA Thermograms of RCA and RCG.

3.5. N_2 Adsorption/Desorption Isotherm of RCA and RCG

The nitrogen adsorption/desorption method (BET method by Breunuer, Emet, and Teller) is very important in the determination of the textural properties of solids. The textural parameters influence the catalytic and adsorptive capacity of the materials in the treatment of liquid pollutants. Based on the International Union of Pure and Applied Chemistry (IUPAC)'s recommended classification system for physical adsorption isotherms, the isotherms of the clays belong to the type IV isotherm (Figure 5) with a hysteresis loop of type H_3 [34]. There are platforms in the region where the $P/P0$ value is higher than the strongly increasing isotherms. These lines indicate that the samples contain macropores more than 50 nm in diameter [35]. Based on the data Table 2, the clay of the region of Fez Meknes (RCA) has a larger surface than the clay of Draa Tafilalet (RCG).even at the level of pore volume and pore diameter, the clay of Agourai is better than that of Geulmima

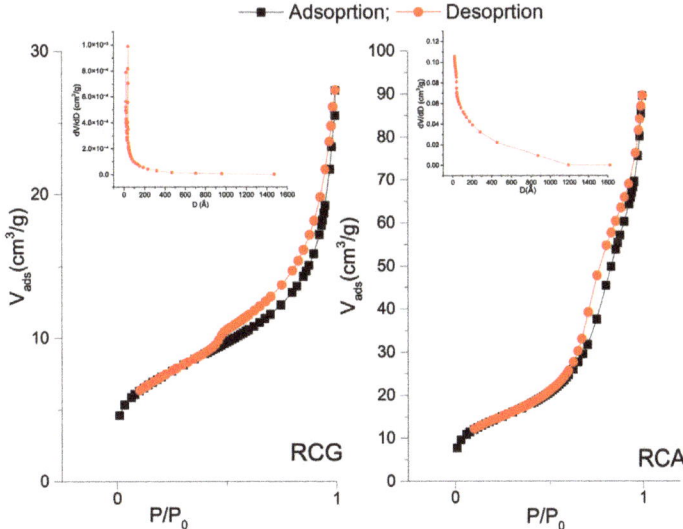

Figure 5. N_2 adsorption/desorption isotherms of RCA and RCG.

Table 2. Textural characteristics of RCA and ACA.

Adsorbent	Proprieties		
	Pore Volume (cm^3 g^{-1})	Specific Surface Area (m^2 g^{-1})	Pore Diameter (Å)
RCA	0.13	51.41	91.35
RCG	0.03	25.31	38.36

3.6. SEM

The SEM micrographic images (Figure 6) of both samples show the porous nature of our materials, which facilitates the penetration of the phenol molecules. Moreover, these images confirm that the structure of the clays is formed by a relatively homogeneous and compact clay matrix. The EDX spectra (Figure 7) show the presence of the main elements such as silica, aluminum, and oxygen, which confirms the validity of the X-ray fluorescence analysis.

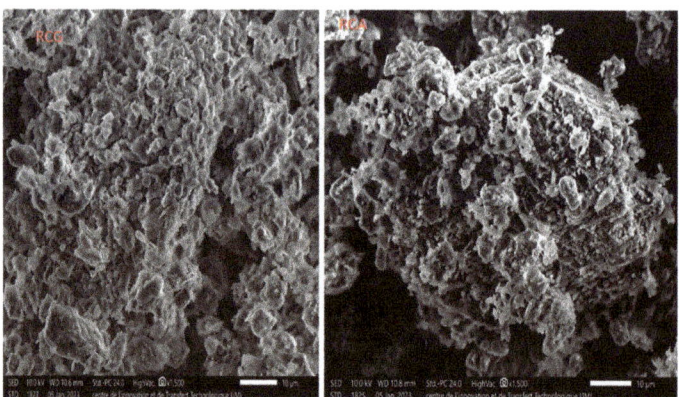

Figure 6. SEM images of RCA and RCG.

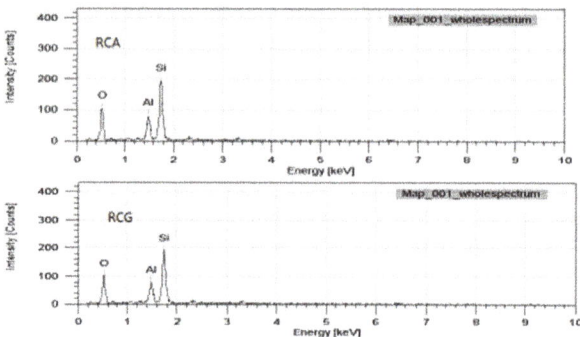

Figure 7. EDX data of RCA and RCG.

3.7. Point of Zero Charge

In the phenomenon of phenol adsorption, the determination of the point of zero charges is important because of the information on this point on the charge of the adsorbent surface and the intervals of change of this charge. The pH of zero charge point pH$_{pzc}$ of the RCG and RCA clays are 7.84 and 8.60, respectively (see Supplementary Materials Figures S1 and S2). Thus, for pH values higher than the pH of the points of null charge of the solids, the surface is negatively charged. At lower pH, the surface is positively charged. The adsorption of phenol is more significant in the case of positively charged surfaces, so we work in the pH< pH$_{pzc}$ of the solids. A difference between the two solids at the pH$_{pzc}$ level can affect the adsorption capacity.

3.8. Determination of CEC

The cation exchange capacity was determined using the method of Delphine Aran [36]. Solids have a very high capacity compared to other types of clays [35]. The CEC of RCA is about 10.6 meq/100 g, while that of RCG is 16.64 meq/100 g. Compared with RCA, RCG has a higher switching capacity. This difference is evidenced by the percentages of elements that make up the two materials.

3.9. pH Effect

The pH of the solution is an important and specific factor in any liquid pollutant adsorption study as it affects the structure of the adsorbent and adsorbate as well as the adsorption mechanism. Therefore, it is logical to know the adsorption efficiency at different pH values to determine the optimal pH for adsorption. The literature gives two cases of the influence of pH on the adsorption of phenol [37]. We can see that in the acidic state, the positive charge on the surface of the adsorbent is dominant, so there is strong static electricity between the adsorbate and the adsorbent on the surface charge. In the ground state, the dramatic drop in adsorption capacity is evidenced by the nature of the predominantly positive charge on the solid surface [38].

For RCG, the adsorption capacity of phenol decreases with a slight increase in pH. The equilibrium adsorption capacity is 2.88 mg. g^{-1} and 2.54 for pH = 4 and 11, respectively. The Agourai clay showed an important decrease in adsorption capacity from pH = 4 to pH = 11, from 1.2 to 0.8 mg g^{-1}, respectively (Figure 8). These results can be explained by the zero-charge pH parameter, which provides information about the dominant charge on the surface and can thus be demonstrated and interpreted in terms of the properties of the materials used. This can be explained by the fact that in the ground state (pH > pH$_{pcn}$), the predominant charge on the adsorbent surface is negative, which reduces the adsorption of the same charged phenoxide. In the acidic state, the positive charge on the surface of the adsorbent is dominant, so there is a relatively high electrostatic attraction between the positive charge on the surface of the adsorbent and the negative charge of the formed phenoxide, which promotes the adsorption. In short, heterogeneous composition leads to different interactions between adsorbate and adsorbent.

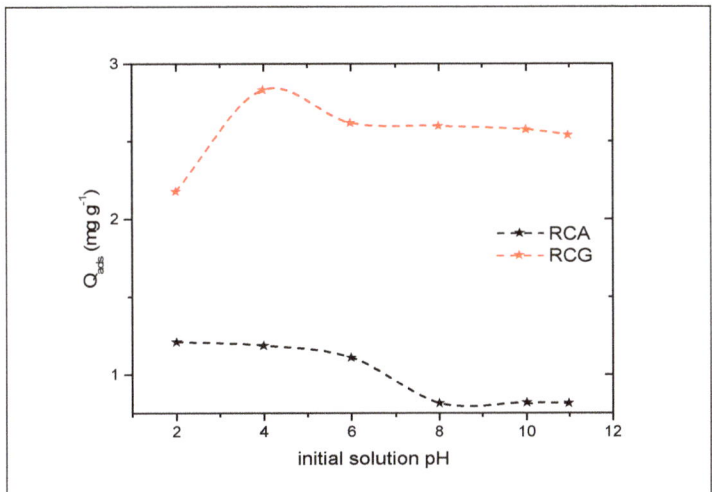

Figure 8. Effect of pH on the phenol adsorption onto RCA and RCG.

3.10. Adsorption of Phenol

3.10.1. Adsorption Kinetics

To evaluate the potential application of the samples in wastewater treatment, the variation in the adsorbed amounts of phenol on the two solids as a function of time and temperature is presented in Figure 9. The adsorbed amount in both solids increases rapidly up to 30 min, and then gradually increases up to 180 min. After this time, it remains almost unchanged at 360 min. The increase in the adsorbed quantity as a function of the temperature can be explained by the effect of the heat on the space between the particles of the solids, a dilatation of the latter imparts fast mobility to the phenol molecules, so an effect can be noticed on the molecules of the phenol by the increase in temperature, which accelerates the fixation on the surface of the clay materials. The amounts of phenol adsorbed by the RCG clay are 1.84, 2.43, and 3.52 mg g^{-1} for temperatures 30 °C, 40 °C, and 50 °C, respectively. While the amount adsorbed by the RCA clay is 1.39, 2.10, and 2.71 mg g^{-1}, respectively, for the same temperatures. The adsorption capacity was better for RCG. This property is dependent on the structural and textural properties of this clay, particularly the zero charge point, which gives an important dominance of positive charges.

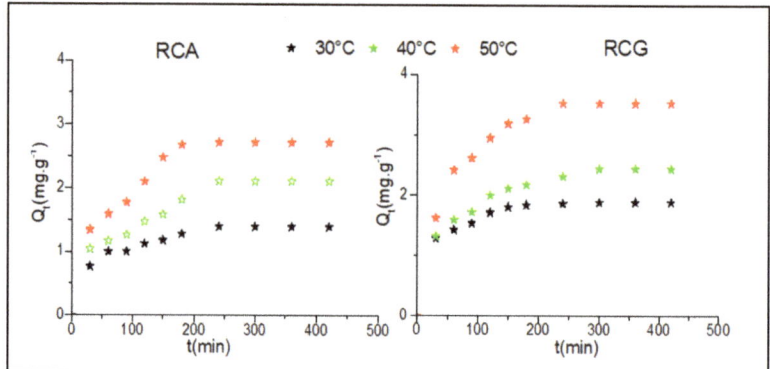

Figure 9. Adsorption kinetics of phenol onto RCA and RCG at different temperatures.

Understanding the adsorption kinetics is necessary to study adsorption because the process can predict the rate of adsorption and explain the mechanism of adsorption. Pseudo-first-order and pseudo-second-order kinetics were used to study the adsorption process of phenol on clay materials. Table 3 lists the parameters of the pseudo-first-order (Equation (2)) and pseudo-second-order (Equation (3)) model equations, and the model representations are shown in Figure 10 [39]. The results show that the adsorption kinetics are described by pseudo-second-order kinetics with a coefficient of determination close to 1 (Table 3).

Table 3. Kinetic parameters of linear and nonlinear modeling of phenol adsorption at different temperatures onto RCA and RCG.

		RCA			RCG		
	Parameters	Temperature (°C)			Temperature (°C)		
		30	40	50	30	40	50
Models	q_{texp} (mg g^{-1})	1.39	2.10	2.71	1.86	2.43	3.52
Pseudo-first-order	k_1 (min^{-1})	0.02	0.01	0.14	0.03	0.01	0.01
	q_e (mg g^{-1})	1.34	2.10	3.73	1.83	2.34	3.48
	R^2	0.95	0.93	0.96	0.96	0.96	0.99
Pseudo-second-order	k_2 (g mg^{-1}·min^{-1})	0.101	0.2278	0.6124	0.4033	0.4918	1.521
	q_e (mg g^{-1})	1.515	2.481	3.17	1.992	2.663	3.99
	R^2	0.9848	0.96018	0.9712	0.9911	0.9895	0.9955

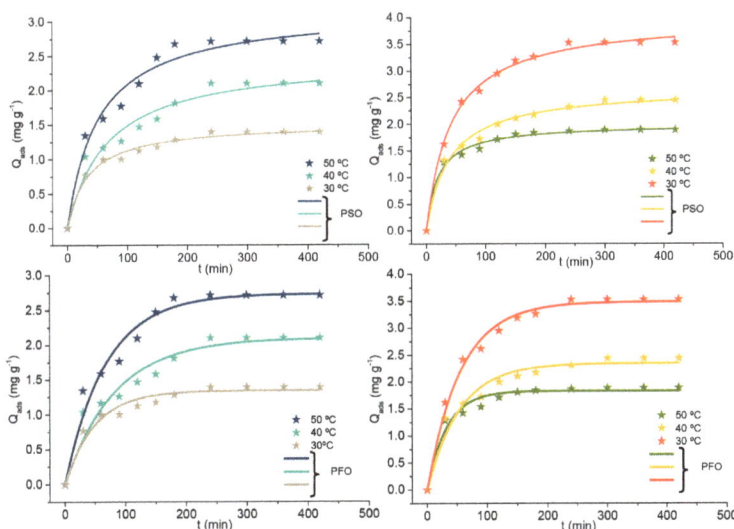

Figure 10. Adsorption kinetics with nonlinear models of pseudo-first-order and pseudo-second-order of phenol (C0 = 5.10^{-4} M) onto RCA and RCG at pH = 4.

3.10.2. Adsorption Isotherms and Physical Statistical Interpretations

The experimental data and the physical statistical prediction are shown in Figure 11. Starting with the experimental data, it was found that both Moroccan clays (RCA and RCG) were able to adsorb the phenol, showing improvements according to the evolution of the system temperature. For the selection of the most suitable Phys-Stat model, the statistical indicators described in Section 2.6 and summarized in Table 4 were employed. Taking into consideration the results obtained from the statistical indicators, the first model to be eliminated is the MM, which corresponded to the Multilayer model. The failure to present a good correlation coefficient (R2), clearly indicates that the phenol molecules are not able to perform multilayer adsorption. The monolayer and dual-layer model present good statistical indicators. However, aiming to obtain the most adequate model, the BIC was used as an evaluation parameter. The BIC indicates that the monolayer model with two different energy sites (MLT) is the most adequate model for describing the adsorption of phenol onto the RCA and RCG clays. In addition to the statistical indicators, the steric parameters obtained for the other models, besides the MLT, do not present coherence. In other words, in cases where the concentration at half-saturation should be increasing, it diminishes or presents fluctuation. Overall the MLT indicates that the different number of phenol molecules (n_1 and n_2) are adsorbed onto two different energetic sites ($-\varepsilon_1$ and $-\varepsilon_2$), which leads to different receptor densities (N_{m1} and N_{m2}). Thus, taking into consideration the MLT, the steric parameters are further discussed aiming to give insight into the phenol adsorption mechanism.

The number of adsorbed molecules per site is also known as the stereographic coefficient, which governs adsorption. The nature of the value also indicated how the phenol molecules are attached to the surface of the RCA and RCG. When the n values are below 1, the molecules are adsorbed in a parallel way. While, for values above 1, the molecules are attached on the surface in a perpendicular or non-parallel fashion [40]. In addition, the anchorage number ($n_a = 1/n$) represents the number of sites occupied by one molecule. The evolution of the number of adsorbed molecules per site according to the temperature and systems is presented in Figure 12. The first notable aspect is that the RCA and RCG clays have different numbers of molecules per site. For the RCA clay, the number of molecules per site is above 1, indicating that all molecules are adsorbed in a perpendicular way to the surface, irrespective of the receptor site. In addition, in all cases, the number of molecules in site 1 is higher than in site 2 ($n_1 > n_2$), meaning that the phenol has a preference for the receptor site 1. In terms of anchor numbers, the estimated variation in phenol molecules is $0.02848 < n_{a1} < 0.03847$ for receptor site 1 and $0.1355 < n_{a2} < 0.6219$ for receptor site 2. This means that the phenol molecules occupy less than one full position in each case. This could be due to affinity or steric effects. In other words, the RCA clay has a deficit of receptor sites with a high affinity for the phenol molecules. The change in the number of molecules per site for the phenol/RCG is depicted in Figure 12B. In this case, the number of molecules per site is 10 times lower in comparison with the phenol/RCA. There are different possibilities to explain this difference between each material: (i) due to the density of receptor sites, (ii) due to textural proprieties, however, this could not be the reason due to the RCA having a higher specific surface area than the RCG, or (iii) the different compositions [41–43]. According to Table 1, the materials present differences in composition, with the RCG clay presenting higher quantities of silicon dioxide (SiO_2). Works in the literature reported that silicon dioxide directly influences the quantities of phenol adsorbed [6,44,45]. In a similar way, it is also possible to estimate the anchorage number for the phenol/RCG system; where n_{a1} ranged from 0.8451 to 1.410 and 0.1824 to 0.8331 for n_{a2}, according to the temperature. This indicates that the phenol molecules tend to bind to more receptor sites of type one as the system temperature increases. As for receptor site type two, the phenol molecules tend to need less than one receptor site per molecule.

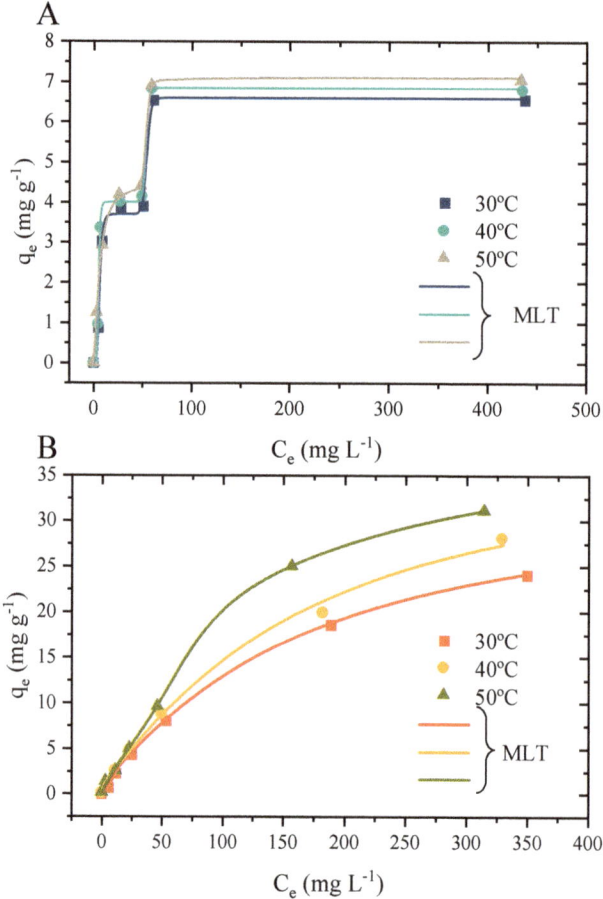

Figure 11. Phenol adsorption isotherms for the RCA (**A**) and RCG (**B**), lines are the model value predictions.

Table 4. Statistical indicators according to the model and adsorbent.

Adsorbent	Model	T (°C)	Statistical Indicators			
			R^2	ARE (%)	MSE $(mg\ g^{-1})^2$	BIC
RCA	MLO	20	0.8946	25.86	1.005	1.953
		30	0.8766	29.38	1.2663	3.573
		40	0.9371	10.50	0.6686	−0.8976
RCG		20	0.9942	11.74	0.7757	0.1421
		30	0.9859	20.46	2.420	8.106
		40	0.9967	22.55	0.7605	0.003962
RCA	MLT	20	0.9979	1.957	0.08037	−19.59
		30	0.9999	0.3973	0.00562	−38.22
		40	0.9995	1.613	0.02302	−28.35
RCG		20	0.9999	0.4556	0.008035	−35.71
		30	0.9999	3.943	0.09776	−18.22
		40	0.9998	5.033	0.1411	−15.65
RCA	DLO	20	0.8939	26.19	1.011	2.000
		30	0.8761	29.59	1.271	3.601
		40	0.9369	10.90	0.6703	−0.880
RCG		20	0.9995	13.23	0.06972	−16.72
		30	0.9966	5.001	0.5792	−1.902
		40	0.9991	12.03	0.2102	−8.998
RCA	DLT	20	0.8939	26.19	1.349	3.946
		30	0.8761	29.59	1.695	5.547
		40	0.9369	10.90	0.8937	1.066
RCG		20	0.9995	13.23	0.09296	−14.78
		30	0.9966	5.001	0.7723	0.0442
		40	0.9991	12.03	0.2802	−7.052
RCA	MM	20	0.7985	28.92	3.840	10.38
		30	−0.2504	44.44	25.66	23.67
		40	−0.2296	28.12	26.14	23.80
RCG		20	−0.2536	83.57	334.9	41.66
		30	−0.3040	83.35	446.7	43.67
		40	−0.1736	83.41	541.9	45.03

The change in receptor density according to the evolution of the temperature and the system, phenol/RCA or phenol/RCG, is depicted in Figure 13. Starting with the RCA, it is possible that as the temperature of the system increases, the density of receptor two increases, while that of receptor one remains almost constant at around 0.5. This indicates that the affinity of receptor two is not dependent on the system temperature and that the increase in the adsorption capacity is due to receptor site two, the density of which increases. This effect can be related to two different causes: (i) a compensation regarding the number of molecules per site, in particular of receptor site one, where it achieved a higher value (above 10); (ii) the adsorbent material tends to change according to the temperature, facilitating the affinity between the phenol molecules and the RCA [46]. The density of the receptor sites for the phenol/RCG was found to observe more typical behavior, with the densities for both receptor sites tending to increase with the temperature. At all temperatures, the density of receptor site one (N_{m1}) is higher than that of receptor site two (N_{m2}), indicating that receptor one has a high affinity and thus, facilitates the adsorption of phenol. In addition, the minor decrease in the number of the adsorbed phenol molecules (n_1) direct reflects the behavior of the receptor density N_{m1}, where the value increases, indicating a higher number of molecules are attached to this site [47].

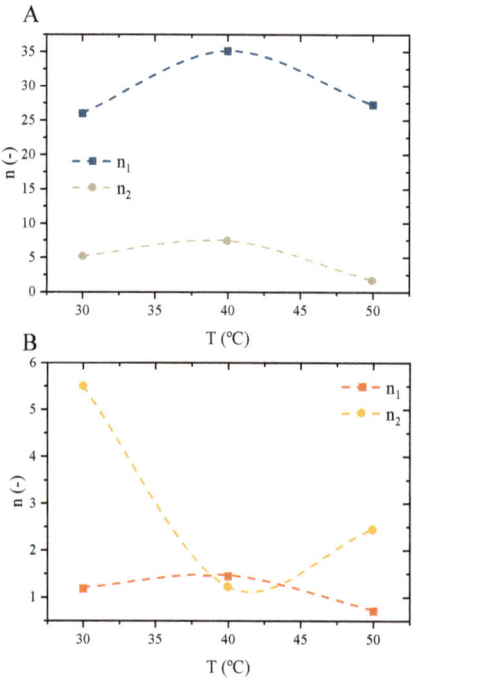

Figure 12. Evolution of the number of molecules per site parameters according to the temperature and system, RCA (**A**) and RCG (**B**).

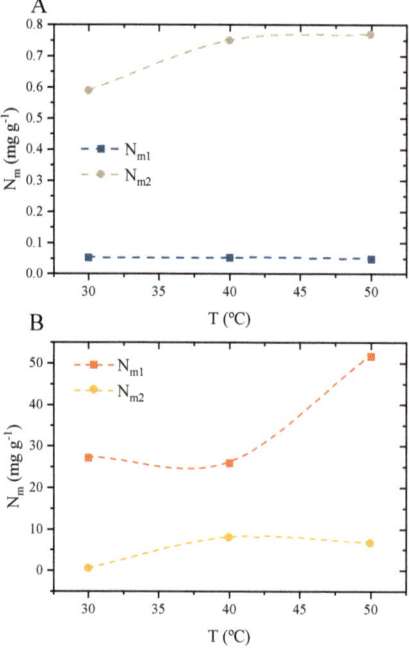

Figure 13. Receptor site density changes according to the temperature and system, RCA (**A**) and RCG (**B**).

From the half-concentration saturation parameters, it is possible to determine the adsorption energy according to the receptor site and system, as shown in Figure 14. The first aspect found is that the adsorption energy is positive for both systems, indicating endothermic adsorption based on physical interactions since the values were below 40 kJ mol^{-1}. For the phenol/RCA, it was found that receptor one has higher adsorption energy than receptor two, $\Delta E_1 > \Delta E_2$. In other words, receptor one has a higher affinity for the phenol. At first glance, this may seem to be in contradiction with the receptor densities, however, this may indicate that lower quantities of the receptor of type one may be present on the surface. For the phenol/RGC system, the energy of receptor one is also higher than receptor two. The minor difference is that, as the temperature evolves, a less them 1% decrease in the energy for the receptor site one occurs. The main explanation for this is the minor change that occurs in the number of molecules for receptor one [47]. In addition, these changes in the adsorption energy do not indicate a change in the nature of the adsorption, thus being endothermic for all the temperatures tackled in this study.

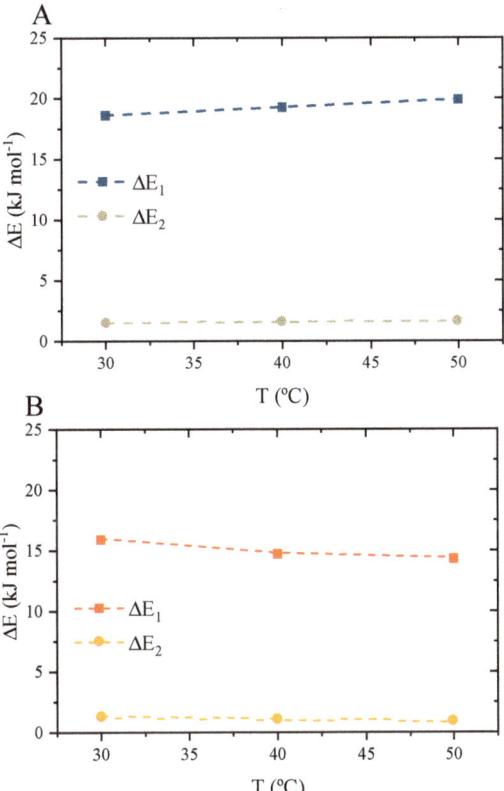

Figure 14. Estimated adsorption energy change according to the temperature and system, RCA (**A**) and RCG (**B**).

3.10.3. Application of the Stat-Phys Model with the Thermodynamic Potential Functions

Thermodynamic potential functions are quantities that are used to describe the state of a thermodynamic system and the energy exchanges that can occur within it. As it is possible to associate the grande canonical partition function to obtain the thermodynamic potential function of the configurational entropy (S_A, kJ mol^{-1} K^{-1}), Gibbs free energy for adsorption (G_A, kJ mol^{-1}), and internal energy of adsorption (E_I, kJ mol^{-1}). The definition of each thermodynamic variable is given by the following Equations [48]:

$$\frac{S_a}{k_B} = \ln(Z_{gc}) - \beta \frac{\partial}{\partial \beta} \ln(Z_{gc}) \tag{18}$$

$$G_a = \mu Q_0 \tag{19}$$

$$E_{int} = \frac{\mu}{\beta}\left(\frac{\partial}{\partial \mu}\ln(Z_{gc})\right) - \frac{\partial}{\partial \beta}\ln(Z_{gc}) \tag{20}$$

where Z_{gc} corresponds to the total grand canonical partition function, μ corresponds to the translational chemical potential of the phenol molecule (kJ mol^{-1}), β corresponds to $1/k_B T$, k_B corresponds to the Boltzmann constant, and T is the temperature of the system.

Taking into consideration that the MLT was the best Stat-Phys model to represent both systems, the thermodynamic parameters can be obtained according to Equations (21)–(23). The derivation of the Equations can be found elsewhere [47,49].

$$\frac{S_a}{k_B} = \begin{cases} N_{m1}\left[\ln\left(1+\left(\frac{C_e}{C_1}\right)^n\right) - \ln\left(1+\left(\frac{C_e}{C_1}\right)^n\right)\frac{\left(\frac{C_e}{C_1}\right)^n}{1+\left(\frac{C_e}{C_1}\right)^n}\right] \\ +N_{m2}\left[\ln\left(1+\left(\frac{C_e}{C_2}\right)^n\right) - \ln\left(1+\left(\frac{C_e}{C_2}\right)^n\right)\frac{\left(\frac{C_e}{C_2}\right)^n}{1+\left(\frac{C_e}{C_2}\right)^n}\right] \end{cases} \tag{21}$$

$$G_a \beta = \ln\left(\frac{C_e}{\left(\frac{2\pi m}{h^2 \beta}\right)^{3/2}}\right)\left(\frac{Q_{m1}}{1+\left(\frac{C_1}{C_e}\right)^n} + \frac{Q_{m2}}{1+\left(\frac{C_2}{C_e}\right)^n}\right) \tag{22}$$

$$E_{int} = \begin{cases} N_{m1}\left[\mu \frac{\left(\frac{C_e}{C_1}\right)^n}{1+\left(\frac{C_e}{C_1}\right)^n} \frac{1}{\beta}\ln\left(1+\left(\frac{C_e}{C_1}\right)^n\right)\frac{\left(\frac{C_e}{C_1}\right)^n}{1+\left(\frac{C_e}{C_1}\right)^n}\right] \\ +N_{m2}\left[\mu \frac{\left(\frac{C_e}{C_2}\right)^n}{1+\left(\frac{C_e}{C_2}\right)^n} - \frac{1}{\beta}\ln\left(1+\left(\frac{C_e}{C_2}\right)^n\right)\frac{\left(\frac{C_e}{C_2}\right)^n}{1+\left(\frac{C_e}{C_2}\right)^n}\right] \end{cases} \tag{23}$$

3.10.4. Interpretations of the Potential Function

The simulations of the change in the configuration entropy according to the phenol equilibrium concentration and the systems are depicted in Figure 15. The simulations take into consideration the maximum concentration equilibrium obtained from the isotherms, meaning around 500 mg L^{-1}, for a fixed value. The first aspect to be observed is that the configuration entropy for the phenol/RCA presents two entropic peaks. The first is at around 5 to 7 mg L^{-1}, which corresponds to the concentration at half-saturation for receptor site two. After that, a second minor entropic peak occurs, corresponding to the receptor site one. After that, the entropic change diminishes until achieving equilibrium. The second aspect is regarding the phenol/RCG system, where, at first glance, the simulation does not show the entropic equilibrium and shows one single entropic peak for all temperatures. Regarding the entropic equilibrium, further simulation indicates that it is only reached after 1500 mg L^{-1}. This indicates that the adsorption could occur beyond the concentration investigated in this work. Regarding the single entropic peak, the main explanation is that receptor sites one and two present close half-saturation concentrations, meaning that the adsorption occurs at the same equilibrium concentration. Last, regarding the temperature effect, the results are in agreement with the endothermic nature, in which the temperature increases both the adsorption capacity and the entropic behavior.

The simulation of the Gibbs free energy according to the equilibrium concentration and temperature system is shown in Figure 16. The first observation is that the Gibbs free energy is negative for all phenol equilibrium concentrations, temperatures, and systems, indicating that the adsorption of phenol is spontaneous at any given point. Regarding the different profiles of each simulation, the explanation goes hand in hand with the entropic behavior. In the first case, phenol/RCA displays a ladder-type profile. This profile is related to the different receptor sites' half-saturation concentrations, where site one is only adsorbed with high phenol concentrations. As for the phenol/RCG, the same effect of not presenting an equilibrium at this concentration indicates that the RCG can further adsorb phenol. The lack of a ladder-type profile also corroborates the entropic change, which presents a single peak, meaning that the adsorption of phenol is spontaneous and occurs in the same way for the two receptor sites.

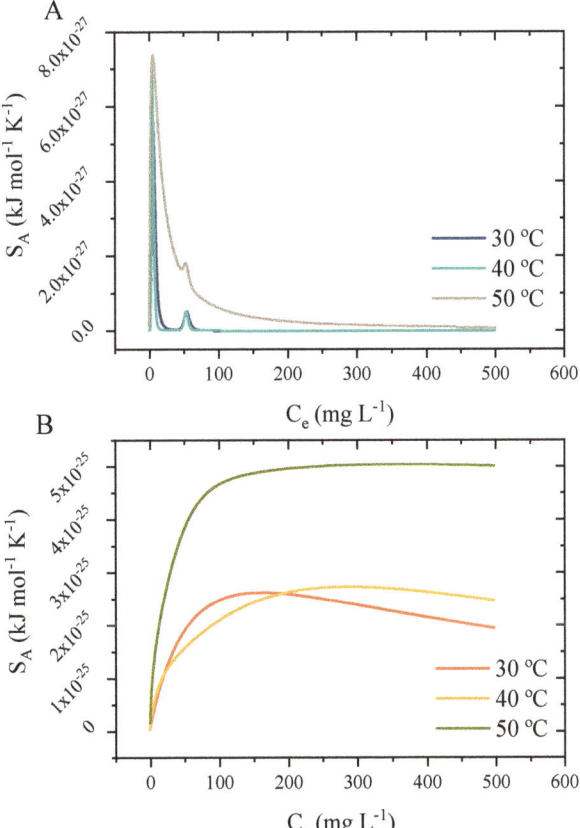

Figure 15. Entropic change according to the phenol equilibrium, temperature, and adsorption system, RCA (**A**) and RCG (**B**).

The simulations for the internal energy according to the phenol equilibrium concentration and system temperature are shown in Figure 17. As expected, the internal energy presents similar behavior to the simulations obtained for the configurational entropy and Gibbs free energy. In other words, the presence of ladders in the phenol/RCA system, due to the energic differences between the receptor sites and the lack of equilibrium for the phenol/RCG system, indicates that more phenol molecules could be adsorbed. Overall, the internal energy tends to increase with temperature, corroborating the results obtained for the experimental isotherms.

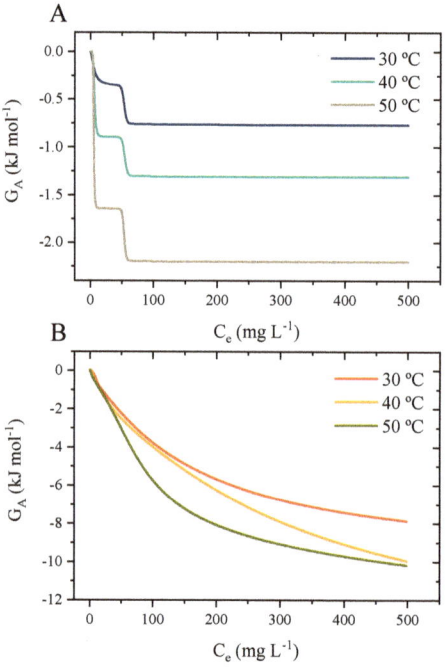

Figure 16. Gibbs free energy changes according to the phenol concentration, temperature, and system, RCA (**A**) and RCG (**B**).

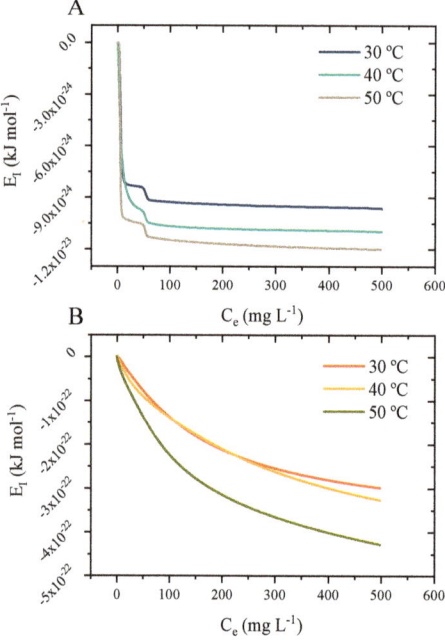

Figure 17. Internal energy changes according to the equilibrium concentration, temperature, and system, RCA (**A**) and RCG (**B**).

3.10.5. Mechanism of Phenol Adsorption

The adsorption process of phenol on clay usually relies on several physicochemical forces occurring at the solid-liquid interface, such as the interaction between phenol molecules and clay functional groups

- Van der Waals force: Dipole-dipole attraction between atoms or molecules through low-level electrical interference. This attractive force is very important for the adsorption of organic substances such as phenol. We are well aware of the inhomogeneity of clay solids.
- Coulomb force: The electrostatic force developed between a charged surface and an opposite charge. Surface charges can result from isostructural substitution or protonation or deprotonation of surface functional groups. Surface charge is determined by changes in the pH of the medium, and the pH at the point of zero charges can help us identify and assign the dominant charge on a clay surface.
- The hydrogen bonds, or the role of H_2O in the adsorption, have intermolecular interactions that occur between hydrogen atoms and electronegative atoms (O, F, S, Cl).

In this work, we try to deepen the study of the mechanism of the adsorption of phenol on clay. For that, structural and textural characterization of the samples was used to determine the nature of the interactions between the adsorbate and the adsorbent either by FTIR or XRD (see the supplement of this work).

To better understand the adsorption mechanism of clay materials, the infrared spectra of phenol and clay were examined. Figures S3 and S5 (Supplementary Materials) show the FTIR spectra of RCG and RCA both before and after phenol adsorption. The internal OH units of the kaolinite structure and the water of hydration of sodium cations are responsible for the characteristic RCG and RCA bands at 3625 and 3420 cm^{-1}, respectively. [50]. With increasing phenol content, the intensity of the RCA and RCG bands (3400–3650 cm^{-1}) increases. These findings imply that phenol penetrates the kaolinite interlayer and forms a hydrogen connection with the water molecules in the cationic hydration spheres. In all samples, the water OH groups' bands at 1638 and 3467 cm^{-1} appear first. Some bands developed in the spectra of the sample after phenol adsorption as opposed to the new peaks not seen in the Agourai clay samples before and after adsorption. While the band at 1584 cm^{-1} is caused by C-O vibrations, the band at 1479 cm^{-1} is caused by the stretching of the aromatic C=C bond. The CH band for Ph(10^{-3} M)/RCG is situated at 1450 cm^{-1} in Figure 9. This alteration in the plane of the phenol CH group is brought on by bending vibrations. Additionally, vibrations below 1030 cm^{-1} that correspond to the Si-O deformation and Si-O-Si stretching modes show band separation, indicating that they are involved in phenolic interactions [51]. Figures S2 and S6 (Supplementary Materials) displays the tilemaps for the RCG and CCG. The DRX analysis suggests that the interlobar space of the solid has been stripped as a result of the insertion of phenol into this space with a large dip angle, as seen by the removal of the quartz line at 20° in the spectrum in the RCG spectrum following phenol adsorption [45]. The distance from the quartz plane may only be extended if this adsorption mechanism is permitted; the average diameter of the phenol molecule is 5, depending on how it enters the interlayer gap. [15,52,53]. In addition, the maximum intensity in the RCG spectrum was reduced after phenol adsorption. In the clay samples collected in the Meknes region, the only thing missing was the lines in the raw clay attributed to calcite. In both diffractograms, there is no movement of lines or the appearance of new lines. Only one effect can be observed, and that is the reduction in peak intensity. After phenol adsorption, neither solid's spectrum displayed any new peaks, supporting the earlier finding of phenol's physical adsorption. According to the findings of XRD, FTIR, and SEM, surface adsorption rather than intercalation were the primary interactions between phenols and clay particles, which happened at the outer surface by electrostatic attraction.

4. Conclusions

The results of this laboratory-scale study demonstrate the utility of using RCA and RCG in the field of remediation of water bodies contaminated with organic pollutants. The adsorption kinetics of phenol on different adsorbents allowed the selection of Geulmima clay as the optimal adsorbent for phenol in an aqueous solution. The parameters under study, the pH value, and the temperature of the medium influence this kinetics. However, the structure and structural characterization of the solid before and after phenol adsorption indicated that the mechanism of the reaction was electrostatic and that hydrogen bonding played an important role in RCG adsorption of phenol, and kinetic modeling showed pseudo-second-order model dynamics. From the physical-statistical modeling, it was found that the phenol adsorption for both cases is well represented by the monolayer model with two types of energy sites, irrespective of the origin of the clay. However, the characteristics of each material direct reflect the steric parameters, as indicated by the model. The RCA has a receptor site with low density but with a high number of molecules and another receptor site that presents a more equilibrate behavior, being able to increase the density with increasing system temperature. As for the RCG, it was found that both receptor sites had similar energy and tend to work together in the increment of the adsorption capacity. Potential thermodynamic functions indicate that the adsorption of phenol is spontaneous and endothermic for all the studied systems, with all the thermodynamic proprieties tending to increase in absolute values with increasing temperature. Furthermore, the thermodynamic analysis indicated that the RCG can further adsorb phenol since the equilibrium is not reached for any variable. Characterization of the solid after phenol adsorption confirmed the physical nature of the process and the type of layer (van der Waals) and the role played by hydrogen bonds in the case of RCG. The latter result suggests that the main interaction between the phenolics and the outer precursor is through electrostatic attraction.

Supplementary Materials: The following supporting information can be downloaded at: https://www.mdpi.com/article/10.3390/w15101881/s1.

Author Contributions: Y.D., D.S.P.F., J.G., T.L., R.O. and Y.B.: Conceptualization, Methodology, Survey, Original Draft Writing, Editing Review; Y.D., D.S.P.F., J.G., T.L., R.O., Y.B., H.M., H.O., A.S. and B.M.: Methodology, Survey, Original draft writing, editing; Y.D., D.S.P.F., Y.B., R.O. and T.L.: Conceptualization, Methodology, Resources, Writing, Editing, Supervision, Acquiring Funding. All authors have read and agreed to the published version of the manuscript.

Funding: This research received no external funding.

Data Availability Statement: Data sharing is not applicable to this article as no datasets were generated or analyzed during the current study.

Acknowledgments: The authors also thank everyone who helped prepare the manuscript.

Conflicts of Interest: We declare that we do not have any commercial or associative interest that represents a conflict of interest in connection with the work submitted.

References

1. Hasan, M.K.; Shahriar, A.; Jim, K.U. Water pollution in Bangladesh and its impact on public health. *Heliyon* **2019**, *5*, e02145. [CrossRef] [PubMed]
2. Ali, S.N.; El-Shafey, E.; Al-Busafi, S.; Al-Lawati, H.A. Adsorption of chlorpheniramine and ibuprofen on surface functionalized activated carbons from deionized water and spiked hospital wastewater. *J. Environ. Chem. Eng.* **2018**, *7*, 102860. [CrossRef]
3. Mohammed, B.B.; Yamni, K.; Tijani, N.; Alrashdi, A.A.; Zouihri, H.; Dehmani, Y.; Chung, I.-M.; Kim, S.-H.; Lgaz, H. Adsorptive removal of phenol using faujasite-type Y zeolite: Adsorption isotherms, kinetics and grand canonical Monte Carlo simulation studies. *J. Mol. Liq.* **2019**, *296*, 111997. [CrossRef]
4. Dehbi, A.; Dehmani, Y.; Omari, H.; Lammini, A.; Elazhari, K.; Abdallaoui, A. Hematite iron oxide nanoparticles (α-Fe$_2$O$_3$): Synthesis and modelling adsorption of malachite green. *J. Environ. Chem. Eng.* **2019**, *8*, 103394. [CrossRef]
5. Dehbi, A.; Dehmani, Y.; Omari, H.; Lammini, A.; Elazhari, K.; Abouarnadasse, S.; Abdallaoui, A. Comparative study of malachite green and phenol adsorption on synthetic hematite iron oxide nanoparticles (α-Fe$_2$O$_3$). *Surf. Interfaces* **2020**, *21*, 100637. [CrossRef]

6. Dehmani, Y.; Dridi, D.; Lamhasni, T.; Abouarnadasse, S.; Chtourou, R.; Lima, E.C. Review of phenol adsorption on transition metal oxides and other adsorbents. *J. Water Process. Eng.* **2022**, *49*, 102965. [CrossRef]
7. Awad, A.M.; Shaikh, S.M.; Jalab, R.; Gulied, M.H.; Nasser, M.S.; Benamor, A.; Adham, S. Adsorption of organic pollutants by natural and modified clays: A comprehensive review. *Sep. Purif. Technol.* **2019**, *228*, 115719. [CrossRef]
8. Tariq, R.; Abatal, M.; Bassam, A. Computational intelligence for empirical modeling and optimization of methylene blue adsorption phenomena using available local zeolites and clay of Morocco. *J. Clean. Prod.* **2022**, *370*, 133517. [CrossRef]
9. Es-Sahbany, H.; El Hachimi, M.; Hsissou, R.; Belfaquir, M.; Nkhili, S.; Loutfi, M.; Elyoubi, M. Adsorption of heavy metal (Cadmium) in synthetic wastewater by the natural clay as a potential adsorbent (Tangier-Tetouan-Al Hoceima—Morocco region). *Mater. Today Proc.* **2021**, *45*, 7299–7305. [CrossRef]
10. Azarkan, S.; Peña, A.; Draoui, K.; Sainz-Díaz, C.I. Adsorption of two fungicides on natural clays of Morocco. *Appl. Clay Sci.* **2016**, *123*, 37–46. [CrossRef]
11. Gładysz-Płaska, A.; Majdan, M.; Pikus, S.; Sternik, D. Simultaneous adsorption of chromium(VI) and phenol on natural red clay modified by HDTMA. *Chem. Eng. J.* **2012**, *179*, 140–150. [CrossRef]
12. Chaari, I.; Fakhfakh, E.; Medhioub, M.; Jamoussi, F. Comparative study on adsorption of cationic and anionic dyes by smectite rich natural clays. *J. Mol. Struct.* **2018**, *1179*, 672–677. [CrossRef]
13. Wang, M.; Hearon, S.E.; Johnson, N.M.; Phillips, T.D. Development of broad-acting clays for the tight adsorption of benzo[a]pyrene and aldicarb. *Appl. Clay Sci.* **2018**, *168*, 196–202. [CrossRef] [PubMed]
14. Kausar, A.; Iqbal, M.; Javed, A.; Aftab, K.; Nazli, Z.; Nawaz, H.; Bhatti, H.N.; Nouren, S. Dyes adsorption using clay and modified clay: A review. *J. Mol. Liq.* **2018**, *256*, 395–407. [CrossRef]
15. Bentahar, Y.; Hurel, C.; Draoui, K.; Khairoun, S.; Marmier, N. Adsorptive properties of Moroccan clays for the removal of arsenic(V) from aqueous solution. *Appl. Clay Sci.* **2016**, *119*, 385–392. [CrossRef]
16. Nabbou, N.; Belhachemi, M.; Boumelik, M.; Merzougui, T.; Lahcene, D.; Harek, Y.; Zorpas, A.A.; Jeguirim, M. Removal of fluoride from groundwater using natural clay (kaolinite): Optimization of adsorption conditions. *Comptes Rendus Chim.* **2018**, *22*, 105–112. [CrossRef]
17. Ouallal, H.; Dehmani, Y.; Moussout, H.; Messaoudi, L.; Azrour, M. Kinetic, isotherm and mechanism investigations of the removal of phenols from water by raw and calcined clays. *Heliyon* **2019**, *5*, e01616. [CrossRef]
18. Khalfaoui, M.; Baouab, M.; Gauthier, R.; Ben Lamine, A. Statistical Physics Modelling of Dye Adsorption on Modified Cotton. *Adsorpt. Sci. Technol.* **2002**, *20*, 17–31. [CrossRef]
19. Knani, S.; Mathlouthi, M.; Ben Lamine, A. Modeling of the Psychophysical Response Curves Using the Grand Canonical Ensemble in Statistical Physics. *Food Biophys.* **2007**, *2*, 183–192. [CrossRef]
20. Khalfaoui, M.; Baouab, M.; Gauthier, R.; Ben Lamine, A. Dye Adsorption by Modified Cotton. Steric and Energetic Interpretations of Model Parameter Behaviours. *Adsorpt. Sci. Technol.* **2002**, *20*, 33–47. [CrossRef]
21. Hua, P.; Sellaoui, L.; Franco, D.; Netto, M.S.; Dotto, G.L.; Bajahzar, A.; Belmabrouk, H.; Bonilla-Petriciolet, A.; Li, Z. Adsorption of acid green and procion red on a magnetic geopolymer based adsorbent: Experiments, characterization and theoretical treatment. *Chem. Eng. J.* **2019**, *383*, 123113. [CrossRef]
22. Zhang, L.; Sellaoui, L.; Franco, D.; Dotto, G.L.; Bajahzar, A.; Belmabrouk, H.; Bonilla-Petriciolet, A.; Oliveira, M.L.; Li, Z. Adsorption of dyes brilliant blue, sunset yellow and tartrazine from aqueous solution on chitosan: Analytical interpretation via multilayer statistical physics model. *Chem. Eng. J.* **2019**, *382*, 122952. [CrossRef]
23. Manni, A.; El, A.; El Amrani, I.; Hassani, E.; El, A.; Sadik, C. Valorization of coffee waste with Moroccan clay to produce a porous red ceramics (class BIII), Boletín La Soc. *Española Cerámica Y Vidr.* **2019**, *58*, 211–220. [CrossRef]
24. Ba Mohammed, B.; Yamni, K.; Tijani, N.; Lee, H.-S.; Dehmani, Y.; El Hamdani, H.; Alrashdi, A.A.; Ramola, S.; Belwal, T.; Lgaz, H. Enhanced removal efficiency of NaY zeolite toward phenol from aqueous solution by modification with nickel (Ni-NaY). *J. Saudi Chem. Soc.* **2021**, *25*, 101224. [CrossRef]
25. Jedli, H.; Brahmi, J.; Hedfi, H.; Mbarek, M.; Bouzgarrou, S.; Slimi, K. Adsorption kinetics and thermodynamics properties of Supercritical CO_2 on activated clay. *J. Pet. Sci. Eng.* **2018**, *166*, 476–481. [CrossRef]
26. Dehmani, Y.; Ed-Dra, A.; Zennouhi, O.; Bouymajane, A.; Filali, F.R.; Nassiri, L.; Abouarnadasse, S. Chemical characterization and adsorption of oil mill wastewater on Moroccan clay in order to be used in the agricultural field. *Heliyon* **2020**, *6*, e03160. [CrossRef]
27. Hadjltaief, H.B.; Sdiri, A.; Ltaief, W.; Da Costa, P.; Gálvez, M.E.; Ben Zina, M. Efficient removal of cadmium and 2-chlorophenol in aqueous systems by natural clay: Adsorption and photo-Fenton degradation processes. *Comptes Rendus Chim.* **2018**, *21*, 253–262. [CrossRef]
28. Ouaddari, H.; Beqqour, D.; Bennazha, J.; El Amrani, I.-E.; Albizane, A.; Solhy, A.; Varma, R.S. Natural Moroccan clays: Comparative study of their application as recyclable catalysts in Knoevenagel condensation. *Sustain. Chem. Pharm.* **2018**, *10*, 1–8. [CrossRef]
29. Bentahar, S.; Dbik, A.; El Khomri, M.; El Messaoudi, N.; Lacherai, A. Adsorption of methylene blue, crystal violet and congo red from binary and ternary systems with natural clay: Kinetic, isotherm, and thermodynamic. *J. Environ. Chem. Eng.* **2017**, *5*, 5921–5932. [CrossRef]
30. Bouna, L.; El Fakir, A.A.; Benlhachemi, A.; Draoui, K.; Villain, S.; Guinneton, F. Physico-chemical characterization of clays from Assa-Zag for valorization in cationic dye methylene blue adsorption. *Mater. Today Proc.* **2019**, *22*, 8–13. [CrossRef]
31. Richards, S.; Bouazza, A. Phenol adsorption in organo-modified basaltic clay and bentonite. *Appl. Clay Sci.* **2007**, *37*, 133–142. [CrossRef]
32. Pawar, R.R.; Lalhmunsiama; Gupta, P.; Sawant, S.Y.; Shahmoradi, B.; Lee, S.-M. Porous synthetic hectorite clay-alginate composite beads for effective adsorption of methylene blue dye from aqueous solution. *Int. J. Biol. Macromol.* **2018**, *114*, 1315–1324. [CrossRef]

33. Gamoudi, S.; Srasra, E. Characterization of Tunisian clay suitable for pharmaceutical and cosmetic applications. *Appl. Clay Sci.* **2017**, *146*, 162–166. [CrossRef]
34. Kragović, M.; Stojmenović, M.; Petrović, J.; Loredo, J.; Pašalić, S.; Nedeljković, A.; Ristović, I. Influence of Alginate Encapsulation on Point of Zero Charge (pHpzc) and Thermodynamic Properties of the Natural and Fe(III)—Modified Zeolite. *Procedia Manuf.* **2019**, *32*, 286–293. [CrossRef]
35. Asuha, S.; Fei, F.; Wurendaodi, W.; Zhao, S.; Wu, H.; Zhuang, X. Activation of kaolinite by a low-temperature chemical method and its effect on methylene blue adsorption. *Powder Technol.* **2019**, *361*, 624–632. [CrossRef]
36. Aran, D.; Maul, A.; Masfaraud, J.-F. A spectrophotometric measurement of soil cation exchange capacity based on cobaltihexamine chloride absorbance. *Comptes Rendus Geosci.* **2008**, *340*, 865–871. [CrossRef]
37. Cheng, W.; Gao, W.; Cui, X.; Hong, J.; Feng, R. Phenol adsorption equilibrium and kinetics on zeolite X/activated. *J. Taiwan Inst. Chem. Eng.* **2016**, *62*, 192–198. [CrossRef]
38. Selim, A.Q.; Sellaoui, L.; Mobarak, M. Statistical physics modeling of phosphate adsorption onto chemically modified carbonaceous clay. *J. Mol. Liq.* **2019**, *279*, 94–107. [CrossRef]
39. Kong, X.; Gao, H.; Song, X.; Deng, Y.; Zhang, Y. Adsorption of phenol on porous carbon from Toona sinensis leaves and its mechanism. *Chem. Phys. Lett.* **2019**, *739*, 137046. [CrossRef]
40. Franco, D.; Piccin, J.S.; Lima, E.C.; Dotto, G.L. Interpretations about methylene blue adsorption by surface modified chitin using the statistical physics treatment. *Adsorption* **2015**, *21*, 557–564. [CrossRef]
41. Oueslati, K.; Naifar, A.; Sakly, A.; Kyzas, G.Z.; Ben Lamine, A. Statistical and physical interpretation of dye adsorption onto low-cost biomass by using simulation methods. *Colloids Surfaces A Physicochem. Eng. Asp.* **2022**, *646*, 128969. [CrossRef]
42. Chen, F.; Zhao, E.; Kim, T.; Wang, J.; Hableel, G.; Reardon, P.J.T.; Ananthakrishna, S.J.; Wang, T.; Arconada-Alvarez, S.; Knowles, J.C.; et al. Organosilica Nanoparticles with an Intrinsic Secondary Amine: An Efficient and Reusable Adsorbent for Dyes. *ACS Appl. Mater. Interfaces* **2017**, *9*, 15566–15576. [CrossRef] [PubMed]
43. Li, Z.; Gómez-Avilés, A.; Sellaoui, L.; Bedia, J.; Bonilla-Petriciolet, A.; Belver, C. Adsorption of ibuprofen on organo-sepiolite and on zeolite/sepiolite heterostructure: Synthesis, characterization and statistical physics modeling. *Chem. Eng. J.* **2019**, *371*, 868–875. [CrossRef]
44. Hank, D.; Azi, Z.; Hocine, S.A.; Chaalal, O.; Hellal, A. Optimization of phenol adsorption onto bentonite by factorial design methodology. *J. Ind. Eng. Chem.* **2014**, *20*, 2256–2263. [CrossRef]
45. Dehmani, Y.; Sellaoui, L.; Alghamdi, Y.; Lainé, J.; Badawi, M.; Amhoud, A.; Bonilla-Petriciolet, A.; Lamhasni, T.; Abouarnadasse, S. Kinetic, thermodynamic and mechanism study of the adsorption of phenol on Moroccan clay. *J. Mol. Liq.* **2020**, *312*, 113383. [CrossRef]
46. Knani, S.; Khalfaoui, M.; Hachicha, M.; Mathlouthi, M.; Ben Lamine, A. Interpretation of psychophysics response curves using statistical physics. *Food Chem.* **2014**, *151*, 487–499. [CrossRef]
47. Franco, D.S.P.; Georgin, J.; Netto, M.S.; Martinello, K.D.B.; Silva, L.F. Preparation of activated carbons from fruit residues for the removal of naproxen (NPX): Analytical interpretation via statistical physical model. *J. Mol. Liq.* **2022**, *356*, 119021. [CrossRef]
48. Sellaoui, L.; Depci, T.; Kul, A.R.; Knani, S.; Ben Lamine, A. A new statistical physics model to interpret the binary adsorption isotherms of lead and zinc on activated carbon. *J. Mol. Liq.* **2016**, *214*, 220–230. [CrossRef]
49. Jedli, H.; Briki, C.; Chrouda, A.; Brahmi, J.; Abassi, A.; Jbara, A.; Slimi, K.; Jemni, A. Experimental and theoretical study of CO_2 adsorption by activated clay using statistical physics modeling. *RSC Adv.* **2019**, *9*, 38454–38463. [CrossRef]
50. Madejová, J. FTIR techniques in clay mineral studies. *Vib. Spectrosc.* **2003**, *31*, 1–10. [CrossRef]
51. Lee, S.G.; Choi, J.I.; Koh, W.; Jang, S.S. Adsorption of β-d-glucose and cellobiose on kaolinite surfaces: Density functional theory (DFT) approach. *Appl. Clay Sci.* **2013**, *71*, 73–81. [CrossRef]
52. Asnaoui, H.; Dehmani, Y.; Khalis, M.; Hachem, E.-K. Adsorption of phenol from aqueous solutions by Na–bentonite: Kinetic, equilibrium and thermodynamic studies. *Int. J. Environ. Anal. Chem.* **2022**, *102*, 3043–3057. [CrossRef]
53. Luo, Z.; Gao, M.; Yang, S.; Yang, Q. Adsorption of phenols on reduced-charge montmorillonites modified by bispyridinium dibromides: Mechanism, kinetics and thermodynamics studies. *Colloids Surf. A Physicochem. Eng. Asp.* **2015**, *482*, 222–230. [CrossRef]

Disclaimer/Publisher's Note: The statements, opinions and data contained in all publications are solely those of the individual author(s) and contributor(s) and not of MDPI and/or the editor(s). MDPI and/or the editor(s) disclaim responsibility for any injury to people or property resulting from any ideas, methods, instructions or products referred to in the content.

Article

In Situ Polyaniline Immobilized ZnO Nanorods for Efficient Adsorptive Detoxification of Cr (VI) from Aquatic System

Fahad A. Alharthi *, Riyadh H. Alshammari and Imran Hasan *

Department of Chemistry, College of Science, King Saud University, Riyadh 11451, Saudi Arabia; ralshammari@ksu.edu.sa
* Correspondence: fharthi@ksu.edu.sa (F.A.A.); iabdulateef@ksu.edu.sa (I.H.)

Abstract: The elimination of toxic heavy metal ions from wastewater has been found to be of great importance in human as well marine animal wellbeing. Among various heavy metals, Cr (VI) has been found to be one of the highly toxic and carcinogenic heavy metals which are found to be dissolved in the water stream, the urgent treatment of which needs to be a priority. The present study demonstrates the fabrication of zinc oxide nanorods (ZnO NRs) and an immobilized polyaniline nanorod (ZnO@PAni NR) composite through an in situ free radical polymerization reactions. The material synthesis and purity were verified by X-ray diffractometer (XRD), Fourier transform infrared (FTIR), scanning electron microscope (SEM), energy dispersive spectroscope (EDS), and transmission electron microscope (TEM). Further, ZnO@PAni NRs were applied as an adsorbent for Cr (VI) in the aquatic system and exhibited a tremendous removal efficiency of 98.76%. The impact of operating parameters such as dose effect and pH on adsorption properties were studied. The uptake mechanism of Cr (VI) by ZnO@PAni was best explained by pseudo-second-order reaction, which suggested that the adsorption of Cr (VI) by the synthesized adsorbent material was processed by chemisorption, i.e., through formation of chemical bonds. The adsorption process proved viable and endothermic thermodynamically, and best supported by a Langmuir model, suggesting a monolayer formation of Cr (VI) on the surface of ZnO@PAni NRs.

Keywords: conducting polymers; wastewater treatment; Cr (VI) management; nanocomposites; zinc oxide

1. Introduction

The combined effects of extensive industrialization, population increase, agricultural activity, and other geological and environmental changes have led to an alarming decline in water quality [1–3]. Industries have become a significant contributor to water pollution by introducing harmful chemicals and heavy metals that have a substantial negative impact on both human and animal health, as well as causing widespread ecological harm [4–6]. Among various heavy metals, Cr (VI) has been recognized as a most harmful pollutant, causing various devastating effects such as the irritation and corrosion of human skin; further, it can lead to rust in the textile and metal finishing industries and can interfere with other processes including electroplating, dyeing, wood preservation, painting, fertilizing, and photography [7–11]. Usually in water streams, chromium exists in two different chemical states: Cr (III) and Cr (VI); Cr (VI) is genotoxic and carcinogenic to both humans and animals as compared to Cr (III), which is less toxic [12,13]. Therefore, based on its genotoxic and carcinogenic effects, it has been necessary to develop types of cost-effective and efficient methods which can remove of Cr (VI) from industrial and municipal wastewater. Scientists and researchers have developed various strategical methods such as sedimentation, precipitation, coagulation, flotation, ion exchange, biological treatment, adsorption, and reverse osmosis to reduce environmental problems [14–17]. Among them, adsorption has been preferred as the most effective and efficient method because of its operational simplicity and environmental friendliness for removing ultra-trace levels of Cr

(VI) from wastewater [18–20]. Although adsorption has been proved to be very effective for heavy metal removal, the efficiency of the method depends upon the development of such types of adsorbent materials that can provide an ample number of surface-active sites to bind the heavy metal ions [21].

Recently, metal-oxide-based nanocomposites have been extensively utilized as adsorbent for the removal of heavy metals owing to their large specific surface area, high porosity, ample adsorption sites, and high removal efficiency [22–24]. Various metal oxide nanoparticles such as Fe_2O_3, Fe_3O_4, TiO_2, SiO_2, CuO, CeO_2, and ZnO have been explored in the literature for Cr (VI) removal [1,25–30]. Among these, the ZnO nanoparticle has been recognized as one of the most propitious materials as an adsorbent for the removal of heavy metal ions because of its low cost, high surface area, and optimized electronic structure [25,31]. However, the degree of agglomeration limits these nanoparticles' efficiency, and to address this issue considerable effort has been made in the past to functionalize the surface of ZnO NPs with some electron-donating functional groups such as -OH, -COOH, -SH, $-NH_2$, etc. [32].

Nowadays, conducting polymers with metal-oxide-based nanocomposite material have become a popular alternative for the sequestration of heavy metals from wastewater because they primarily offer more interfacial surface area for the adhesion of heavy metal ions, are simple to synthesize, and are cost-effective [33]. Polyaniline (PAni) is a highly conducting polymer which has excellent conducting properties and electronic features [34]. PAni has been recognized as one of the cost-effective conductive supports that improve the adsorptive properties of these hybrid nanocomposite materials towards heavy metals through the provision of electron-donating amine and imine groups on the surface of the material [35]. The functional groups and electronic properties in PAni generally involve electrostatic attractions and π–π interactions in the adsorption process which synergistically add on to the efficiency of ZnO NPs [36]. The addition of fillers such as natural clays, metal oxide nanoparticles, etc., can improve the stability of the polymer matrix [37].

Table 1 comprises some of the specific methods for synthesis of ZnO NPs using various modes and their advantages and disadvantages.

Table 1. Synthesis methods of ZnO nanoparticles.

Method	Precursor	Solvent	Features	Advantages/Cost	Disadvantages	References
Chemical coprecipitation	Zinc acetate	Double distilled water	30–60 nm of nanorod	Low energy input/low cost	High cost of precursors	[38]
Microwave decomposition	Zinc acetate dehydrate	1-Butyl-3-methylimidazolium bis (trifluoromethyl-sulfonyl) imide [bmim][NTf2]	37–47 nm Sphere	Industrial-scale production/low cost	Parameter control	[39]
Hydrothermal process	Zinc acetate dihydrate	Polyvinylpyrrolidone (PVP)	50–200 nm of nanorod	Uncomplicated equipment/low cost	Nanoparticle stability	[40]
Wet chemical Method	Zinc nitrate hexahydrate	Sodium hydroxide (NaOH) as precursors and soluble starch as stabilizing agent	20–30 nm Acicular	Easy parameter tailoring, low cost	Nanoparticle stability	[41]
Sol–gel method	Zinc nitrate	Distilled water and gelatin as substrate	30–60 nm Circular and hexagonal	Inexpensive and easy to handle chemical reagents		[42]
Solvothermal	Zinc acetate dihydrate	Polyethylene glycol, absolute ethanol	10–20 nm Quasi-spherical	Industrial-scale production	High cost of precursors	[43]
Micro-emulsion	Zn (AOT)2	Heptane, diethyl oxalate, chloroform, methanol	10–20 nm Quasi-spherical	Easy parameter tailoring	Surfactant use	[44]
Sono-chemical	zinc nitrate hexahydrate	Potassium hydroxide, cetyltrimethylammonium bromide	200–400 nm flakes	Industrial-scale production	Parameter control	[45]

In the present study, our research group has synthesized ZnO@PAni NRs using an in-situ synthesis approach. Furthermore, adsorption properties of the prepared ZnO@PAni NRs were investigated for the removal of Cr (VI) from an aqueous solution.

2. Experimental Section

2.1. Chemicals

Zinc nitrate hexahydrate (Zn (NO$_3$)$_2$.6H$_2$O, 98% RG), sodium hydroxide pallets (97%, ACS grade), and hydrochloric acid (HCl, 37% ACS grade) were purchased from Sigma Aldrich (St. Luis, MO, USA). Aniline (monomer, 99.5%) and ammonium per sulphate (APS, (NH$_4$)$_2$S$_2$O$_8$, 98% RG) were supplied by Alpha Aesar (Somerville, MA, USA). All the chemicals and reagents were used as received without any further purification.

2.2. Synthesis of Zinc Oxide Nanorods (ZnO NRs)

The ZnO NRs were synthesized by the one-pot chemical coprecipitation method reported elsewhere, with some modification [46]. A 0.2 M solution of zinc acetate was prepared in ethanol and under magnetic stirring to attain homogeneity. A 15 mL solution of 0.1 M KOH was added dropwise until the pH of the reaction reached 9–10 and left on magnetic stirring for 12 h. After the stimulated time, the product was collected through centrifuge, washed with deionized water and absolute alcohol in order to remove any unreacted species, dehydrated in a hot-air oven under 90 °C temperature for 3 h, and calcined at 500 °C for another 3 h.

2.3. Synthesis of ZnO Immobilized PAni NRs

A method of oxidative free radical polymerization was utilized for the synthesis of the material reported elsewhere [47]. In a conical flask, a 2% (w/v) colloidal solution of ZnO was taken and sonicated for 30 min at 25 °C in an HCl solvent system. After sonication, a solution of 10% (v/v) aniline monomer in 0.1 M HCl solution was added to the dispersed colloidal system and magnetically mixed thoroughly to attain a homogeneous condition. After achieving homogeneity, a solid powder of ammonium persulfate of approximately 12.5 g was added to the colloidal mixture prepared above and the reaction was left on ice bath for 12 h. After the stimulated time, the reaction was stopped by adding an excess amount of HCl, and a green-colored precipitate was obtained, which was filtered and washed with deionized water several times to remove unreacted species. The material was dehydrated in a hot-air oven at 80 °C for 4 h.

2.4. Apparatus Used for Characterization

For recording XRD patterns in the current investigation, a powder X-ray diffractometer (XRD) from Rigaku, Tokyo, Japan, was used. Scanning electron microscopy (SEM) photographs of the produced materials were taken using a Zeiss microscope (Jena, Germany). On an Oxford EDX (MA, USA) apparatus coupled to an SEM, research using the energy dispersive X-ray spectroscopy (EDX) technique was carried out. On a Tecnai G2, F30 apparatus, pictures of the acquired samples under a transmission electron microscope (TEM) were taken. TGA analysis was carried out on Perkin Elmer model STA 6000 under an N2 environment. We used a BRUKER spectrometer to record the Fourier transform infrared (FTIR) spectra of the prepared samples. BET analysis was performed on a BET Surface Area Analyzer quanta chrome, Autosorb iQ2. The remaining concentration of Cr (VI) ions in the solution after adsorption was determined by using an atomic absorption spectrometer (AAS) GBC-908AA.

2.5. Adsorption of Cr (VI) Experiments

The adsorption experiments were performed according to the batch method. An amount of 0.02 g of ZnO@PAni NRs was dispersed in 20 mL of 50 mg/L Cr (VI) solution and was placed in sa 50 mL Erlenmeyer flask and shaken into a thermostat water bath at 100 rpm at variable temperature values such as 298 K, 308 K, and 318 K. The effects of various reaction variables such as contact time (50–200 min), pH of the medium (2–7), and adsorbent dose (12–28 mg) were optimized. The adsorption capacity of the synthesized material was evaluated by using Equation (1):

$$q_e = \left(\frac{C_0 - C_e}{C_0}\right) \times 100 \tag{1}$$

where q_e is the amount of Cr (VI) adsorbed per gram of the adsorbent at equilibrium, C_0 and C_e are the initial and final concentration of Cr (VI) ions, respectively, V is the volume of the metal ion solution taken, and W is the mass of the adsorbent taken (g).

The pH of the solution can easily influence the migration of charge on the surfaces of the adsorbent. Therefore, adsorption experiments were performed by varying the pH of the 50 mg/L Cr (VI) solution 15 mg adsorbent dose. The amount of adsorbent has a significant impact on how well metal ions in the solution are able to be absorbed. The effect of various adsorbent doses was optimized for improved adsorption capacity. A variation of adsorbent dose from 12 to 28 mg was used in 20 mL of 50 mg/L aqueous solution of Cr (VI) to investigate the impact of the adsorbent dose on the removal of Cr (VI).

2.6. Effects of Variable Nanorod Size, Initial Cr (VI) Concentration, and Individual Constituents on Adsorption

The effect of varying nanorod size on the adsorption capacity of ZnO@PAni towards Cr (VI) was observed by synthesizing the material with nanorods of various sizes such as 10.5, 19, 32.6, and 52.15 nm. Experiments were performed by taking 20 mL of 50 ppm Cr (VI) solution with 13 mg of the adsorbent dose. The adsorption efficiency of the synthesized material ZnO@PAni was observed towards the variable Cr (VI) concentration from 10 mg/L to 100 mg/L at an optimized reaction condition. Simultaneously the effects of individual constituents, viz. ZnO NRs, PAni, and ZnO@PAni NRs, on Cr (VI) adsorption were also studied at optimized reaction conditions.

2.7. Adsorption Isotherms

The equilibrium attained after the uptake of the metal ion by the adsorbent surface was studied by adsorption isotherms. The present work involves the nonlinear regression analysis of two isotherm models, namely, the Langmuir [48] and Freundlich [49] models, to study the adsorption properties of ZnO@PAni towards Cr (VI).

2.7.1. Langmuir Isotherm

According to the Langmuir model, a reversible chemical equilibrium happens between the ZnO@PAni surface and Cr (VI) ions with a validity of monolayer formation on a surface with a limited number of identical sites, which is nonlinearly represented by Equation (2).

$$q_e = \frac{q_m K_L C_e}{1 + K_L C_e} \tag{2}$$

2.7.2. Freundlich Isotherm

Equation (3) represents the empirical linear equation pertaining to the Freundlich model, which is based on the idea of a heterogenous surface and reversible multilayer adsorption. The model is nonlinearly represented by Equation (3):

$$q_e = K_F C_e^{1/n} \tag{3}$$

Herein, n is the degree of favorability of adsorption reaction, e.g., if n > 1, then adsorption is favorable, while if n < 1, then adsorption is non-favorable. K_F (mg/g) is the Freundlich adsorption capacity.

2.8. Adsorption Kinetics

2.8.1. Pseudo-First Order

The adsorption rate is directly proportional to the difference between the equilibrium adsorption capacity and the adsorption capacity at any given time t, according to the pseudo-first-order kinetics model, which is based on the premise that adsorption is regulated by diffusion steps [50],

$$q_t = q_e\left(1 - e^{-k_1 t}\right) \quad (4)$$

Herein, q_e = amount of Cr (VI) adsorbed on ZnO@PAni (mg/g) at equilibrium, and q_t = amount of Cr (VI) adsorbed (mg/g) at time t.

2.8.2. Pseudo-Second Order

The following equation provides the linear equation for pseudo-second-order kinetics [51],

$$q_t = \frac{k_2 q_e^2 t}{1 + k_2 q_e t} \quad (5)$$

where k_2 = pseudo-second-order rate constant, q_e = amount of Cr (VI) adsorbed on ZnO@PAni (mg/g) at equilibrium, and q_t = amount of adsorbed (mg/g) of Cr (VI) at time t.

2.8.3. Elovich Model

The Elovich equation may be used to understand the sorption kinetics as follows if the process is a chemisorption on highly heterogeneous sorbents [52]:

$$q_t = \frac{1}{\beta}\ln(\alpha\beta t + 1) \quad (6)$$

Herein, α = initial adsorption rate (mg/g·min), β = desorption constant (g/mg), and q_t = adsorption capacity at any time t (mg/g).

2.8.4. Intraparticle Diffusion Model

In a batch reactor system, pore and intraparticle diffusion are frequently rate-limiting when an adsorbent is translocated from the solution into the solid phase of absorbents [53]. Intraparticle diffusion was investigated using the Weber and Morris equation, as given below:

$$q_t = K_{int} \times t^{0.5} + C \quad (7)$$

In Equation (12), q_t = amount of Cr (VI) adsorbed (mg/g) at time t, K_{int} = intraparticle diffusion constant (mg/g·min$^{0.5}$), and t = time.

2.9. Adsorption Thermodynamics

The Gibbs equation and the van't Hoff equation were used to compute thermodynamic parameters such as the Gibbs free energy change (G°), enthalpy change (H°), and entropy change (S°) in order to support our assertion that the adsorption process is endothermic. The equilibrium constant using the Langmuir constant (K_L) is given by Equation (8) [54]

$$K_{eq} = \frac{K_L \times M_A}{\gamma_e} \quad (8)$$

where K_L is the Langmuir constant (L/mg), M_A is the molar mass of Cr (VI) in mg, and γ_e is the activity of adsorbate, i.e., Cr (VI). For non-ionic species or dilute solutions, $\gamma_e = 1$, so Equation (8) reduces to Equation (9) [55];

$$K_{eq} = K_L \times M_A \quad (9)$$

The equilibrium constant is related to Gibbs free energy by Equation (10);

$$\Delta G = -RT \ln K_{eq} \tag{10}$$

since

$$\Delta G = \Delta H - T\Delta S \tag{11}$$

Using Equations (10) and (11), we obtained the following:

$$\ln K_{eq} = -\frac{\Delta H}{RT} + \frac{\Delta S}{R} \tag{12}$$

R = gas constant, K_{eq} = equilibrium constant, and T = temperature of solution. The values of ΔH and ΔS are determined from the slope and intercept of a plot of ln K_{eq} as a function of 1/T. The following Equation (11) can be used to compute the free energy change (ΔG) of the adsorption reaction.

2.10. Determination of Point of Zero Charge and Zeta Potential

The point of zero charge (pHpzc) signifies the state of a material when the net electrical charge on its surface becomes zero; at this stage, adsorption takes place through an ion exchange process [56]. To determine the pHpzc, experiments were conducted in a batch mode by taking 20 mL of 0.1 M KCl solution with 50 mg of adsorbent with 10 replicates. The pH of all the 10 samples was adjusted from 1 to 10 using 0.1 M HCl and NaOH solutions. After 24 h, the supernatants were collected using centrifuge and employed to measure the ζ potential of the solution. A curv6e of ζ potential vs. pH was plotted and is given in Figure S1. The ζ potential was positive till pH 4 and negative from pH 5 to 10. The value of pHpzc was found to be 4.2, which signifies the positive surface of the adsorbent below pH 4.2 and negative surface above pH 4.2.

3. Results and Discussion

3.1. Characterization of ZnO@PAni

Surface morphological properties of the ZnO, PAni, and ZnO@PAni were examined by SEM analysis, and elemental composition was confirmed by EDX. Figure 1a demonstrates the recorded SEM image of PAni which shows flakelike surface structures while the ZnO nanoparticles exhibited a nanowire-type-shaped surface with agglomerations, as presented in Figure 1b. Thus, the SEM investigations suggested that ZnO and PAni possess different surface morphologies as described above. Therefore, it was an interesting thought to utilize the surface properties of both ZnO and PAni to synthesize a composite ZnO@PAni with improved adsorption efficiency. The SEM results of ZnO@PAni indicated that ZnO nanowires are strongly attached on the PAni flakes (Figure 1c). The elemental composition of the prepared ZnO@PAni NRs has been examined by EDX studies, and Figure 1d exhibited the presence of C, N, Zn, and O elements, which revealed the formation of ZnO@PAni NRs with good phase purity. Figure 1e represents the SEM image of ZnO@PAni NRs after the adsorption of Cr (VI), which exhibits a surface covered with small dots on ZnO NRs and PAni flakes, which is due to adsorption of water molecules alongside Cr (VI) ions. Figure 1f represents the EDX spectra of Cr-adsorbed ZnO@PAni NRs, which represent the presence of elements such as C, O, Zn, and Cr. The absence of an N atom from the EDX spectra suggests the involvement of amine and imine groups in the binding of Cr (VI) from the aqueous solution.

Figure 1. SEM image of (**a**) PAni, (**b**) ZnO nanowires, and (**c**) ZnO@PAni; (**d**) EDX spectra of ZnO@PAni; (**e**) SEM image of ZnO@PAni after Cr (VI) adsorption; and (**f**) EDX spectra of ZnO@PAni-Cr (VI) adsorbed.

The morphological properties of the prepared ZnO@PAni were further authenticated by recording a TEM image. The TEM image of the synthesized ZnO@PAni NRs showed that ZnO nanowires are immobilized with PAni flakes (Figure 2a). Therefore, this confirms the formation of ZnO@PAni NRs using an in situ synthetic method. The Gaussian distribution profiles for average particle size of ZnO@PAni NRs are presented in Figure 2b, which represents the average particle size as 22 nm.

Fourier transform infrared spectroscopy (FTIR) was further used to analyze the type of functional groups formed during processing of the nanocomposite material involving ZnO, PAni, and ZnO@PAni composite and the results are given in Figure 3.

FTIR spectra of ZnO exhibited the presence of various bands at 3443, 1629, 1022, and 443 cm^{-1}. The peak at 3443, 1629 cm^{-1} (stretching and bending vibrational modes of hydroxyl compounds (–OH groups)), and the characteristic peaks at 1022 and 443 cm^{-1} of Zn-O bonds authenticates the formation of ZnO nanorods [57,58]. The FTIR spectrum of PAni exhibits an absorption band at ~829 cm^{-1} (C–H bonding out of plane bending in the benzenoid ring), 515 cm^{-1} (C–N–C bonding mode of aromatic ring), 698 cm^{-1} (C–C and C–H bonding mode of aromatic ring), 1036, 1296, and 1499 cm^{-1} can be attributed to the C–N stretching of the benzenoid ring, and 1576 cm^{-1} may be due to the C–N stretching of the quinoid ring [59]. The FTIR spectrum of ZnO@PAni showed similar characteristics to the band of PAni. However, slight displacement in the peak was observed, which may be due to the presence of ZnO on the surface of PAni. The band at 3355 cm^{-1} is due to the presence of hydrogen bonding formed between N-H bonds of PAni and oxygen of ZnO NRs [60].

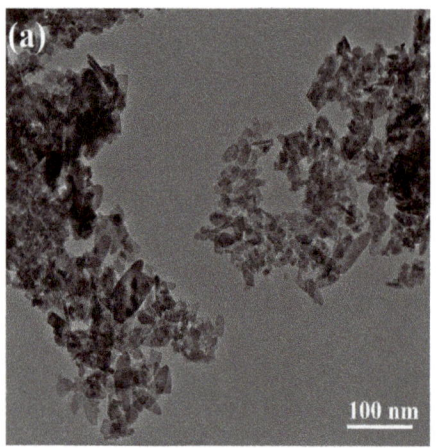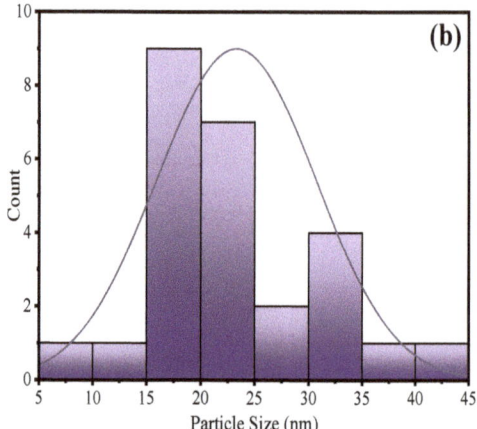

Figure 2. (**a**) TEM image of ZnO@PAni NRs, and (**b**) corresponding Gaussian distribution profiles for average particle size.

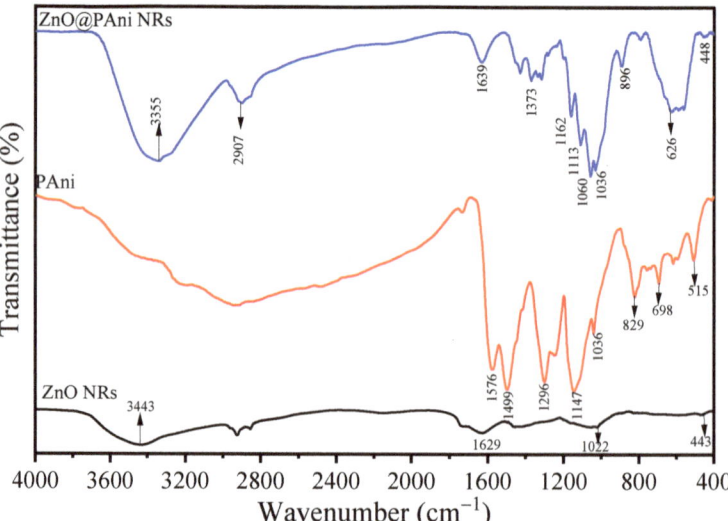

Figure 3. FTIR spectra of ZnO (black line) PAni (red line) and ZnO@PAni NRs (blue line).

Furthermore, recorded XRD data of the ZnO, PAni, and ZnO/PAni are presented in Figure 4. The characteristic peaks of ZnO were observed in the recorded XRD pattern of ZnO, and presence of (011), (100), (002), (101), (020), (102), (302), (110), (103), and (112) diffraction planes confirmed the formation of ZnO corresponding to JCPDS No. 36–1451. The strong diffraction peaks indicated the good crystalline nature of the prepared ZnO. The XRD of PAni has been displayed in Figure 4, and a broad diffraction peak appeared at 25.5°. This is the characteristic peak of PAni and suggests the presence of the (200) diffraction plane in the prepared PAni. The XRD results for ZnO@PAni demonstrated the presence of the (200) diffraction plane of PAni and authenticated the formation of ZnO@PAni [47,61].

However, crystallinity of the ZnO@PAni was reduced due to the amorphous nature of PAni (Figure 4).

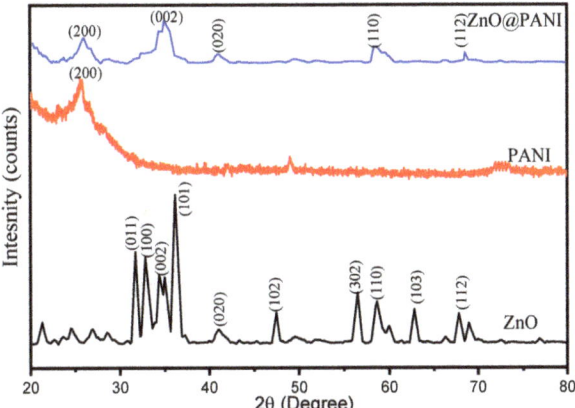

Figure 4. XRD spectra of ZnO (black line) PAni (red line) and ZnO@PAni NRs (blue line).

The thermal stability of ZnO@PAni and PAni were also investigated, and the obtained results are shown in Figure 5. The weight loss for PAni around 124 °C may be attributed to the evaporation of water. The weight loss for PAni at ~700–725 °C was found to be 57.46%. In the case of ZnO@PAni, the weight loss after 700 °C was found to be 44.69%. This shows that weight loss in ZnO@PAni is less compared to pristine PAni. Thus, the introduction of stable ZnO enhances the thermal stability of ZnO@PAni (Figure 5) [62]. The specific surface area of adsorbent plays a vital role, and it is necessary to study the surface area of ZnO@PAni. Brunauer–Emmett–Teller (BET) investigations were carried out to examine the specific surface area of the prepared ZnO@PAni. The nitrogen adsorption–desorption isotherm of ZnO@PAni has been presented in Figure 6a. The BET studies showed that ZnO@PAni has a high surface area of 113.5 m^2/g. The pore width distribution curve of ZnO@PAni is shown in Figure 6b. The average pore width of the ZnO@PAni was found to be 7 nm.

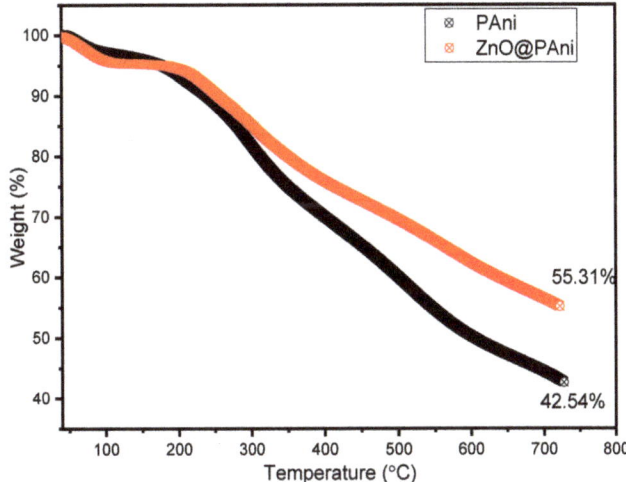

Figure 5. Thermogravimetric (TGA) profile for PAni (black line) and ZnO@PAni (red line).

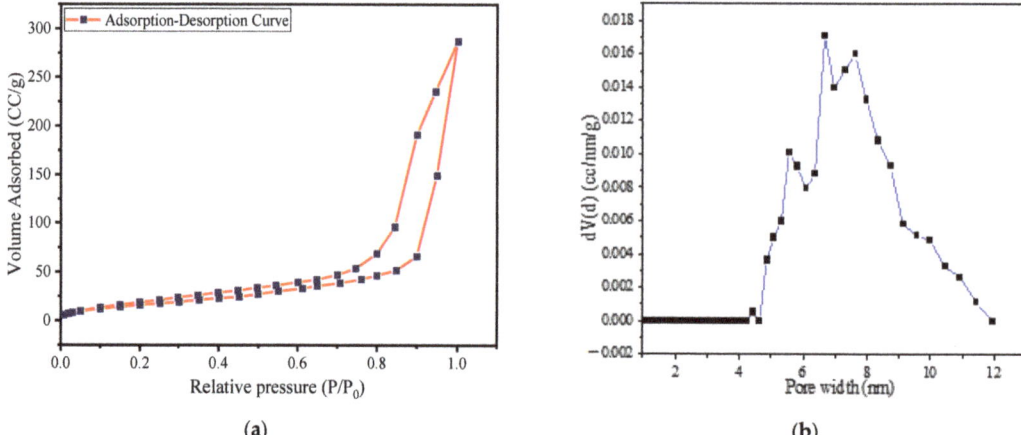

Figure 6. (a) N_2 adsorption–desorption curve, and (b) corresponding BJH pore size distribution curve for ZnO@PAni NRs.

3.2. Adsorption Studies

The adsorption properties of synthesized ZnO@PAni NRs were explored towards Cr (VI) from an aqueous system, the effect of various reaction variables such as pH, adsorbent dose, and contact time were observed, and the results are given in Figure 7a–c. Figure 7a shows the 2D contour plot for the simultaneous interaction of contact time and pH on the adsorption capacity of ZnO@PAni NRs. According to the observations, ZnO@PAni exhibited maximum adsorption capacity for Cr (VI) at pH 2, which continues to decrease with the further increase in pH value. Different forms of Cr (VI) such as CrO_4^{2-} are more prominent in neutral pH, whereas $HCrO_4^-$ is prominent in an acidic medium, which are the key factors for the adsorption of Cr (VI) on the ZnO@PAni surface [63]. Regarding a pH less than 4 (pH_{pzc} = 4.2, Figure 7d), the adsorbent surface becomes positively charged due to protonation in the presence of an excess of H^+ ions in the solution, and at lower pH, the $HCrO_4^-$ form of Cr (VI) ions predominate. Higher removal efficiency resulted from the strong electrostatic interaction between the positively charged adsorbent surface and the negatively charged $HCrO_4^-$ ions. With increasing pH, deprotonation of the surface of ZnO@PAni can occur due to decreasing H^+ ions [64].

Therefore, at higher pH values, there is less interaction between Cr (VI) ions and adsorbent surface, which leads to lesser adsorption capacity (Figure 7a). The simultaneous effect of contact time (min) on the adsorption of Cr (VI) was also observed in a time span of 50–200 min. Figure 7a shows that >35 mg/g adsorption capacity was observed at 120 min of contact time, and after that a negligible variation was observed with further increase in time, suggesting the occupation of all the surface-active sites or saturation of the adsorbent surface by Cr (VI) ions. Therefore, 120 min was chosen as the optimized time for further adsorption experiments.

The effect of various adsorbent doses was optimized for improved adsorption capacity. The results presented in Figure 7b suggest that the adsorption capacity of ZnO@PAni towards Cr (VI) decreases with an increase in adsorbent dose. The optimized amount of 13 mg is a more suitable adsorbent for Cr (VI) removal, which shows maximum adsorbent capacity. Initially with a lower amount of adsorbent dose, the higher number of surface-active sites are available to bind with Cr (VI) and as a result a high value of adsorbent capacity is achieved. At a higher adsorbent dose, due to the agglomeration of nanoparticles, a smaller number of surface-active sites are available and as a result a lower value of adsorbent capacity occurs [65]. Furthermore, we have also investigated the adsorption capacity by varying the pH and adsorbent dose, and the observations indicated that the highest adsorption capacity of more than 50 mg/g for 13 mg ZnO@PAni in an aqueous solution of pH 2 (Figure 7c) was achieved. Therefore, it was concluded that ZnO@PAni has the potential to adsorb the Cr (VI) effectively. To further confirm the adsorption of Cr (VI) by ZnO@PAni, we have recorded a SEM image of the ZnO@PAni surface after Cr (VI) adsorption. Figure 1e shows the recorded SEM image of ZnO@PAni with Cr (VI) adsorption. The Cr (VI) particles are seen on the surface of ZnO@PAni. The presence of Cr (VI) on ZnO@PAni was also further authenticated by EDX. Figure 1f shows the obtained EDX data of Cr (VI)-adsorbed ZnO@PAni. The EDX data reveal the presence of Cr, C, Zn, and O elements. Thus, it is confirmed that Cr (VI) particles are present on the ZnO@PAni surface. This confirms the adsorption properties of ZnO@PAni nanocomposite.

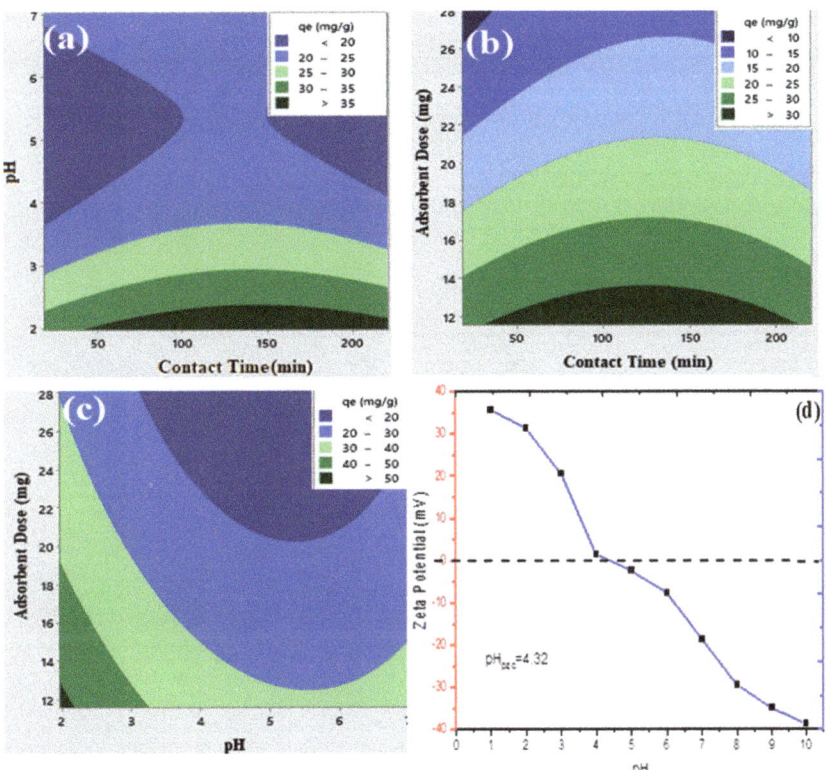

Figure 7. The 2D contour plots for optimization of process variable: (**a**) pH vs. contact time, (**b**) adsorbent dose vs. contact time, and (**c**) pH vs. adsorbent dose for removal of (50 mg/L) Cr (VI) by ZnO@PAni; (**d**) pH vs. zeta potential curve for the determination of the point of zero charge (pH$_{pzc}$).

3.3. Effect of Varying Nanorod Size, Initial Cr (VI) Concentration, and Individual Adsorbent Material

The effect of varying nanorod size on the adsorption capacity of ZnO@PAni towards Cr (VI) was observed, and the obtained results are summarized in Figure 8a. The observations demonstrated that with the increase in nanorod size from 10.5 to 20 nm, adsorption capacity of ZnO@PAni increased from 25.56 to 37.93 mg/g, while beyond 20 nm, a gradual decrease in adsorption capacity was observed, which might be due to the lower surface area possessed by larger-size nanorods [66]. Initially, at 10.5 nm nanorod size, there may be a possibility of stacking of nanorods due to small size and hence a lower adsorption capacity was observed. As the particle size increased to 19 nm, the possibility of agglomeration decreased, and as a result, nanorods were distributed throughout the aqueous media and thus a high adsorption capacity was achieved. At a higher particle size, due to low surface area, a smaller number of surface-active sites were available and hence a decrease in adsorption capacity was constantly observed. The adsorption efficiency of the synthesized material ZnO@PAni was observed towards the variable Cr (VI) concentration and the obtained results are summarized in Figure 8b. The observations and results suggested that with gradual increase in Cr (VI) concentration from 10 mg/L to 100 mg/L, the adsorption capacity of ZnO@PAni also increases. This can be explained as the adsorbent consists of plenty of surface-active sites that can easily accommodate Cr (VI) ions, even at a 100 mg/L concentration. Figure 8c represents the adsorption capacity bars with respect to the synthesized material (ZnO@PAni) and its constituent materials, including ZnO and PAni. It was observed that the composite formed by assembling ZnO and PAni results in higher adsorption capacity as compared to its individual constituents. This is due to the synergistic effect provided by -OH groups on the surface of ZnO and -NH$_2$ groups on PAni that provide plenty of charge density to bind the positively charged Cr (VI) [67,68].

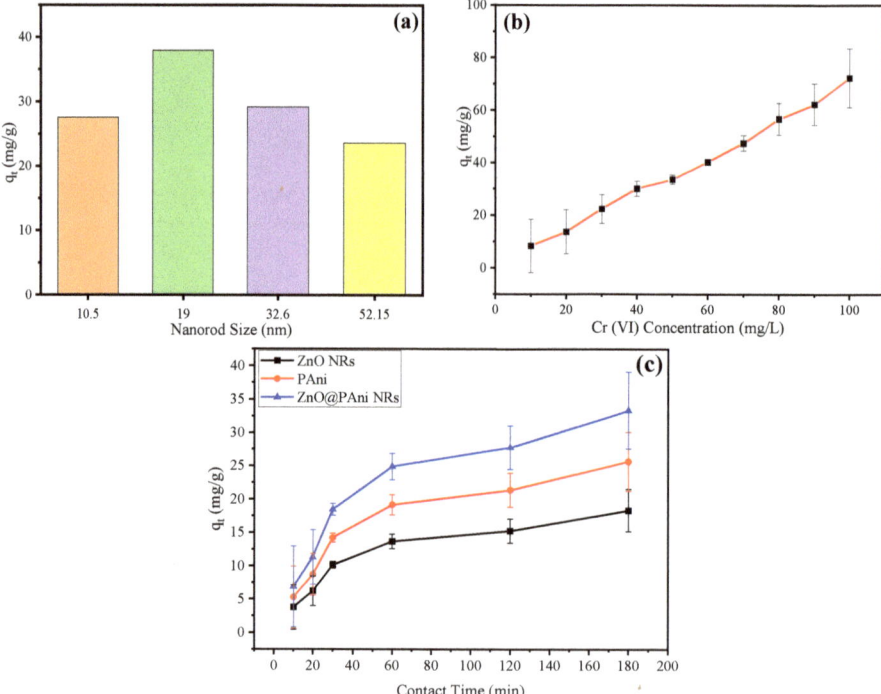

Figure 8. Variation of adsorption capacity of ZnO@PAni with respect to (**a**) particle size, (**b**) initial Cr (VI) concentration, and (**c**) the synthesized material and its constituents.

3.4. Adsorption Isotherm Studies

The equilibrium data obtained after experiments were applied to the nonlinear models of Langmuir and Freundlich, and the results are given in Table 2 and Figure 9a,b. Table 2 shows that with a low value of chi square (χ^2 = 3.41) and high value of R^2 (0.99), the Langmuir model was found to be the best model for explaining the adsorption of Cr (VI) on ZnO@PAni. The maximum monolayer adsorption capacity was found to be 142.27 mg/g at 298 K, 219.18 mg/g at 308 K, and 310.47 mg/g at 318 K. The Langmuir constant K_L was found to be 0.412, 0.767, and 0.849 L/mg at temperatures from 298 to 318 K. The significance of the KL value also establishes the affinity of adsorbent towards adsorbate, and in this case, as the temperature increases, the value of K_L also increases, suggesting a high affinity of ZnO@PAni NRs towards Cr (VI) and a high value of adsorption capacity achieved at high temperature. The Freundlich adsorption parameters were calculated from the nonlinear plot between q_e versus ln C_e presented in Figure 9b and given in Table 2, which shows that the values of n are >1 for all temperature ranges, indicating good adsorption. The values of n being greater than 1 suggest the existence of a substantial interaction between the ZnO@PAni surface and Cr (VI) ions and the gradual increase in the value of 1 with temperature suggests high favorability of adsorption at high temperature. The obtained results through nonlinear regression for Langmuir and Freundlich models also suggest the data are statistically valid for explaining the adsorption behavior of ZnO@PAni NRs towards Cr (VI) [69,70].

Table 2. Adsorption isotherm parameters for the removal of (100 mg/L) Cr (VI) on ZnO@PAni NRs at 298 K, 308 K, and 318 K.

Model	Parameters	298 K	308 K	318 K
Langmuir	q_m (mg/g)	142.27	219.18	310.47
	K_L (L/mg)	0.412	0.767	0.849
	R^2	0.99	0.99	0.99
	χ^2	4.13	3.55	3.41
Freundlich	K_F (mg/g) (mg/L)$^{1/n}$	109.79	111.17	121.44
	n	1.84	2.52	3.42
	R^2	0.98	0.97	0.96
	χ^2	21.61	19.04	15.13

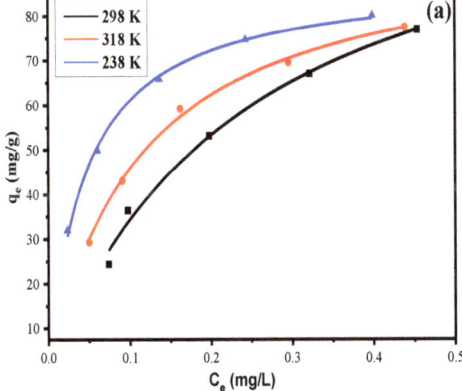
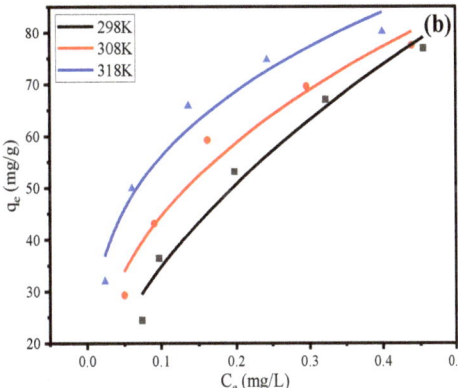

Figure 9. Nonlinear regression fitted adsorption isotherms of (**a**) Langmuir and (**b**) Freundlich using 20 mL of 100 mg/L Cr (VI) solution at pH 2 and 13 mg of adsorbent dose at 298–318 K temperature range.

3.5. Adsorption Kinetics

The kinetic data obtained from the kinetic studies are listed in Table 3, and graphs are given in Figure 10a,b. The high value of correlation coefficients $R^2 = 0.99$ and low value of $\chi^2 = 0.36$ characterized the pseudo-second-order kinetic model best describing the sorption rate of Cr (VI) on the surface of ZnO@PAni. According to the pseudo-second-order kinetic model, chemical adsorption may be the rate-limiting phase, and there may be entitlement of valence forces through the electron deficient Cr (VI) orbitals and electron rich -NH$_2$ and -OH groups, which can involve a complex reaction involving chemical bond formation and diffusion reactions simultaneously [71]. Moreover, the obtained q_e (calculated values) for PSO are found to be ore closer to the q_e (experimental values). According to The Weber–Morris equation, if the adsorption rate is controlled by the intraparticle diffusion process, then the plot between qt and $t^{0.5}$ will be a straight line. Figure 10b represents the intraparticle diffusion plot between qt and $t^{0.5}$ which showed a multilinear plot which does not pass through the origin, suggesting the involvement of boundary-layer diffusion in Cr (VI) adsorption. The first linear portion with a k_{p1} value of 5.02 represents the rapid external diffusion of Cr (VI) to the ZnO@PAni surface and second portion with k_{p2} value of 1.48 corresponds to the intraparticle diffusion effect.

Table 3. Kinetic parameters for the removal of 50 mg/L Cr (VI) on ZnO@PAni NRs at optimized reaction condition.

Model	Parameters	318 K
Pseudo-first order	$q_{e,exp}$ (mg/g)	35.69
	$q_{e,cal}$ (mg/g) k_1 (1/min)	41.36
	R^2	0.98
	χ^2	1.63
Pseudo-second order	$q_{e,exp}$ (mg/g)	35.69
	$q_{e,cal}$ (mg/g)k_2 (g/mg × min)	36.27
	R^2	0.99
	χ^2	0.36
Elovich	α (mg/g × min)	1.64
	β (g/mg)	0.089
	R^2	0.96
	χ^2	2.36
Intraparticle diffusion	k_{p1} (mg/g × min$^{1/2}$)	5.02
	k_{p2} (mg/g × min$^{1/2}$)	1.48
	R_1^2	0.98
	R_2^2	0.96
	χ^2	1.38

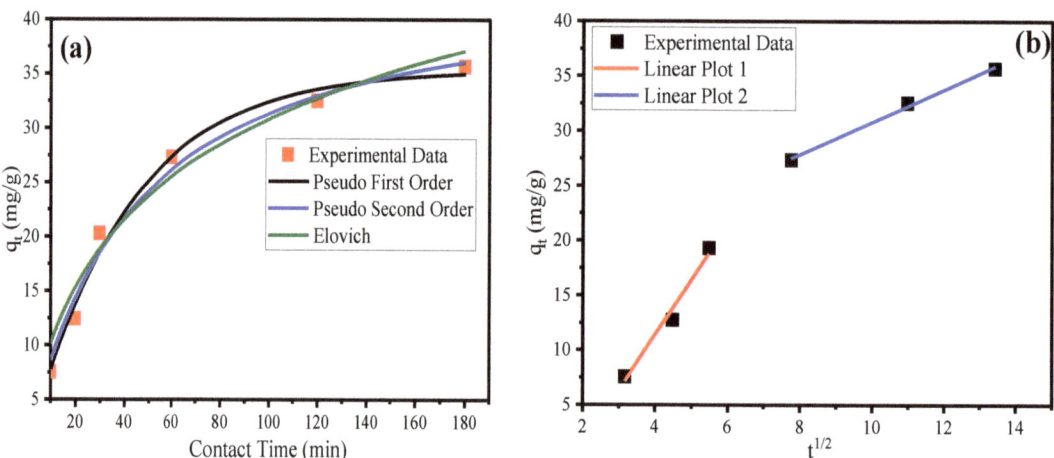

Figure 10. (a) Nonlinear fitted kinetic curves for pseudo-first order, pseudo-second order, and Elovich model, and (b) linear plot of intraparticle diffusion.

3.6. Adsorption Thermodynamics

Table 4 and Figure 11 provide a list of the thermodynamic characteristics related to the adsorption of Cr (VI) on the ZnO@PAni surface.

Table 4. Thermodynamic parameters for the removal of Cr (VI) on ZnO@PAni NRs at 298 K, 308 K, and 318 K.

Temperature (K)	ΔG° (kJ/mol)	ΔH° (kJ/mol)	ΔS° (kJ/mol·K)
298	−23.79		
308	−25.67	32.01	0.187
318	−27.54		

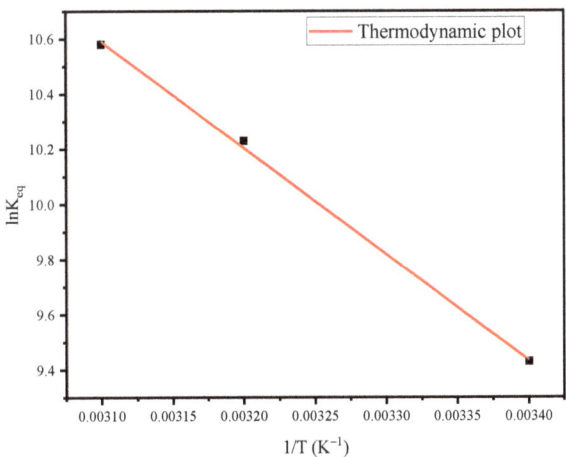

Figure 11. Adsorption thermodynamics curve for evaluation of enthalpy and entropy of the reaction.

Table 4 shows that the value of ΔH° is found to be 32.01 kJ/mol, which suggests that the adsorption of Cr (VI) on ZnO@PAni is endothermic in nature. Owing to this explanation, the adsorption reaction is highly favorable at a high temperature, i.e., 318 K, and less favorable at 298 K. The entropy ΔS° was found to be 0.187 kJ/mol × K, which is also positive, suggesting a randomness created in the system. The Gibbs free energy value ΔG° was found to be −23.79 kJ/mol at 298 K, −25.67 kJ/mol at 308 K, and −27.54 kJ/mol at 318 K, which suggests the feasible nature of the adsorption reaction and increase in negative magnitude with temperature, suggesting high feasibility at high temperature [72].

3.7. Adsorption Mechanism

From the kinetic and isotherm studies, it was observed that both adsorbent and adsorbate govern the adsorption process and as the time progresses, more and more portions of surface-active sites on the adsorbent surface are occupied accordingly. The pseudo-second-order model revealed the primary rate-controlling step as the adherence of Cr (VI) on the surface of ZnO@PAni. Additionally, the intraparticle diffusion also suggested the involvement of boundary-layer diffusion in Cr (VI) adsorption. The first step is ascribed as the mass transfer of Cr (VI) to the adsorbent surface, and the second step is attributed as the diffusion process through the inner nanopores of adsorbent. From the FTIR and SEM-EDX analysis, it was confirmed that the adsorbent surface is rich with electron-donating functional groups such as -OH and $-NH_2$ groups, which are involved in the adsorption process by the following Equations (13) and (14). Figure 1f shows the obtained EDX data of Cr (VI)-adsorbed ZnO@PAni. The EDS data reveal the presence of Cr, C, Zn, and O elements. Thus, it is confirmed that Cr (VI) particles are present on the ZnO@PAni surface. This confirms the adsorption properties of ZnO@PAni nanocomposite.

$$R\text{-}OH + HCrO_4^- + H^+ \; R\text{-}(OH_2)^+/HCrO_4^- \tag{13}$$

$$R-NH_2 + HCrO_4^- + H^+ \: R-(NH_3)^+/HCrO_4^- \tag{14}$$

3.8. Comparison with the Literature

The adsorption studies of the synthesized material ZnO@PAni were compared with other ZnO-based nanocomposite materials reported in the literature and are listed in Table 5. It was observed from the comparison data that modification of ZnO with PAni results in enhanced adsorption capacity towards Cr (VI) as compared to other materials.

Table 5. Comparison of the adsorption performance of present study with literature.

Adsorbent	Cr (VI) Removal Capacity (mg/g)	pH	Contact Time (min)	References
CuO–ZnO–C	201.56	2	600	[68]
ZnO-MWCNTs	152.2	2	120	[73]
rGO-ZnO	25.45	3	240	[74]
Ag/ZnO-AC	4.17	2.5	1200	[75]
PA/TSC-ZnO	90.83	4	250	[76]
GO ZnO-ZnFe$_2$O$_4$	109.89	4	120	[77]
GO-CS@MOF [Zn(BDC)(DMF)]	144.92	2	360	[78]
ZnO@PAni NRs	310.47	2	120	Present study

4. Conclusions

In summary, a multifunctional nanocomposite material using PAni and ZnO nanorods was synthesized through a free radical oxidative polymerization reaction. Various spectroscopic and morphological studies confirmed the successful formation of the material. The FTIR studies suggested the formation of H bonding between ZnO and -NH$_2$ groups of PAni. The XRD results for ZnO@PAni demonstrated the presence of the (200) diffraction plane of PAni with characteristic peaks of ZnO. The TEM image of the synthesized ZnO/PAni composite showed that ZnO nanowires are linked to the flakelike PAni with an average particle size of 22 nm. The synergistic effect of solid ZnO NRs in the polymer matrix results in high thermal stability of PAni. The BET studies showed that ZnO@PAni has a high surface area of 113.5 m^2/g with average pore width of 7 nm. The material as an adsorbent showed high affinity towards Cr (VI) from an aqueous solution at optimized reaction conditions of pH 2 and 13 mg adsorbent dose. The adsorption reaction was highly pH dependent and the obtained data were best explained by the Langmuir model, resulting in maximum monolayer adsorption capacity (q_m), as 142.27 mg/g at 298 K, 219.18 mg/g at 308 K, and 310.47 mg/g at 318 K were obtained. The kinetic studies suggested the uptake of Cr (VI) by ZnO@PAni NRs was best supported by a pseudo-second-order model, indicating a chemical bond formation between Cr (VI) and the adsorbent surface. In future, we plan to utilize the adsorptive and optical properties of the material towards Cr (VI)-polluted wastewater, splitting the simultaneous hydrogen production reaction under UV-Vis light radiations. Thus, the synthesized nanomaterial could emerge as cost-effective and highly efficient towards sustainable wastewater management and energy production.

Supplementary Materials: The following supporting information can be downloaded at: https://www.mdpi.com/article/10.3390/w15101949/s1, Figure S1: Point of zero charge (pHpzc) analysis using Zeta potential; Figure S2: Non-linear regression for (a) Langmuir (b) Freundlich and (c) Temkin isotherms at 298 to 318 K temperature. Table S1: Adsorption isotherm parameters for the removal of (100 mg/L) Cr (VI) on ZnO@PAni NRs at 298 K, 308 K and 318 K obtained by non-linear regression analysis.

Author Contributions: Conceptualization, F.A.A.; Methodology, F.A.A.; Software, F.A.A., R.H.A. and I.H.; Validation, I.H.; Formal analysis, R.H.A.; Investigation, R.H.A.; Resources, R.H.A.; Data curation, I.H.; Writing—original draft, R.H.A.; Writing—review & editing, I.H.; Supervision, F.A.A.; Project administration, F.A.A. All authors have read and agreed to the published version of the manuscript.

Funding: This research received no external funding.

Data Availability Statement: Data is contained within the article.

Acknowledgments: The authors extend their thanks to Researchers Supporting Project (Ref: RSP2023R442), King Saud University, Riyadh, Saudi Arabia.

Conflicts of Interest: The authors declare no conflict of interest.

References

1. Barad, J.M.; Kohli, H.P.; Chakraborty, M. Adsorption of hexavalent chromium from aqueous stream by maghemite nanoparticles synthesized by the microemulsion method. *Energy Nexus* **2021**, *5*, 100035. [CrossRef]
2. Nematollahzadeh, A.; Seraj, S.; Mirzayi, B. Catecholamine coated maghemite nanoparticles for the environmental remediation: Hexavalent chromium ions removal. *Chem. Eng. J.* **2015**, *277*, 21–29. [CrossRef]
3. Marti, N.; Bouzas, A.; Seco, A.; Ferrer, J. Struvite precipitation assessment in anaerobic digestion processes. *Chem. Eng. J.* **2008**, *141*, 67–74. [CrossRef]
4. Chai, W.S.; Cheun, J.Y.; Kumar, P.S.; Mubashir, M.; Majeed, Z.; Banat, F.; Ho, S.-H.; Show, P.L. A review on conventional and novel materials towards heavy metal adsorption in wastewater treatment application. *J. Clean. Prod.* **2021**, *296*, 126589. [CrossRef]
5. Imessaoudene, A.; Cheikh, S.; Bollinger, J.C.; Belkhiri, L.; Tiri, A.; Bouzaza, A.; Mouni, L. Zeolite Waste Characterization and Use as Low-Cost, Ecofriendly, and Sustainable Material for Malachite Green and Methylene Blue Dyes Removal: Box;Behnken Design, Kinetics, and Thermodynamics. *Appl. Sci.* **2022**, *12*, 7587. [CrossRef]
6. Mammar, A.C.; Mouni, L.; Bollinger, J.-C.; Belkhiri, L.; Bouzaza, A.; Assadi, A.A.; Belkacemi, H. Modeling and optimization of process parameters in elucidating the adsorption mechanism of Gallic acid on activated carbon prepared from date stones. *Sep. Sci. Technol.* **2019**, *55*, 3113–3125. [CrossRef]
7. Bayuo, J. An extensive review on chromium (vi) removal using natural and agricultural wastes materials as alternative biosorbents. *J. Environ. Health Sci. Eng.* **2021**, *19*, 1193–1207. [CrossRef]
8. Al-Qodah, Z.; Dweiri, R.; Khader, M.; Al-Sabbagh, S.; Al-Shannag, M.; Qasrawi, S.; Al-Halawani, M. Processing and characterization of magnetic composites of activated carbon, fly ash, and beach sand as adsorbents for Cr (VI) removal. *Case Stud. Chem. Environ. Eng.* **2023**, *7*, 100333. [CrossRef]
9. Al-Waqfi, M.S.E.-A.; Al-Qodah, Z. An iron rotating disk for the elimination of hexavalent chromium ion from industrial wastewaters: Putting it to work. *Desalination Water Treat* **2022**, *270*, 83–91. [CrossRef]
10. Zhang, H.; Li, P.; Wang, Z.; Zhang, X.; Zheng, S.; Zhang, Y. In Situ synthesis of γ-AlOOH and synchronous adsorption separation of V(V) from highly concentrated Cr (VI) multiplex complex solutions. *ACS Sustainable Chem. Eng.* **2017**, *5*, 6674–6681. [CrossRef]
11. Hu, Q.; Paudyal, H.; Zhao, J.; Huo, F.; Inoue, K.; Liu, H. Adsorptive recovery of vanadium(V) from chromium(VI)-containing effluent by Zr(IV)-loaded orange juice residue. *Chem. Eng. J.* **2014**, *248*, 79–88. [CrossRef]
12. Sun, F.; Liu, M.; Yuan, B.; He, J.; Wu, P.; Liu, C.; Jiang, W. Separation of vanadium and chromium by selective adsorption by titanium-based microspheres. *Chem. Eng. J.* **2022**, *450*, 138039. [CrossRef]
13. Rajapaksha, A.U.; Selvasembian, R.; Ashiq, A.; Gunarathne, V.; Ekanayake, A.; Perera, V.; Wijesekera, H.; Mia, S.; Ahmad, M.; Vithanage, M.; et al. A systematic review on adsorptive removal of hexavalent chromium from aqueous solutions: Recent advances. *Sci. Total. Environ.* **2022**, *809*, 152055. [CrossRef]
14. Fan, Y.; Wang, X.; Wang, M. Separation and recovery of chromium and vanadium from vanadium-containing chromate solution by ion exchange. *Hydrometallurgy* **2013**, *136*, 31–35. [CrossRef]
15. Yang, K.; Zhang, X.; Tian, X.; Yang, Y.; Chen, Y. Leaching of vanadium from chromium residue. *Hydrometallurgy* **2010**, *103*, 7–11. [CrossRef]
16. He, K.; Chen, P.; Yuan, B.; Sun, F.; He, J.; Wu, P.; Liu, C.; Jiang, W. Removing trace chromium from high concentration vanadium solution by photoreduction deposition with Ti–Zr solid solution. *Sep. Purif. Technol.* **2022**, *290*, 120855. [CrossRef]
17. Imessaoudene, A.; Cheikh, S.; Hadadi, A.; Hamri, N.; Bollinger, J.-C.; Amrane, A.; Tahraoui, H.; Manseri, A.; Mouni, L. Adsorption Performance of Zeolite for the Removal of Congo Red Dye: Factorial Design Experiments, Kinetic, and Equilibrium Studies. *Separations* **2023**, *10*, 57. [CrossRef]
18. Bouchelkia, N.; Tahraoui, H.; Amrane, A.; Belkacemi, H.; Bollinger, J.-C.; Bouzaza, A.; Zoukel, A.; Zhang, J.; Mouni, L. Jujube stones based highly efficient activated carbon for methylene blue adsorption: Kinetics and isotherms modeling, thermodynamics and mechanism study, optimization via response surface methodology and machine learning approaches. *Process. Saf. Environ. Prot.* **2023**, *170*, 513–535. [CrossRef]
19. El Messaoudi, N.; El Mouden, A.; Fernine, Y.; El Khomri, M.; Bouich, A.; Faska, N.; Ciğeroğlu, Z.; Américo-Pinheiro, J.H.P.; Jada, A.; Lacherai, A. Green synthesis of Ag2O nanoparticles using Punica granatum leaf extract for sulfamethoxazole antibiotic adsorption: Characterization, experimental study, modeling, and DFT calculation. *Environ. Sci. Pollut. Res.* **2022**, 1–18. [CrossRef]

20. El Mouden, A.; El Guerraf, A.; El Messaoudi, N.; Haounati, R.; El Fakir, A.A.; Lacherai, A. Date Stone Functionalized with 3-Aminopropyltriethoxysilane as a Potential Biosorbent for Heavy Metal Ions Removal from Aqueous Solution. *Chem. Afr.* **2022**, *5*, 745–759. [CrossRef]
21. El Mouden, A.; El Messaoudi, N.; El Guerraf, A.; Bouich, A.; Mehmeti, V.; Lacherai, A.; Jada, A.; Américo-Pinheiro, J.H.P. Removal of cadmium and lead ions from aqueous solutions by novel dolomite-quartz@Fe3O4 nanocomposite fabricated as nanoadsorbent. *Environ. Res.* **2023**, *225*, 115606. [CrossRef] [PubMed]
22. Gupta, K.; Joshi, P.; Gusain, R.; Khatri, O.P. Recent advances in adsorptive removal of heavy metal and metalloid ions by metal oxide-based nanomaterials. *Coord. Chem. Rev.* **2021**, *445*, 214100. [CrossRef]
23. Wadhawan, S.; Jain, A.; Nayyar, J.; Mehta, S.K. Role of nanomaterials as adsorbents in heavy metal ion removal from waste water: A review. *J. Water Process. Eng.* **2020**, *33*, 101038. [CrossRef]
24. De Jesus, R.A.; de Assis, G.C.; de Oliveira, R.J.; Costa, J.A.S.; da Silva, C.M.P.; Bilal, M.; Iqbal, H.M.; Ferreira, L.F.R.; Figueiredo, R.T. Environmental remediation potentialities of metal and metal oxide nanoparticles: Mechanistic biosynthesis, influencing factors, and application standpoint. *Environ. Technol. Innov.* **2021**, *24*, 101851. [CrossRef]
25. Mittal, A.; Ahmad, R.; Hasan, I. Iron oxide-impregnated dextrin nanocomposite: Synthesis and its application for the biosorption of Cr(VI) ions from aqueous solution. *Desalination Water Treat.* **2015**, *57*, 15133–15145. [CrossRef]
26. Liu, M.; Yin, W.; Qian, F.-J.; Zhao, T.-L.; Yao, Q.-Z.; Fu, S.-Q.; Zhou, G.-T. A novel synthesis of porous TiO$_2$ nanotubes and sequential application to dye contaminant removal and Cr(VI) visible light catalytic reduction. *J. Environ. Chem. Eng.* **2020**, *8*, 104061. [CrossRef]
27. Sharma, P.; Prakash, J.; Palai, T.; Kaushal, R. Surface functionalization of bamboo leave mediated synthesized SiO$_2$ nanoparti-cles: Study of adsorption mechanism, isotherms and enhanced adsorption capacity for removal of Cr (VI) from aqueous solution. *Environ. Res.* **2022**, *214*, 113761. [CrossRef]
28. Wang, S.Z.; Zheng, M.; Zhang, X.; Zhuo, M.P.; Zhou, Q.Q.; Zheng, M.; Liao, L.S. Fine synthesis of hierarchical CuO/Cu(OH)$_2$ urchin-like nanoparticles for efficient removal of Cr(VI). *J. Alloys Compd.* **2021**, *884*, 161052. [CrossRef]
29. Sahu, U.K.; Zhang, Y.; Huang, W.; Ma, H.; Mandal, S.; Sahu, S.; Sahu, M.K.; Patel, R.K.; Pu, S. Nanoceria-loaded tea waste as bio-sorbent for Cr(VI) removal. *Mater. Chem. Phys.* **2022**, *290*, 126563. [CrossRef]
30. Onwudiwe, D.C.; Oyewo, O.A.; Ogunjinmi, O.E.; Ojelere, O. Hexavalent chromium reduction by ZnO, SnO$_2$ and ZnO-SnO$_2$ synthesized using biosurfactants from extract of Solanum macrocarpon. *South Afr. J. Chem. Eng.* **2021**, *38*, 21–33. [CrossRef]
31. Pandey, M.; Tripathi, B.D. Synthesis, characterization and application of zinc oxide nano particles for removal of hexavalent chromium. *Res. Chem. Intermed.* **2017**, *43*, 121–140. [CrossRef]
32. Rout, D.R.; Jena, H.M. Enhanced Cr(VI) adsorption using ZnO decorated graphene composite: Batch and continuous studies. *J. Taiwan Inst. Chem. Eng.* **2022**, *140*, 104534. [CrossRef]
33. Ruotolo, L.; Gubulin, J. Chromium(VI) reduction using conducting polymer films. *React. Funct. Polym.* **2005**, *62*, 141–151. [CrossRef]
34. Dognani, G.; Hadi, P.; Ma, H.; Cabrera, F.C.; Job, A.; Agostini, D.L.; Hsiao, B.S. Effective chromium removal from water by polyaniline-coated electrospun adsorbent membrane. *Chem. Eng. J.* **2019**, *372*, 341–351. [CrossRef]
35. Herath, A.; Reid, C.; Perez, F.; Pittman, C.U.; Mlsna, T.E. Biochar-supported polyaniline hybrid for aqueous chromium and nitrate adsorption. *J. Environ. Manag.* **2021**, *296*, 113186. [CrossRef]
36. Muhammad, A.; Shah, A.U.H.A.; Bilal, S. Effective Adsorption of Hexavalent Chromium and Divalent Nickel Ions from Water through Polyaniline, Iron Oxide, and Their Composites. *Appl. Sci.* **2020**, *10*, 2882. [CrossRef]
37. Mohammad, N.; Atassi, Y. Enhancement of removal efficiency of heavy metal ions by polyaniline deposition on electrospun polyacrylonitrile membranes. *Water Sci. Eng.* **2021**, *14*, 129–138. [CrossRef]
38. Mahmood, N.B.; Saeed, F.R.; Gbashi, K.R.; Mahmood, U.S. Synthesis and characterization of zinc oxide nanoparticles via ox-alate co-precipitation method. *Mater. Lett. X* **2022**, *13*, 100126.
39. Hasanpoor, M.; Aliofkhazraei, M.; Delavari, H.J.P.M.S. Microwave-assisted Synthesis of Zinc Oxide Nanoparticles. *Proc. Mater. Sci.* **2015**, *11*, 320–325. [CrossRef]
40. Gerbreders, V.; Krasovska, M.; Sledevskis, E.; Gerbreders, A.; Mihailova, I.; Tamanis, E.; Ogurcovs, A. Hydrothermal synthesis of ZnO nanostructures with controllable morphology change. *Crystengcomm* **2020**, *22*, 1346–1358. [CrossRef]
41. Li, J.; Wu, Z.; Bao, Y.; Chen, Y.; Huang, C.; Li, N.; He, S.; Chen, Z. Wet chemical synthesis of ZnO nanocoating on the surface of bamboo timber with improved mould-resistance. *J. Saudi Chem. Soc.* **2017**, *21*, 920–928. [CrossRef]
42. Al Abdullah, K.; Awad, S.; Zaraket, J.; Salame, C. Synthesis of ZnO Nanopowders by Using Sol-Gel and Studying Their Struc-tural and Electrical Properties at Different Temperature. *Energy Procedia* **2017**, *119*, 565–570. [CrossRef]
43. Wang, Y.; Yang, C.; Liu, Y.; Fan, Y.; Dang, F.; Qiu, Y.; Zhou, H.; Wang, W.; Liu, Y. Solvothermal Synthesis of ZnO Nanoparticles for Photocatalytic Degradation of Methyl Orange and *p*-Nitrophenol. *Water* **2021**, *13*, 3224. [CrossRef]
44. Pineda-Reyes, A.M.; Olvera, M.D.L.L. Synthesis of ZnO nanoparticles from water-in-oil (w/o) microemulsions. *Mater. Chem. Phys.* **2018**, *203*, 141–147. [CrossRef]
45. Banerjee, P.; Chakrabarti, S.; Maitra, S.; Dutta, B.K. Zinc oxide nano-particles—Sonochemical synthesis, characterization and application for photo-remediation of heavy metal. *Ultrason. Sonochem.* **2012**, *19*, 85–93. [CrossRef] [PubMed]
46. Chandraiahgari, C.R.; De Bellis, G.; Ballirano, P.; Balijepalli, S.K.; Kaciulis, S.; Caneve, L.; Sarto, F.; Sarto, M.S. Synthesis and characterization of ZnO nanorods with a narrow size distribution. *RSC Adv.* **2015**, *5*, 49861–49870. [CrossRef]

47. Mostafaei, A.; Zolriasatein, A. Synthesis and characterization of conducting polyaniline nanocomposites containing ZnO nanorods. *Prog. Nat. Sci. Mater. Int.* **2012**, *22*, 273–280. [CrossRef]
48. Langmuir, I. The adsorption of gases on plane surfaces of glass, mica and platinum. *J. Am. Chem. Soc.* **1918**, *40*, 1361–1403. [CrossRef]
49. Freundlich, H. Über Die Adsorption in Lösungen. *Z. Für Phys. Chem.* **1907**, *57U*, 385–470. [CrossRef]
50. Lagergren, S.K. On the Theory of So-Called Adsorption of Materials. *R. Swed. Acad. Sci.* **1898**, *24*, 1–13.
51. Ho, Y.S.; McKay, G. Pseudo-second order model for sorption processes. *Process Biochem.* **1999**, *34*, 451–465. [CrossRef]
52. Aharoni, C.; Ungarish, M. Kinetics of Activated Chemisorption. Part 1.—The Non-Elovichian Part of the Isotherm. *J. Chem. Soc. Faraday Trans. 1 Phys. Chem. Condens. Phases* **1976**, *72*, 400–408. [CrossRef]
53. Weber, W.J., Jr.; Morris, J.C. Kinetics of Adsorption on Carbon from Solution. *J. Sanit. Eng. Div.* **1963**, *89*, 31–59. [CrossRef]
54. Ghosal, P.S.; Gupta, A.K. Determination of thermodynamic parameters from Langmuir isotherm constant-revisited. *J. Mol. Liq.* **2017**, *225*, 137–146. [CrossRef]
55. Raghav, S.; Kumar, D. Adsorption Equilibrium, Kinetics, and Thermodynamic Studies of Fluoride Adsorbed by Tetrametallic Oxide Adsorbent. *J. Chem. Eng. Data* **2018**, *63*, 1682–1697. [CrossRef]
56. Primo, J.D.O.; Bittencourt, C.; Acosta, S.; Sierra-Castillo, A.; Colomer, J.-F.; Jaerger, S.; Teixeira, V.C.; Anaissi, F.J. Synthesis of Zinc Oxide Nanoparticles by Ecofriendly Routes: Adsorbent for Copper Removal from Wastewater. *Front. Chem.* **2020**, *8*, 1100. [CrossRef]
57. Khan, M.; Naqvi, A.H.; Ahmad, M. Comparative study of the cytotoxic and genotoxic potentials of zinc oxide and titanium dioxide nanoparticles. *Toxicol. Rep.* **2015**, *2*, 765–774. [CrossRef]
58. Taghizadeh, S.-M.; Lal, N.; Ebrahiminezhad, A.; Moeini, F.; Seifan, M.; Ghasemi, Y.; Berenjian, A. Green and Economic Fabrication of Zinc Oxide (ZnO) Nanorods as a Broadband UV Blocker and Antimicrobial Agent. *Nanomaterials* **2020**, *10*, 530. [CrossRef]
59. Saravanan, S.; Joseph Mathai, C.; Anantharaman, M.; Venkatachalam, S.; Prabhakaran, P. Investigations on the electrical and structural properties of polyaniline doped with camphor sulphonic acid. *J. Phys. Chem. Solids* **2006**, *67*, 1496–1501. [CrossRef]
60. Bera, S.; Khan, H.; Biswas, I.; Jana, S. Polyaniline hybridized surface defective ZnO nanorods with long-term stable photoelectrochemical activity. *Appl. Surf. Sci.* **2016**, *383*, 165–176. [CrossRef]
61. Mutlaq, S.; Albiss, B.; Al-Nabulsi, A.A.; Jaradat, Z.W.; Olaimat, A.N.; Khalifeh, M.S.; Holley, R.A. Con-ductometric Immunosensor for *Escherichia coli* O157:H7 Detection Based on Polyaniline/Zinc Oxide (PANI/ZnO) Nano-composite. *Polymers* **2021**, *13*, 3288. [CrossRef] [PubMed]
62. Mohsen, R.M.; Morsi, S.M.M.; Selim, M.M.; Ghoneim, A.M.; El-Sherif, H.M. Electrical, thermal, morphological, and antibacterial studies of synthesized polyaniline/zinc oxide nanocomposites. *Polym. Bull.* **2018**, *76*, 6135–6160. [CrossRef]
63. Yusuff, A.S. Optimization of adsorption of Cr(VI) from aqueous solution by *Leucaena leucocephala* seed shell activated carbon using design of experiment. *Appl. Water Sci.* **2018**, *8*, 232. [CrossRef]
64. Rouhaninezhad, A.A.; Hojati, S.; Masir, M.N. Adsorption of Cr (VI) onto micro- and nanoparticles of palygorskite in aqueous solutions: Effects of pH and humic acid. *Ecotoxicol. Environ. Saf.* **2020**, *206*, 111247. [CrossRef] [PubMed]
65. Olad, A.; Farshi Azhar, F. study on the adsorption of chromium (VI) from aqueous solutions on the algi-nate-montmorillonite/polyaniline nanocomposite. *Desalination Water Treat* **2014**, *52*, 2548–2559. [CrossRef]
66. Aliabadi, M.; Khazaei, I.; Fakhraee, H.; Mousavian, M.T.H. Hexavalent chromium removal from aqueous solutions by using low-cost biological wastes: Equilibrium and kinetic studies. *Int. J. Environ. Sci. Technol.* **2012**, *9*, 319–326. [CrossRef]
67. Amaku, J.F.; Ogundare, S.A.; Akpomie, K.G.; Ngwu, C.M.; Conradie, J. Sequestered uptake of chromium (VI) by *Irvingia gab-onensis* stem bark extract anchored silica gel. *Biomass Convers Biorefin.* **2021**, 1–13. [CrossRef]
68. Prajapati, A.K.; Mondal, M.K. Novel green strategy for CuO–ZnO–C nanocomposites fabrication using marigold (*Tagetes* spp.) flower petals extract with and without CTAB treatment for adsorption of Cr(VI) and Congo red dye. *J. Environ. Manag.* **2021**, *290*, 112615. [CrossRef]
69. Tran, H.N.; You, S.-J.; Hosseini-Bandegharaei, A.; Chao, H.-P. Mistakes and inconsistencies regarding adsorption of contaminants from aqueous solutions: A critical review. *Water Res.* **2017**, *120*, 88–116. [CrossRef]
70. Bollinger, J.-C.; Tran, H.N.; Lima, E.C. Comments on "removal of methylene blue dye from aqueous solution using citric acid modified apricot stone". *Chem. Eng. Commun.* **2022**, 1–6. [CrossRef]
71. Tran, H.N. Applying Linear Forms of Pseudo-Second-Order Kinetic Model for Feasibly Identifying Errors in the Initial Periods of Time-Dependent Adsorption Datasets. *Water* **2023**, *15*, 1231. [CrossRef]
72. Salvestrini, S.; Ambrosone, L.; Kopinke, F.-D. Some mistakes and misinterpretations in the analysis of thermodynamic adsorption data. *J. Mol. Liq.* **2022**, *352*, 118762. [CrossRef]
73. Murali, A.; Sarswat, P.K.; Free, M.L. Adsorption-coupled reduction mechanism in ZnO-Functionalized MWCNTs nanocom-posite for Cr (VI) removal and improved anti-photocorrosion for photocatalytic reduction. *J. Alloys Compd.* **2020**, *843*, 155835. [CrossRef]
74. Naseem, T.; Bibi, F.; Arif, S.; Waseem, M.; Haq, S.; Azra, M.N.; Liblik, T.; Zekker, I. Adsorption and Kinetics Studies of Cr (VI) by Graphene Oxide and Reduced Graphene Oxide-Zinc Oxide Nanocomposite. *Molecules* **2022**, *27*, 7152. [CrossRef]
75. Taha, A.; Da'Na, E.; Hassanin, H.A. Modified activated carbon loaded with bio-synthesized Ag/ZnO nanocomposite and its application for the removal of Cr (VI) ions from aqueous solution. *Surf. Interfaces* **2021**, *23*, 100828. [CrossRef]
76. Dinari, M.; Haghighi, A. Ultrasound-assisted synthesis of nanocomposites based on aromatic polyamide and modified ZnO nanoparticle for removal of toxic Cr(VI) from water. *Ultrason. Sonochem.* **2018**, *41*, 75–84. [CrossRef]

77. Sahoo, S.K.; Hota, G. Amine-Functionalized GO Decorated with ZnO-ZnFe2O4 Nanomaterials for Remediation of Cr(VI) from Water. *ACS Appl. Nano Mater.* **2019**, *2*, 983–996. [CrossRef]
78. Samuel, M.S.; Subramaniyan, V.; Bhattacharya, J.; Parthiban, C.; Chand, S.; Singh, N.P. A GO-CS@MOF [Zn(BDC)(DMF)] material for the adsorption of chromium(VI) ions from aqueous solution. *Compos. Part B Eng.* **2018**, *152*, 116–125. [CrossRef]

Disclaimer/Publisher's Note: The statements, opinions and data contained in all publications are solely those of the individual author(s) and contributor(s) and not of MDPI and/or the editor(s). MDPI and/or the editor(s) disclaim responsibility for any injury to people or property resulting from any ideas, methods, instructions or products referred to in the content.

Article

Algae and Hydrophytes as Potential Plants for Bioremediation of Heavy Metals from Industrial Wastewater

Naila Farid [1], Amin Ullah [1,*], Sara Khan [1], Sadia Butt [2], Amir Zeb Khan [3], Zobia Afsheen [1], Hamed A. El-Serehy [4], Humaira Yasmin [5,6], Tehreem Ayaz [7] and Qurban Ali [8,*]

1. Department of Health and Biological Sciences, Abasyn University Peshawar, Peshawar 25000, Pakistan
2. Department of Microbiology, Shaheed Benazirbhutoo Women University Peshawar, Peshawar 25000, Pakistan
3. Pak-Austria Fachhochschule: Institute of Applied Sciences and Technology (IAST), Haripur 22721, Pakistan; amerzeb1986@gmail.com
4. Department of Zoology, College of Science, King Saud University, Riyadh 11451, Saudi Arabia
5. Department of Infectious Disease, Faculty of Medicine, South Kensington Campus, Imperial College, London SW7 2BX, UK
6. Department of Biosciences, COMSATS University, Islamabad 45550, Pakistan
7. School of Resources and Environmental Engineering, East China University of Science and Technology, Shanghai 200237, China; tehreemayaz17301@yahoo.com
8. Department of Plant Breeding and Genetics, University of the Punjab, Lahore 54599, Pakistan
* Correspondence: aminbiotech7@gmail.com (A.U.); qurbanali.pbg@pu.edu.pk (Q.A.)

Abstract: Aquatic bodies contaminated by heavy metals (HMs) are one of the leading issues due to rapidly growing industries. The remediation of using algae and hydrophytes acts as an environmentally friendly and cost effective. This study was performed to investigate the pollution load, especially HMs, in the wastewater of the Gadoon Industrial Estate and to utilize the hydrophytes (*Typha latifolia* (TL) and *Eicchornia crassipes* (EI)) and algae (*Zygnema pectiantum* (ZP) and *Spyrogyra* species (SS)) as bioremediators. The wastewater was obtained and assessed for physiochemical parameters before treating with the selected species. The pot experiment was performed for 40 days. Then the wastewater samples and selected species were obtained from each pot to analyze the metal removal efficiency and assess for metal concentrations using atomic absorption spectrophotometry. The dissolved oxygen (DO; 114 mg/L), total suspended solids (TSS; 89.30 mg/L), electrical conductivity (EC; 6.35 mS/cm), chemical oxygen demand (COD) (236 mg/L), biological oxygen demand (BOD; 143 mg/L), and total dissolved solids (TDS; 559.67 mg/L), pH (6.85) were analyzed. The HMs were noted as Zn (5.73 mg/L) and Cu (7.13 mg/L). The wastewater was then treated with the species, and significant reductions were detected in physiochemical characteristics of the wastewater such as DO (13.15–62.20%), TSS (9.18–67.99%), EC (74.01–91.18%), COD (25.84–73.30%), BOD (21.67–73.42%), and TDS (14.02–95.93%). The hydrophytes and algae removed up to 82.19% of the Zn and 85.13% of the Cu from the wastewater. The study revealed that the hydrophytes and algae significantly decreased the HM levels in the wastewater ($p \leq 0.05$). The study found TL, EI, ZP, and SS as the best hyper accumulative species for Zn and Cu removal from wastewater. The HMs were removed in the order of Cu > Zn. The most efficient removal for Cu was found by *Typha latifolia* and Zn by *Zygnema pectiantum*. It was concluded that bioremediation is an environmentally friendly and cost-effective technique that can be used for the treatment of wastewater due to the efficiency of algae and hydrophytes species in terms of HM removal.

Keywords: bioremediation; hydrophytes; algae; heavy metals; wastewater

Citation: Farid, N.; Ullah, A.; Khan, S.; Butt, S.; Khan, A.Z.; Afsheen, Z.; El-Serehy, H.A.; Yasmin, H.; Ayaz, T.; Ali, Q. Algae and Hydrophytes as Potential Plants for Bioremediation of Heavy Metals from Industrial Wastewater. *Water* **2023**, *15*, 2142. https://doi.org/10.3390/w15122142

Academic Editor: Hai Nguyen Tran

Received: 7 April 2023
Revised: 29 May 2023
Accepted: 29 May 2023
Published: 6 June 2023

Copyright: © 2023 by the authors. Licensee MDPI, Basel, Switzerland. This article is an open access article distributed under the terms and conditions of the Creative Commons Attribution (CC BY) license (https://creativecommons.org/licenses/by/4.0/).

1. Introduction

Industrial wastewater (IWW) is considered one of the significant sources of aquatic contamination [1]. Due to the important use of heavy metals (HMs) in industrial processes, different poisonous HMs are widely used in many industrial activities such as mining, smelting, plating, fertilizer manufacturing, the textile, chemical, plastics, and pigments industries, etc. [2,3]. Heavy-metal-contaminated water is a serious issue due to the non-degradable nature of HMs and their tendency to accumulate in natural water bodies [4–6]. When the metals are discharged in the water bodies, they settle down and enter the aquatic bodies, leading to severe health problems and even causing the death of aquatic organisms [7].

The conventional methods used for metal removal from wastewater include electrochemical methods (electrocoagulation, electroflotation, and electrodeposition), coagulation–flocculation, floatation, membrane filtration, ion exchange, and chemical precipitation [8]. However, these methods have the drawbacks of maintenance costs, expensive operation, and secondary contamination due to the formation of toxic sludge. For landfill leachate treatment, constructed wetlands may be regarded as more environmentally friendly and sustainable solutions. Constructed wetlands have been used to remove pollution from various wastewater streams. In constructed wetlands, wastewater is remediated by physical (e.g., filtration and sedimentation) and chemical processes (e.g., adsorption and precipitation), as well as biological processes (e.g., uptake from the water column and root zone and microbial degradation) [9]. Constructed wetlands have been used to treat a wide range of storm water runoff, waste streams, landfill leachate, agricultural drainage including mine drainage, domestic wastewater, as well as industrial effluents [9,10]. Plants induce physicochemical and biological processes to remediate metals from wastewater effluents [11], which help in its treatment [12]. Hydrophytes such as Pistia stratiotes [13], Lemna gibba [14], and *Eichhornia crassipes* [12] are free-floating and known for pollution absorption particularly for HMs in polluted water. Similarly, other classes of emergent species such as *Typha latifolia* [15] have the capability of metal accumulation in higher concentration in shoots and roots [16]. According to the study of Khan et al. [17] regional species of hydrophytes are adapted to local climatic conditions and appropriate to use in constructed wetlands for the treatment of wastewater. So, one of the most effective and best substitutes for the conventional methods is bioremediation, which uses biological materials to convert toxic metals into less harmful substances [18,19]. Bioremediation is a cost-effective and environmentally friendly method in the revitalization of the environment [20,21]. There are two main approaches to bioremediation. Phytoremediation (plant-centered) is a technique that uses either genetically modified or raw plants to restore polluted water and land sources [22]. Some hydrophytes such as water cabbage (*Pistia stratiotes*), duckweed (*Lemna gibba*) [14], and water hyacinth (*Eichhornia crassipes*) [12] are floating hydrophytes famous for pollution absorption capability mostly for metals in IWW.

Phycoremediation is a valuable technique used to remediate metal-polluted water. The algae efficiently remove metals from wastewater [23,24]. In living algal cells, metal absorption occurs through binding with the intracellular ligands and the cell surface. The metal-bonded ligands accumulate further by active biological transport [25]. The metal removal mechanisms rely on various functional groups such as hydroxyls, amines, and carboxylates forming complexes with metal [26] and decreasing metal concentrations in the treated water. Earlier, studies have verified that algal species improve the wastewater quality and IWW by metal absorption [24,27,28]. Moreover, the microalgal biomass after bioremediation can be processed further to produce biochar, biodiesel, value-added products, and metal extraction, representing the process as environmentally sustainable and cost effective. Algae gained the attention of researchers as a potential applicant for biodiesel production among different feedstock. Microalgae cultivation can use non-agricultural land, suggesting that the land requirement for algae biodiesel production is low [29]. Algae can utilize CO_2 as a carbon source for oil (i.e., biodiesel) and biomass production. Algae have rapid growth potential and have oil contents up to 50% dry weight of biomass [30]. The waste biomass can be converted into biochar, which plays a crucial role in protecting and repairing the environment [31].

In Pakistan, it is alarming that most industries and cities lack facilities for wastewater treatment. Huge industrial effluents are being directly released to surface water, resulting in severe pollution. Due to organic loads and toxic materials such as HMs, the IWW forms a main source of aquatic pollution in Pakistan. It causes water to be unsuitable for drinking, irrigation, or any other use. HM pollution has become a severe risk to living organisms in the environment. It can cause different health issues such as lung insufficiency, nervous disorders, bone injury, cancer, teratogenic and embryotoxic effects, and hypertension in humans. To deal with this issue, a variety of methods can be utilized, but these methods are expensive and badly affect the environment. In developing countries such as Pakistan, these methods cannot be used to remove HMs. Therefore, it is expected that hydrophytes and algal species may contribute in reclaiming the wastewater discharged from the Gadoon Industrial Estate, Pakistan. This study aims to assess the physiochemical parameters of the wastewater collected from the Gadoon Industrial Estate and the removal efficiency of the HMs and to compare the HM removal efficiency of algae and hydrophytes.

2. Materials and Methods

2.1. Algae and Hydrophyte Collection and Transplantation

In the study, the algal species *spirogyra spp.* and *Zygnema pectinatum* were collected from the Khiyali River, Peshawar. The new buds of *Typha latifolia* (cattail) and *Eichhornia crassipes* (water hyacinth) were collected from the ponds in Sabi village, District Peshawar. The species were first washed with tap and distilled water. The algae were then observed and identified by the procedure adopted by [32]. To remove dust and sand, the selected species were washed thoroughly and acclimatized and transplanted into IWW for 40 days at room temperature (18–20 °C). Figure 1 shows the collection of hydrophytes and algae.

Figure 1. Collection sites of selected hydrophytes and algae.

2.2. Collection and Analysis of IWW

The IWW was obtained from the Gadoon Industrial Estate (34°7′8″ N 72°38′45″ E) using the procedure mentioned by Ayaz et al. [15]. The wastewater sample was obtained during the month of May with average temperatures ranging from 78 to 103 F° and an average rainfall of 3.3 inches. The Gadoon Industrial Estate has many different industrial units such as the steel, chemical, soap and oil, textile, marble, and ghee industries. The study area consists topographically of flood deposits and uneven natural and ground drains and contains widespread gravel deposits with shingle beds and clay and silt stratifications, generally flat in the west and south of the industrial estate. The solid waste substances were removed [32]. Then the IWW was analyzed for physiochemical characteristics such as temperature, TDS, electrical conductivity (EC), pH, TSS, DO, BOD, COD, and metals (Cu and Zn). The metals were then analyzed by atomic absorption spectrophotometry in the Central Resource Laboratory, University of Peshawar.

2.3. Experimental Design

The pot experiment was performed to observe the efficiency of metal (Cu and Zn) removal by individual hydrophytes and algal species (Figure 2). For this experiment, four clean pots were first rinsed using double-deionized distilled water and 10% HNO_3 (diluted nitric acid). The treatment pots for hydrophytes were then marked as TL (containing 50 g *Typha latifolia*) and EI (containing 50 g *Eichhornia crassipes*) and were provided with 20 L of IWW. Similarly, the algal pots were named as ZP (containing 15 g *Zygnema pectinatum*) and SS (containing 15 g *Spirogyra* spp.) provided with 5 L of wastewater sample for transplantation. The hydrophytes were transplanted in sediments containing IWW. All four pots were placed at room temperature (27 °C) under natural light/dark 14:10 conditions for 40 days. After experimentation, species and water samples were obtained from all containers and were analyzed for different physiochemical characteristics and metals (Cu and Zn).

Figure 2. Experimental pots at initial and final stage.

2.4. Sampling and Analysis of Sediment

For the transplantation of the hydrophytes, clean sediment was utilized in each pot (5 kg). Before experimentation, triplicate samples of the sediment were obtained and analyzed from TL and EI containers. The samples were observed for HMs (Zn and Cu) and various physiochemical characteristics. Temperature, EC, and pH were determined in the sediment samples (10 g) using an EC meter (InoLab level-1) and a pH Meter (Model: pH-208). Using the Australian standard procedure adopted by Ayaz et al. [15], the soil moisture content (MC) was determined, whereas the TOM (total organic matter) was observed using the method adopted by Walkley and Black [33]. On the ignition method, through weight loss, the TOC (total organic carbon) was calculated [34]. For the MC, TOM, and TOC, 50 g of sediment sample was used. In sediment, (0.5 g) metals were extracted using a wet digestion procedure [35] and quantified by atomic absorption spectrophotometry (AAS-700 PerkinElmer: Norwalk, CT, USA) in the Central Resource Laboratory, University of Peshawar.

Water samples were collected from the TL, EI, ZP, and SS containers at the end of the experiment. The samples were examined for physicochemical parameters and HMs (Zn and Cu). A quantity of 50 mL of water samples were taken, and the physicochemical characteristics such as EC, pH, and temperature were observed using the EC meter (InoLab-WTB GmbH; Weilheim, Germany) and the pH Meter (Model: pH-208), respectively. The TDS and TSS were determined for the water sample (10 mL) using the gravimetric method. The DO, BOD$_5$, and COD were observed for the water sample (200 mL titrametrically by standard procedure adopted by the American Public Health Association [36]). The samples were also examined using atomic absorption spectrophotometry for HMs (Zn and Cu).

2.5. Sampling, Preparation, and Analysis of Hydrophyte

After experimentation, samples of TL, EI, ZP, and SS were collected in triplicate and adequately labeled to observe the metal removal efficiency of selected species. The root and aerial parts of the TL and EI were processed separately for further analysis using standard methods [17]. The harvested species samples were washed adequately to remove dust, clay, sand, and particles. The samples were then dried at 70 °C in an oven and then ground and stored in polythene bags for wet digestion [35]. After wet digestion, the extracted samples were quantified for heavy metals using atomic absorption spectrophotometry.

2.6. Formula

2.6.1. Bioconcentration Factor

The bioconcentration factor (*BCF*) of algal and hydrophyte efficiency for metal accumulation from the wastewater samples was determined using Equation (1) [37,38].

$$BCF\ (\%) = \frac{C_{algae}}{C_{water}} \times 100 \qquad (1)$$

where C_{algae} refers to the metal concentration in the algae and C_{water} is the metal concentration in the water.

2.6.2. Translocation Factor

The translocation factor (TF) was used to measure the metals translocated from wastewater and accumulated in the hydrophyte's tissues using Equation (2) as below:

$$TCF(\%) = \frac{C_{aerial}}{C_{root}} \times 100 \qquad (2)$$

C_{aerial} refers to the metals accumulated in the aerial parts, and C_{root} refers to the metal concentrations in the root part.

2.6.3. Bioaccumulation Measurement

The bioaccumulation capacity (*q*) was determined using Equation (3) as given below [39].

$$q = \frac{C_i - C_f}{M} \times V \qquad (3)$$

C_i is the initial metal concentration, C_f is the final metal concentration, M is the amount of algal or hydrophyte dry biomass (g), and V is the water volume in (L).

2.6.4. Bioremoval Efficiency

The bioremoval efficiency (%) was measured using Equation (4) [24].

$$R = \frac{C_i - C_f}{C_i} \times 100 \qquad (4)$$

R refers to the removal percentage, C_i is the initial metal concentration in wastewater, and C_f is the final metal concentration in wastewater.

2.7. Statistical Analysis

The statistical analysis was carried out as follows: Microsoft Excel and the statistical package for social sciences (SPSS) was used. ANOVA and t-test were applied for the significance between the variables of the parameters.

3. Results and Discussion

3.1. Physicochemical Characteristics of Sediment Sample

Table 1 summarizes the physicochemical analysis of sediments. The results showed lower concentrations of Zn and Cu (5.30 and 19.20 mg/kg) below the maximum permissible limit set by Pak-EPA (Pakistan Environmental Protection Agency, 2008) before starting the experiment. However, the levels of Zn and Cu raised to 7.47 and 20.82 mg/kg in the pots, respectively, ensuring the sedimentation and presence of metals from the water samples. Mass balancing suggested that the HMs were primarily deposited in sediments and hydrophytes. This character was also observed in the research studies reported by Hadad et al. [40] and Mays and Edwards [41].

Table 1. Physiochemical parameters of sediment sample.

Parameters (n = 3)	Samples *	Mean ± SD
pH	Initial	8.09 ± 0.03
	Final	8.02 ± 0.03
EC (mS/cm)	Initial	341.5 ± 2.14
	Final	94.0 ± 2.85
MC (mg/m^3)	Initial	0.62 ± 0.03
	Final	0.85 ± 0.013
TOM (mg/kg)	Initial	3.24 ± 0.03
	Final	3.81 ± 0.03
TOC (mg/kg)	Initial	1.78 ± 0.013
	Final	2.39 ± 2.011
Zn (mg/kg)	Initial	5.30 ± 3.17
	Final	7.47 ± 2.70
Cu (mg/kg)	Initial	19.07 ± 4.01
	Final	20.82 ± 4.96

* Initial: samples at the initial stage of the experiment. Final: samples at the final stage of the experiment.

3.2. Characteristics and Pollution Load of IWW

The physiochemical parameters' values of IWW are given in Supplementary Materials Table S1. The IWW collected from all four containers, including TL (treatment container for *Typha latifolia*), EI (treatment container for *Eicchornia crassipes*), ZP (treatment container for *Zygnema pectinatum*), and SS (treatment container for *Spyrogyra* spp.), was observed before and after the treatment or experimentation for parameters such as EC, pH, TSS, BOD, temperature, TDS, DO, and COD.

The pH (6.85) observed for IWW was within the maximum permissible limit set by Pak-EPA. The EC, DO, BOD, COD, TSS, and TDS values determined for IWW were 6.35 mS/cm, 114 mg/L, 143 mg/L, 236 mg/L, 89.30 mg/L, and 559.67 mg/L. The TSS and TDS values were analyzed below the Pak-EPA limit (150 mg/L and 3500 mg/L, respectively). However, the COD and BOD values exceeded the limits (150 mg/L and 80 mg/L, respectively) as set by Pak-EPA. The concentrations of Zn (5.73 mg/L) and Cu (7.13 mg/L) in the IWW exceeded the maximum permissible limit of Pak-EPA.

Previously, similar findings (pH 7.6) were reported by Fito et al. [42], who carried out a study on the wastewater of the sugar industry. The EC results agreed with the results

(320 S/cm) of Asia and Akporhonor [43], who analyzed the physicochemical characteristics of IWW from the rubber industry. The present TSS value was primarily similar to the findings (43 mg/L) reported by Shamshad et al. [24]. Hossain et al. [44] investigated the physical and chemical parameters of the IWW discharged from various industries (Bangladesh) and observed similar findings (EC: 2.64 mS/cm). Similarly, the present values conformed with the EC (0.24–5.04 mS/cm), DO (341 mg/L), BOD (143 mg/L), and COD values of the IWW in a research study reported by Ayaz et al. [15], a phytoremediation study performed on the Hayatabad Industrial Estate. However, the results were different from the EC (149.1–881.3 mS/cm), TSS (2470 mg/L), and COD (1231 mg/L) values observed by Aniyikaiye et al. [45] in a study conducted on the IWW released from the Nigerian paint industries. The present results were much different from the DO (1.83 mg/L) and BOD (25 mg/L) values reported on the phycoremediation of the wastewater of the Hayatabad Industrial Estate performed by Khan et al. [46]. Similarly, Singh et al. [47] carried out a similar study on the textile industry of Ludhiana, India, using wastewater collected from different dyeing mills, and higher COD (3050 mg/L) and BOD values (790 mg/L) were reported. Similarly, the TSS finding (397.5 mg/L) of the research study of Abrha et al. [48] on the IWW obtained from Ethiopian beverage industries was much different from this study's findings. In another study performed on treatments for removing pharmaceutical wastes from wastewater, Badawy et al. [49] reported high TDS values. Higher values of TDS (7072 mg/L) in the textile industry wastewater have been reported by Paul et al. [50]. The differences in the findings can be accredited to the differences in sampling sites and study areas.

3.3. Effect of Hydrophytes and Algal Species on Physicochemical Parameters

The effects of hydrophytes and algae on various characteristics of the water samples are shown in the Supplementary Materials Table S1. The results revealed that *Typha latifolia* had significantly ($p < 0.05$) reduced EC (91.18%), TSS (50.94%), TDS (14.02%), DO (13.15%), BOD (21.67%), and COD (25.84%) during 40 days. In a recent study, the effect of *Eicchornia crassipes* was found lower on EC (77.48%) and TSS (9.18%) and higher on TDS (95.93%), DO (60.52%), BOD (73.42%), and COD (73.30%) than *Typha latifolia*. Similarly, the algal species *Zygnema pectinatum* had also significantly ($p < 0.05$) decreased the pollution of IWW such as EC (74.01%), TSS (63.04%), TDS (75.84%), DO (18.42%), BOD (38.46%), and COD (38.55%) during 40 days. In a recent study, the effect of *Spyrogyra* spp. was found to be lower on TDS (73.51%), BOD (29.37%), and COD (26.27%) and higher on EC (80.31%), TSS (67.99%), and DO (62.20%) than *Zygnema pectinatum* (Supplementary Materials Table S1). The results are consistent with the results reported by Khan et al. [46], the research conducted for the remediation of the IWW of the Hayatabad Industrial Estate by algae (*Oedogonium westi*, *Cladophora glomerata*, *Zygnema insigne*, and *Vaucheria debaryana*) who recorded a significant decrease in the COD (30.7%), EC (85.9%), TDS (79%), and BOD (52.4%). However, these findings were different from the findings reported by Sharma et al. [51] in their study on *Chlorella minutissima*, and significant reductions in the COD (80.5%), TDS (94.4%), and BOD (93%) of the IWW were detected. Similarly, the results of this study were inconsistent with the findings of Okpozu et al. [52], whose study reported on the bioremediation of cassava IWW by *Desmodes musarmatus* and detected a reduction in the COD by 92% and the BOD by 87%. The differences in the results could be linked to the wastewater characteristics, species of algae, and nutrients. Okpozu et al. [52] transplanted the algae in Bold's basal medium and hydrolyzed cassava wastewater.

3.4. Effects of Selected Species on Heavy Metals

3.4.1. Zinc (Zn)

The average or mean of the Zinc concentration at the initial stage of the IWW was recorded as 5.73 mg/L. The mean Zn concentrations observed in the water samples collected after 40 days of the experiment from TL, EI, ZP, and SS were 1.63, 2.17, 1.02, and 2.24 mg/L, respectively (Figure 3). The efficiency of the Zn removal was noted as ranging from

60.90 to 82.19%, where 71.55, 62.12, 82.19, and 60.90% was removed by TL, EI, ZP, and SS, respectively, as shown in Table S1. The t-test results showed that TL, EI, ZP, and SS significantly ($p \leq 0.05$) decreased the Zn level in the final stage water samples.

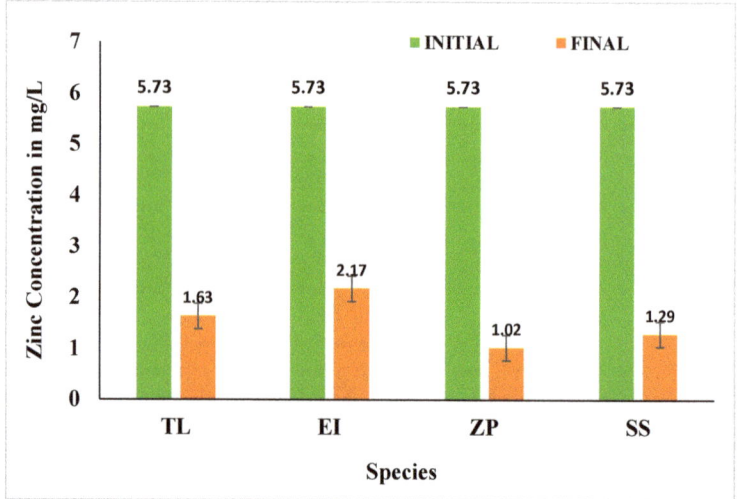

Figure 3. Concentration of zinc in water samples.

The Zn accumulation in the hydrophyte species differs in selected species, and the Zn level was observed to be the maximum in the underground parts of the hydrophyte species compared with the upper or aerial parts. The aerial parts of the TL and EI uptake were 71.23 and 33.61 mg/kg, respectively. The maximum uptake by the aerial tissues of the hydrophyte species was determined (71.23 mg/kg) in TL, and the minimum uptake (33.61 mg/kg) was determined in EI. The Zn uptakes by the root tissue of TL and EI were discovered to be 105.41 and 59.05 mg/kg, respectively (Figure 4). The Zn concentration recorded in the algal biomass after 40 days of experimentation is shown in Figure 5.

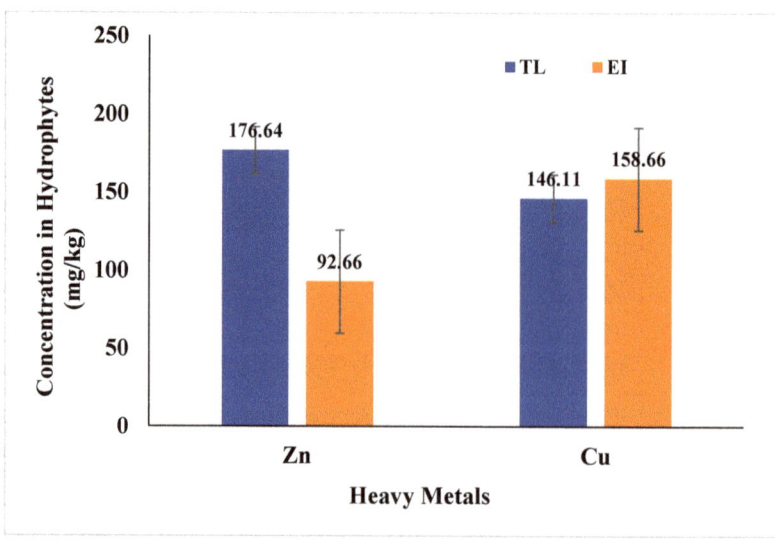

Figure 4. Heavy metal concentrations in hydrophytes.

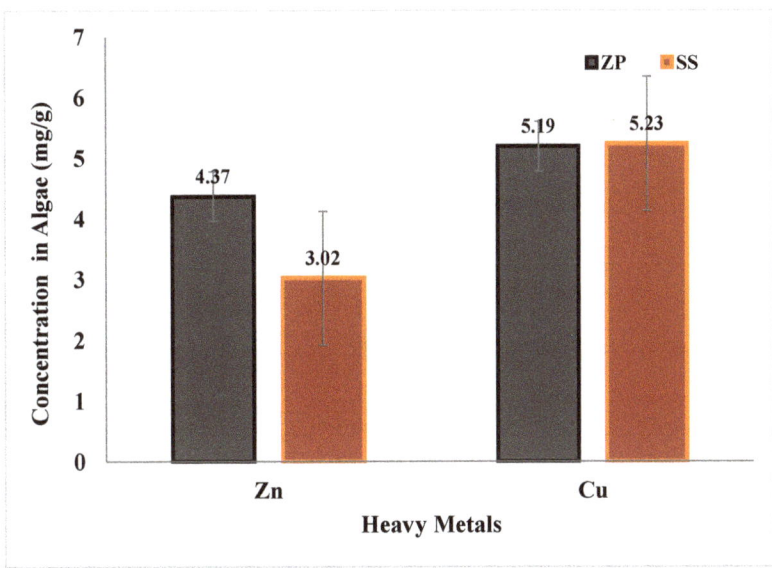

Figure 5. Heavy metal concentrations in algal biomass.

The translocation factors or bioconcentrations factor determined for TL, EI, ZP, and SS for Zn were 67.57, 56.91, 76.26, and 52.70%, respectively (Figure 6). The ZP was recorded for the highest and the SS was recorded for the lowest bioconcentration factor, as shown in Figure 5. The bioaccumulation capacities (q) of TL, EI, ZP, and SS for Zn were 1.64, 1.42, 1.57, and 1.16 mg/g. TL was observed as having the highest and SS was observed as having the lowest bioaccumulation capacity (Figure 7). The bioremoval efficiencies recorded for TL, EI, ZP, and SS were 71.55, 62.12, 82.19, and 60.90%, respectively. The highest Zn bioremoval efficiency was determined for ZP, and lowest was determined for SS (Supplementary Materials Table S1).

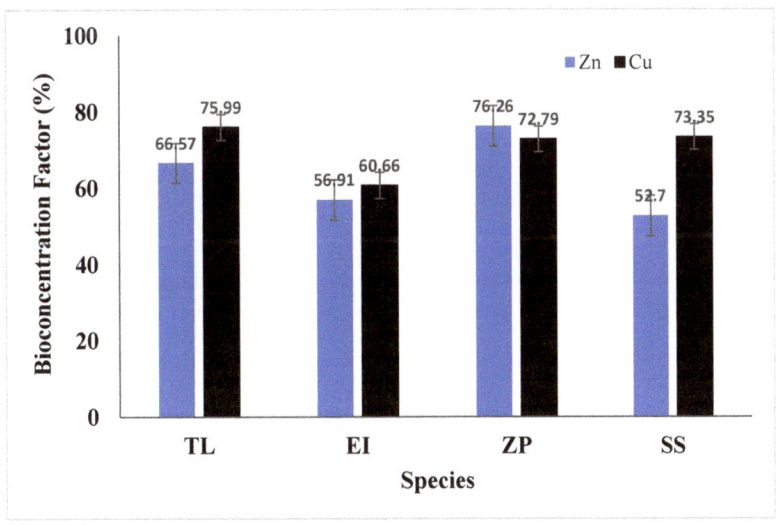

Figure 6. Bioconcentration factor (BCF) (%) or translocation factor of heavy metals in species.

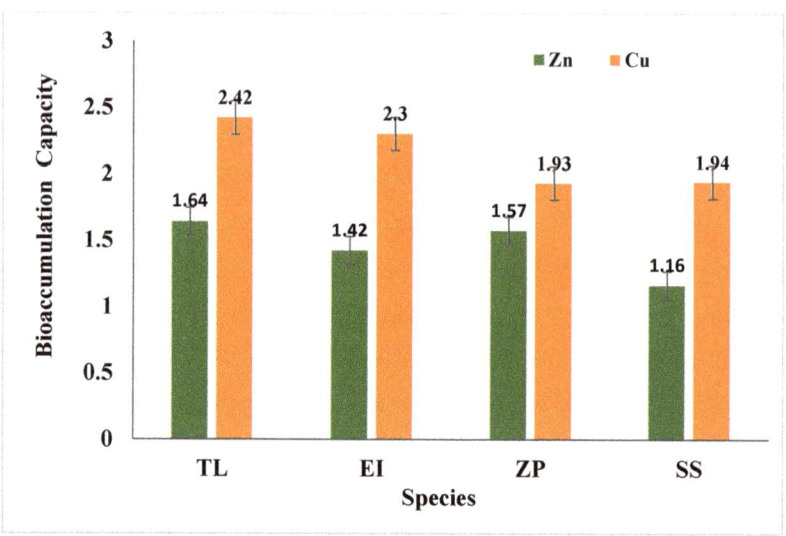

Figure 7. Bioaccumulation of heavy metals in hydrophytes and algae.

However, the algal bioaccumulation values did not support the results (0.745–1.286 mg/kg) of Khan et al. [46]. The bioaccumulation capacity variation may have been due to the different algal species used for bioremediation or the differences in metal concentrations in the IWW collected. The results obtained from the hydrophytes agreed with the results (98.8–99.3%) of the research study presented by Kanagy et al. [53]. Moreover, the findings were much higher than those obtained from the constructed wetland remediating IWW in the central Mediterranean [54]. The present results were much high than those in the literature (48.3%) reported by Khan et al. [17]. The t-test showed that the Zn level in the IWW samples was more reduced ($p \leq 0.05$) than the initial samples of TL, EI, ZP, and SS, suggesting the efficient removal of Zn by selected hydrophytes from the IWW.

The comparative study of the bioremoval efficiency, bioaccumulation capacity, and bioconcentration factor observed for hydrophyte treatments (TL and EC) revealed that *Typha latifolia* was the most effective species for the remediation of Zn as compared with *Eicchornia crassipes*. Our study's findings aligned with the literature presented by Ayaz et al. [15], revealing that *Typha latifolia* was more effective in metal removal from IWW than *Eicchornia crassipes*. Similarly, the comparative study of removal efficiency, bioaccumulation capacity, and bioconcentration factor observed for algal treatments by ZP and SS revealed that *Zygnema pectinatum* was the most effective species for Zn remediation as compared with *Spyrogyra* species.

3.4.2. Copper (Cu)

The mean of copper concentration in the initial stage IWW was 7.13 mg/L. The mean Cu concentrations observed in the water samples collected from TL, EI, ZP, and SS were 1.06, 1.37, 1.34, and 1.29 mg/L, respectively, as shown in Figure 8. The maximum Cu level was observed in EI samples, and the lowest value was observed in TL for the final point samples. Efficiencies of Cu removal were noted that ranged from 80.78 to 85.13%, where 85.13, 80.78, 81.20, and 81.90% were removed by TL, EI, ZP, and SS, respectively (Table S1). In a previous study conducted by Mane and Bhosle [55], higher percentages of bioremoval for Cu (89.6%) and Zn (81.53%) were observed by living *Spirogyra* sp. In other phycoremediation studies executed on wastewater, Uddin et al. [56] concluded that 92% of the Cu removal efficiency was observed for *Spirogyra* on ninety minutes of treatments. The t-test results showed that TL, EI, ZP, and SS significantly ($p \leq 0.05$) decreased the Cu level in the final stage water samples compared with the initial stage.

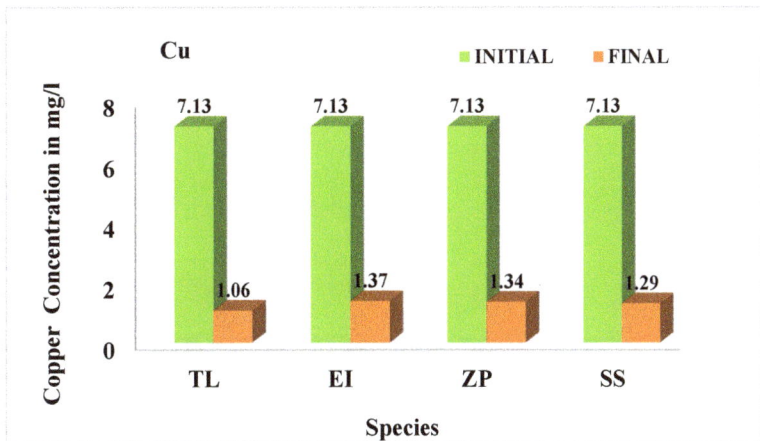

Figure 8. The Cu concentration in water.

The Cu bioaccumulation in selected hydrophytes species differed in various species, and the Cu level was observed to be maximum in the roots of the hydrophyte as compared with the aerial tissue. The upper or aerial parts of the TL and EI took up 63.09 and 59.91 mg/kg of Cu, respectively (Figure 3). The maximum uptake by the aerial parts (63.09 mg/kg) was determined in TL, and the minimum uptake (59.91 mg/kg) was determined in EI. The Cu absorption by the roots of TL and EI was determined to be 83.02 and 98.75 mg/kg, respectively. The lowest uptake by the root (83.02 mg/kg) was recorded in TL, and the highest absorption (98.75 mg/kg) was recorded in EI. The Cu uptakes detected in ZP and SS were 5.19 and 5.23 mg/g, respectively (Figure 5). The highest Cu uptake (5.23 mg/g) was observed in SS, and the lowest uptake (5.19 mg/g) was observed in ZP.

The study results indicated the highest uptake of Cu by EI, followed by TL. The analysis showed that Cu concentrations differed in various species. The Cu concentrations were observed to be higher in the roots as compared with the aerial parts of the hydrophytes. The findings of the study agreed with the literature on hydrophytes reported by Kumari and Tripathi [57], Victor et al. [58], and Galal et al. [59].

The translocation factor or bioconcentration factors determined for TL, EI, ZP, and SS for Cu were 75.99, 60.66, 72.79, and 73.35%, respectively. TL was recorded as having the highest and EI was recorded as having the lowest bioconcentration factor (Figure 6). Figure 7 shows the bioaccumulation capacities (q) of TL, EI, ZP, and SS for Cu, which were 2.42, 2.30, 1.93, and 1.94 mg/g. TL was observed as having the highest and ZP was observed as having the lowest bioaccumulation capacity (Figure 7). The bioremoval efficiencies recorded for TL, EI, ZP, and SS were 85.13, 80.78, 81.20, and 81.90%, respectively. The highest Cu bioremoval efficiency was determined for TL, and lowest was determined for EI (Supplementary Materials Table S1).

The comparative study of copper bioremoval efficiency, bioaccumulation capacity, and bioconcentration factors observed for hydrophyte treatment using TL and EI revealed that *Typha latifolia* was the most effective species for the remediation of Cu as compared with *Eicchornia crassipes*. Our study's findings agreed with the study presented by Ayaz et al. [15], showing that *Typha latifolia* more effectively removed Cu from IWW compared with *Eicchornia crassipes*. Similarly, comparing the copper bioremoval efficiency, bioaccumulation capacity, and bioconcentration factors observed for algal treatments revealed that *Spyrogyra* species were the most effective species for Cu remediation as compared with *Zygnema pectinatum*.

4. Conclusions

The pot experiments concluded that hydrophytes and algae have a key role in the bioremediation of IWW and efficiently reduce metal level in industrial wastewater (IWW). *Zygnema pectinatum* (ZP) had the best Zn removal efficiency, and *Typha Latifolia* was the best bioremediator for Cu. The species removed 85.13% of Cu and 82.19% of Zn. The results recommend that *Typha latifolia*, *Eicchornia crassipes*, *Zygnema pectinatum*, and *Spyrogyra* spp. are the best hyper accumulative species for Zn and Cu from IWW. The heavy metal (Zn and Cu) concentrations of IWW were significantly reduced ($p \leq 0.05$) after the treatments, which clearly showed the significance of hydrophytes and algae in the removal of metals. Moreover, the selected species survived in the conditions of stress triggered by the metal concentrations. Therefore, this property can be encouraging evidence for hydrophytes and algae to be utilized for remediation. So, the phytoremediation and phycoremediation are inexpensive, environmentally friendly treatment processes that can be utilized for the remediation of IWW contaminated with metals.

Supplementary Materials: The following supporting information can be downloaded at https://www.mdpi.com/article/10.3390/w15122142/s1: Table S1: Physiochemical parameters of wastewater samples.

Author Contributions: Conceptualization, N.F., A.U. and H.Y.; Methodology, A.U. and S.K.; Software, T.A., S.K., A.Z.K. and S.B.; Validation, N.F. and H.A.E.-S.; Formal analysis, A.U., Q.A. and S.B.; Investigation, A.U., A.Z.K., Z.A. and H.Y.; Resources, A.U., Z.A. and S.K.; Data curation, N.F. and A.U.; Writing—original draft, N.F. and S.K.; Writing—review & editing, Q.A.; Supervision, A.U.; Project administration, H.A.E.-S.; Funding acquisition, H.A.E.-S. All authors have read and agreed to the published version of the manuscript.

Funding: Researchers Supporting Project Number (RSP2023R19) of King Saud University, Riyadh, Saudi Arabia.

Data Availability Statement: The datasets generated during the current study are available from the corresponding author on reasonable request.

Acknowledgments: The authors would also like to extend their sincere appreciation to the Researchers Supporting Project Number (RSP2023R19), King Saud University, Riyadh, Saudi Arabia.

Conflicts of Interest: The authors declare no conflict of interest.

References

1. Goutam, S.P.; Saxena, G.; Singh, V.; Yadav, A.K.; Bharagava, R.N. Green synthesis of TiO$_2$ nanoparticles using leaf extract of *Jatropha curcas* L. for photocatalytic degradation of tannery wastewater. *Chem. Eng. J.* **2018**, *336*, 386–396. [CrossRef]
2. Thompson, C.O.; Ndukwe, A.O.; Asadu, C.O. Application of activated biomass waste as an adsorbent for the removal of lead (II) ion from wastewater. *Emerg. Contam.* **2020**, *6*, 259–267. [CrossRef]
3. Tokay, B.; Akpınar, I. A comparative study of heavy metals removal using agricultural waste biosorbents. *Bioresour. Technol. Rep.* **2021**, *15*, 100719. [CrossRef]
4. Chung, S.; Chung, J.; Lee, Y.W. Use of TiO$_2$ media by sputtering and peroxy gel method for wastewater reuse in the electronics industry. *Sci. Adv. Mater.* **2019**, *11*, 1547–1555. [CrossRef]
5. Wang, X.L.; Zhang, Y. Synthesis and Characterization of Zeolite A Obtained from Coal Gangue for the Adsorption of F–in Wastewater. *Sci. Adv. Mater.* **2019**, *11*, 277–282. [CrossRef]
6. Yuan, Y.; Niu, B.; Yu, Q.; Guo, X.; Guo, Z.; Wen, J.; Wang, N. Photoinduced multiple effects to enhance uranium extraction from natural seawater by black phosphorus nanosheets. *Angew. Chem.* **2020**, *132*, 1236–1243. [CrossRef]
7. Sheela, T.; Sadasivam, S.K. Dye degradation potential and its degradative enzymes synthesis of Bacillus cereus SKB12 isolated from a textile industrial effluent. *J. Appl. Biol. Biotechnol.* **2020**, *8*, 42–46.
8. Saleh, T.A.; Mustaqeem, M.; Khaled, M. Water treatment technologies in removing heavy metal ions from wastewater: A review. *Environ. Nanotechnol. Monit. Manag.* **2022**, *17*, 100617. [CrossRef]
9. Bakhshoodeh, R.; Alavi, N.; Oldham, C.; Santos, R.M.; Babaei, A.A.; Vymazal, J.; Paydary, P. Constructed wetlands for landfill leachate treatment: A review. *Ecol. Eng.* **2020**, *146*, 105725. [CrossRef]
10. Yin, T.; Chen, H.; Reinhard, M.; Yi, X.; He, Y.; Gin, K.Y.H. Perfluoroalkyl and polyfluoroalkyl substances removal in a full-scale tropical constructed wetland system treating landfill leachate. *Water Res.* **2017**, *125*, 418–426. [CrossRef] [PubMed]
11. DalCorso, G.; Fasani, E.; Manara, A.; Visioli, G.; Furini, A. Heavy metal pollutions: State of the art and innovation in phytoremediation. *Int. J. Mol. Sci.* **2019**, *20*, 3412. [CrossRef] [PubMed]

12. Khan, H.N.; Faisal, M. Phytoremediation of industrial wastewater by hydrophytes. *Phytoremediation Manag. Environ. Contam.* **2018**, *6*, 179–200. [CrossRef]
13. Prajapati, S.K.; Meravi, N.; Singh, S. Phytoremediation of Chromium and Cobalt using Pistia stratiotes: A sustainable approach. *Proc. Int. Acad. Ecol. Environ. Sci.* **2012**, *2*, 136.
14. Demim, S.; Drouiche, N.; Aouabed, A.; Benayad, T.; Couderchet, M.; Semsari, S. Study of heavy metal removal from heavy metal mixture using the CCD method. *J. Ind. Eng. Chem.* **2014**, *20*, 512–520. [CrossRef]
15. Ayaz, T.; Khan, S.; Khan, A.Z.; Lei, M.; Alam, M. Remediation of industrial wastewater using four hydrophyte species: A comparison of individual (pot experiments) and mix plants (constructed wetland). *J. Environ. Manag.* **2020**, *255*, 109833. [CrossRef]
16. Rezania, S.; Taib, S.M.; Din, M.F.M.; Dahalan, F.A.; Kamyab, H. Comprehensive review on phytotechnology: Heavy metals removal by diverse aquatic plants species from wastewater. *J. Hazard. Mater.* **2012**, *318*, 587–599. [CrossRef]
17. Khan, S.; Ahmad, I.; Shah, M.T.; Rehman, S.; Khaliq, A. Use of constructed wetland for the removal of heavy metals from industrial wastewater. *J. Environ. Manag.* **2009**, *90*, 3451–3457. [CrossRef]
18. Ndeddy Aka, R.J.; Babalola, O.O. Effect of bacterial inoculation of strains of Pseudomonas aeruginosa, Alcaligenesfeacalis and Bacillus subtilis on germination, growth and heavy metal (Cd, Cr, and Ni) uptake of Brassica juncea. *Int. J. Phytoremediat.* **2016**, *18*, 200–209. [CrossRef]
19. Akhtar, F.Z.; Archana, K.M.; Krishnaswamy, V.G.; Rajagopal, R. Remediation of heavy metals (Cr, Zn) using physical, chemical and biological methods: A novel approach. *SN Appl. Sci.* **2020**, *2*, 1–14. [CrossRef]
20. Katarzyna, H.; Baum, C. Application of microorganisms in bioremediation of environment from heavy metals. In *Environmental Deterioration and Human Health*; Springer: Dordrecht, The Netherlands, 2014; pp. 215–227. [CrossRef]
21. Sharma, S.; Malaviya, P. Bioremediation of tannery wastewater by Aspergillus flavus SPFT2. *Int. J. Curr. Microbiol. Appl. Sci.* **2016**, *5*, 137–143. [CrossRef]
22. Parmar, S.; Singh, V. Phytoremediation approaches for heavy metal pollution: A review. *J. Plant Sci. Res.* **2015**, *2*, 1–8.
23. Ajayan, K.V.; Selvaraju, M.; Unnikannan, P.; Sruthi, P. Phycoremediation of tannery wastewater using microalgae Scenedesmus species. *Int. J. Phytoremediation* **2015**, *17*, 907–916. [CrossRef]
24. Shamshad, I.; Khan, S.; Waqas, M.; Ahmad, N.; Ur-Rehman, K.; Khan, K. Removal and bioaccumulation of heavy metals from aqueous solutions using freshwater algae. *Water Sci. Technol.* **2022**, *71*, 38–44. [CrossRef]
25. Kiran, B.; Thanasekaran, K. Copper biosorption on Lyngbyaputealis: Application of response surface methodology (RSM). *Int. Biodeterior. Biodegrad.* **2011**, *65*, 840–845. [CrossRef]
26. Piotrowska-Niczyporuk, A.; Bajguz, A.; Talarek, M.; Bralska, M.; Zambrzycka, E. The effect of lead on the growth, content of primary metabolites, and antioxidant response of green alga *Acutodesmus obliquus* (Chlorophyceae). *Environ. Sci. Pollut. Res.* **2015**, *22*, 19112–19123. [CrossRef]
27. Li, H.; Ye, Z.H.; Wei, Z.J.; Wong, M.H. Root porosity and radial oxygen loss related to arsenic tolerance and uptake in wetland plants. *Environ. Pollut.* **2011**, *159*, 30–37. [CrossRef] [PubMed]
28. Liang, Z.; Liu, Y.; Ge, F.; Liu, N.; Wong, M. A pH-dependent enhancement effect of co-cultured Bacillus licheniformis on nutrient removal by Chlorella vulgaris. *Ecol. Eng.* **2015**, *75*, 258–263. [CrossRef]
29. Li, G.; Zhang, J.; Li, H.; Hu, R.; Yao, X.; Liu, Y.; Lyu, T. Towards high-quality biodiesel production from microalgae using original and anaerobically-digested livestock wastewater. *Chemosphere* **2021**, *273*, 128578. [CrossRef]
30. Chen, J.; Li, J.; Dong, W.; Zhang, X.; Tyagi, R.D.; Drogui, P.; Surampalli, R.Y. The potential of microalgae in biodiesel production. *Renew. Sustain. Energy Rev.* **2018**, *90*, 336–346. [CrossRef]
31. Nie, J.; Sun, Y.; Zhou, Y.; Kumar, M.; Usman, M.; Li, J.; Tsang, D.C. Bioremediation of water containing pesticides by microalgae: Mechanisms, methods, and prospects for future research. *Sci. Total Environ.* **2020**, *707*, 136080. [CrossRef]
32. Shamshad, I.; Khan, S.; Waqas, M.; Asma, M.; Nawab, J.; Gul, N.; Li, G. Heavy metal uptake capacity of fresh water algae (*Oedogonium westti*) from aqueous solution: A mesocosm research. *Int. J. Phytoremediat.* **2016**, *18*, 393–398. [CrossRef] [PubMed]
33. Walkley, A.; Black, I.A. An examination of the Degtjareff method for determining soil organic matter 2016, and a proposed modification of the chromic acid titration method. *Soil Sci.* **1934**, *37*, 29–38. [CrossRef]
34. Dean, W.E. Determination of carbonate and organic matter in calcareous sediments and sedimentary rocks by loss on ignition; comparison with other methods. *J. Sediment. Res.* **1974**, *44*, 242–248. [CrossRef]
35. Khan, S.; Rehman, S.; Khan, A.Z.; Khan, M.A.; Shah, M.T. Soil and vegetables enrichment with heavy metals from geological sources in Gilgit, northern Pakistan. *Ecotoxicol. Environ. Saf.* **2010**, *73*, 1820–1827. [CrossRef]
36. American Public Health Association. *APHA Compendium of Methods for the Microbiological Examination*, 3rd ed.; American Public Health Association: Washington, DC, USA, 1992; pp. 105–119, 325–367, 371–415, 451–469, 637–658.
37. Basílico, G.; Faggi, A.; de Cabo, L. Tolerance to metals in two species of fabaceae grown in riverbank sediments polluted with chromium 2020, copper, and lead. In *Phytoremediation*; Springer: Cham, Switzerland, 2018; pp. 169–178. [CrossRef]
38. Wilson, B.; Pyatt, F.B. Heavy metal bioaccumulation by the important food plant, *Olea europaea* L.; in an ancient metalliferous polluted area of Cyprus. *Bull. Environ. Contam. Toxicol.* **2007**, *78*, 390–394. [CrossRef]
39. Flouty, R.; Estephane, G. Bioaccumulation and biosorption of copper and lead by a unicellular algae *Chlamydomonas reinhardtii* in single and binary metal systems: A comparative study. *J. Environ. Manag.* **2012**, *111*, 106–114. [CrossRef]
40. Hadad, H.R.; Maine, M.A.; Bonetto, C.A. Macrophyte growth in a pilot-scale constructed wetland for industrial wastewater treatment. *Chemosphere* **2006**, *63*, 1744–1753. [CrossRef]

41. Mays, P.A.; Edwards, G.S. Comparison of heavy metal accumulation in a natural wetland and constructed wetlands receiving acid mine drainage. *Ecol. Eng.* **2001**, *16*, 487–500. [CrossRef]
42. Fito, J.; Tefera, N.; Kloos, H.; Van Hulle, S.W. Physicochemical properties of the sugar industry and ethanol distillery wastewater and their impact on the environment. *Sugar Tech.* **2019**, *21*, 265–277. [CrossRef]
43. Asia, I.O.; Akporhonor, E.E. Characterization and physicochemical treatment of wastewater from rubber processing factory. *Int. J. Phys. Sci.* **2007**, *2*, 061–067.
44. Hossain, M.B.; Islam, M.N.; Alam, M.S.; Hossen, M.Z. Industrialisation scenario at Sreepur of Gazipur, Bangladesh and physico-chemical properties of wastewater discharged from industries. *Asian J. Environ. Ecol.* **2019**, 1–14. [CrossRef]
45. Aniyikaiye, T.E.; Oluseyi, T.; Odiyo, J.O.; Edokpayi, J.N. Physico-chemical analysis of wastewater discharge from selected paint industries in Lagos, Nigeria. *Int. J. Environ. Res. Public Health* **2019**, *16*, 1235. [CrossRef] [PubMed]
46. Khan, S.; Shamshad, I.; Waqas, M.; Nawab, J.; Ming, L. Remediating industrial wastewater containing potentially toxic elements with four freshwater algae. *Ecol. Eng.* **2017**, *102*, 536–541. [CrossRef]
47. Singh, D.; Singh, V.; Agnihotri, A.K. Study of textile effluent in and around Ludhiana district in Punjab, India. *Int. J. Environ. Sci.* **2013**, *3*, 1271.
48. Abrha, B.H.; Chen, Y. Analysis of physico-chemical characteristics of effluents from beverage industry in Ethiopia. *J. Geosci. Environ. Prot.* **2017**, *5*, 172–182. [CrossRef]
49. Badawy, M.I.; Wahaab, R.A.; El-Kalliny, A.S. Fenton-biological treatment processes for the removal of some pharmaceuticals from industrial wastewater. *J. Hazard. Mater.* **2009**, *167*, 567–574. [CrossRef]
50. Paul, S.A.; Chavan, S.K.; Khambe, S.D. Studies on characterization of textile industrial waste water in Solapur city. *Int. J. Chem. Sci.* **2012**, *10*, 635–642.
51. Sharma, G.K.; Khan, S.A.; Shrivastava, M.; Gupta, N.; Kumar, S.; Malav, L.C.; Dubey, S.K. Bioremediation of sewage wastewater through microalgae (*Chlorella minutissima*). *Indian J. Agric. Sci.* **2020**, *90*, 2024–2028. [CrossRef]
52. Okpozu, O.O.; Ogbonna, I.O.; Ikwebe, J.; Ogbonna, J.C. Phycoremediation of cassava wastewater by *Desmodesmus armatus* and the concomitant accumulation of lipids for biodiesel production. *Bioresour. Technol. Rep.* **2019**, *7*, 100255. [CrossRef]
53. Kanagy, L.E.; Johnson, B.M.; Castle, J.W.; Rodgers, J.H., Jr. Design and performance of a pilot-scale constructed wetland treatment system for natural gas storage produced water. *Bioresour. Technol.* **2008**, *99*, 1877–1885. [CrossRef]
54. Terzakis, S.; Fountoulakis, M.S.; Georgaki, I.; Albantakis, D.; Sabathianakis, I.; Karathanasis, A.D.; Manios, T. Constructed wetlands treating highway runoff in the central Mediterranean region. *Chemosphere* **2008**, *72*, 141–149. [CrossRef]
55. Mane, P.C.; Bhosle, A.B. Bioremoval of some metals by living algae *Spirogyra* sp. and *Spirullina* sp. from aqueous solution. *Int. J. Environ. Resour.* **2012**, *6*, 571–576.
56. Uddin, A.; Lall, A.M.; Rao, K.P.; John, S.A.; Chattree, A. Phycoremediation of Pb, Cd, and of Cu by *Spirogyra cummins* from wastewater. *J. Entomol. Zool. Stud.* **2019**, *7*, 1042–1046.
57. Kumari, M.; Tripathi, B.D. Efficiency of *Phragmites australis* and *Typha latifolia* for heavy metal removal from wastewater. *Ecotoxicol. Environ. Saf.* **2015**, *112*, 80–86. [CrossRef] [PubMed]
58. Victor, K.K.; Ladji, M.; Adjiri, A.O.; Cyrille, Y.D.A.; Sanogo, T.A. Bioaccumulation of heavy metals from wastewaters (Pb, Zn, Cd, Cu and Cr) in water hyacinth (*Eichhornia crassipes*) and water lettuce (*Pistia stratiotes*). *Int. J. Chem. Tech. Res.* **2016**, *9*, 189–195.
59. Galal, T.M.; Eid, E.M.; Dakhil, M.A.; Hassan, L.M. Bioaccumulation and rhizofiltration potential of *Pistia stratiotes* L. for mitigating water pollution in the Egyptian wetlands. *Int. J. Phytoremediation* **2018**, *20*, 440–447. [CrossRef] [PubMed]

Disclaimer/Publisher's Note: The statements, opinions and data contained in all publications are solely those of the individual author(s) and contributor(s) and not of MDPI and/or the editor(s). MDPI and/or the editor(s) disclaim responsibility for any injury to people or property resulting from any ideas, methods, instructions or products referred to in the content.

Article

Adsorption-Reduction of Cr(VI) with Magnetic Fe-C-N Composites

Xu Liu [1,2,†], Huilai Liu [1,2,†], Kangping Cui [1], Zhengliang Dai [3], Bei Wang [3], Rohan Weerasooriya [2,4] and Xing Chen [1,2,*]

1. Key Laboratory of Nanominerals and Pollution Control of Higher Education Institutes, School of Resources and Environmental Engineering, Hefei University of Technology, Hefei 230009, China; akira0519@sina.com (X.L.); liuhuilai1996@163.com (H.L.); cuikangping@hfut.edu.cn (K.C.)
2. Key Laboratory of Aerospace Structural Parts Forming Technology and Equipment of Anhui Province, Institute of Industry and Equipment Technology, Hefei University of Technology, Hefei 230009, China; rohan.we@nifs.ac.lk
3. Anqing Changhong Chemical Co., Ltd., Anqing 246002, China; zl_dai@163.com (Z.D.); 13855694589@163.com (B.W.)
4. National Centre for Water Quality Research, National Institute of Fundamental Studies, Hantana, Kandy 20000, Sri Lanka
* Correspondence: xingchen@hfut.edu.cn; Tel./Fax: +86-551-62902634
† These authors contributed equally to this work.

Citation: Liu, X.; Liu, H.; Cui, K.; Dai, Z.; Wang, B.; Weerasooriya, R.; Chen, X. Adsorption-Reduction of Cr(VI) with Magnetic Fe-C-N Composites. Water 2023, 15, 2290. https://doi.org/10.3390/w15122290

Academic Editor: Hai Nguyen Tran

Received: 9 May 2023
Revised: 9 June 2023
Accepted: 12 June 2023
Published: 19 June 2023

Copyright: © 2023 by the authors. Licensee MDPI, Basel, Switzerland. This article is an open access article distributed under the terms and conditions of the Creative Commons Attribution (CC BY) license (https://creativecommons.org/licenses/by/4.0/).

Abstract: In this study, the iron-based carbon composite (hereafter FCN-x, x = 0, 400, 500, and 600 calcination) was synthesized by a simple high-temperature pyrolysis method using iron-containing sludge coagulant generated from wastewater treatment settling ponds in chemical plants. The FCN-x was used for the adsorptive reduction of aqueous phase Cr(VI) effectively. The FCN-x was characterized by scanning electron microscopy (SEM), X-ray diffraction (XRD), Fourier infrared spectrometer (FT-IR), X-ray photoelectron spectrometer (XPS), and Brunauer-Emmett-Teller theory (BET). FCN-x adsorption of Cr(VI) was examined in batch experiments using CrO_4^{2-} as a function of physicochemical parameters. The chemical kinetics of Cr(VI) adsorption by FCN-500 were modeled by 1st and 2nd order empirical pseudo kinetics. Based on these experiments, FCN-500 has been selected for further studies on Cr(VI) adsorptive reduction. The maximum Cr(VI) adsorption by FCN-500 was 52.63 mg/g showing the highest removal efficiency. The Cr(VI) adsorption by the FCN-500 was quantified by the Langmuir isotherm. XPS result confirmed the reduction of Cr(VI) to Cr(III) by the FCN-500. The iron-based carbon composites have high reusability and application potential in water treatment. The electroplating wastewater with 117 mg/L Cr(VI) was treated with FCN-500, and 99.93% Cr(VI) was removed within 120 min, which is lower than the national chromium emission standard of the People's Republic of China. This work illustrates the value-added role of sludge generated from dye chemical plants to ensure environmental sustainability.

Keywords: magnetic Fe-C-N composite; chromium; adsorption; reduction; coagulation sludge; pyrolyzation

1. Introduction

The exponential growth of industrial production plants with a concomitant discharge of hazardous wastes into the environment has required urgent attention by regulatory agencies. Electroplating, tanning, painting, printing, textiles, and other industries produce voluminous chromium-laden wastewater discharges [1–3]. As a heavy metal, chromium will cause irreversible damage to human health and the ecological environment [4,5]. In water, mainly trivalent and hexavalent chromium states are common [6,7]. Because Cr(VI) has higher mobility than Cr(III), it is considered to have strong toxicity [8]. Chromium(VI) compounds are potent human carcinogens and are highly genotoxic in bacterial and mammalian cells [9]. Several methods have been developed to remove Cr(VI) in water-based adsorption, biological processes, membrane separation, and advanced water treatment [10–12]. Among them, methods based on adsorption have been widely used due to their low cost, convenient sampling, high removal efficiency, and easy implementation steps [13,14].

The common adsorbents used for chromium ions in water treatment are carbon-based materials, metal-organic frameworks, biomasses, metal oxides, conductive polymers, nanocomposites, and magnetic nanoparticles [5,15–18]. Among them, methods based on magnetic nanoparticles are attractive because of their simple preparation, low cost, and easy separation and recovery [19–22]. Magnetic Fe_3O_4 without any modification has been used to purify wastewater containing chromium (VI) [23]. However, the bare Fe_3O_4 is unstable and dissolved in complex environments, thus reducing surface-active sites [24,25]. Different carriers are proposed to fix the bare magnetic Fe_3O_4 to enhance its stability [26]. Wang et al. (2021) reported sludge as a carrier for Fe_3O_4 to produce sludge based magnetic Fe_3O_4 microspheres. The sludge based magnetic Fe_3O_4 microspheres showed a 118 mg/g maximum adsorption capacity at 323 K. Besides, sludge-based Fe_3O_4 microspheres are environmentally benign and possesses excellent capabilities for magnetic separation and reduction of Cr(VI) → Cr(III) [27]. Zhao et al. (2016) synthesized magnetic nanocomposites by a single-step reaction using GO, diethylenetriamine, and Fe_3O_4. AMGO showed efficient adsorption of Cr(VI) and the maximum adsorption amount of Cr(VI) was 123.4 mg/g [28]. Although magnetic nanoparticles have achieved excellent results in Cr(VI) adsorption, the high cost of magnetic nanomaterials often limits their use in environmental remediation, thus requiring more effective and economically feasible production of magnetic adsorbents [29].

Dye chemical wastewater which contains a variety of organic compounds, such as azo, indigo, anthraquinone, aromatic methane, etc., is very complicated [30]. In the past few decades, iron-based chemicals have been widely used as effective coagulants for the purification of dye chemical wastewater and as catalysts for advanced oxidation processes, resulting in the generation of large amounts of sludge containing iron, anthraquinones, benzenes, naphthalenes, and other substances [31,32]. If this sludge is directly discharged or disposed of as waste, it will not only easily cause secondary pollution to the environment, but also be a waste of resources. In this study, the iron-containing coagulant sludge produced in industrial wastewater of a dye chemical plant was calcined at different temperatures under inert conditions. The iron-based carbon material was obtained, which was applied in the Cr(VI) removal experiment. SEM, FTIR, XPS, XRD, and so on were used to characterize the surface structure, morphology, and chemical composition of iron-based carbon materials. The Cr(VI) adsorption on FCN-x was examined as a function of pH, initial adsorbent concentration, equilibrium time, and adsorbate loadings. Besides, an Cr(VI) adsorption mechanism on FCN-500 was developed by coupling kinetic and equilibrium data and spectroscopic measurements. Finally, FCN-500 was used to treat chromium-containing wastewater from the electroplating wastewater plant with successful results.

2. Materials and Methods

2.1. Chemicals and Reagents

Analytical grade sulfuric acid (H_2SO_4), sodium hydroxide (NaOH), hydrochloric acid (HCl), potassium dichromate ($K_2Cr_2O_7$), phosphoric acid (H_3PO_4), diphenylcarbazide ($C_{13}H_{14}N_{14}O$), acetone (C_3H_6O) were obtained from Sigma-Aldrich (Burlington, MA, USA) and used as received. Iron-containing coagulant sludge was collected from the sedimentation tank of Changhong Chemical Co., Ltd. (Anqing, China).

2.2. Preparation of Adsorbent

The iron-bearing sludge came from a chemical plant wastewater, treatment sedimentation pond in Anqing City. Firstly, the iron-containing coagulant sludge was placed in an oven at 80 °C before being ground into a powder. The weighted amount of dried composites was calcined in a tubular furnace, heated to 400 °C, 500 °C, and 600 °C at a rate of 2.4 °C/min under nitrogen purging, and kept at these conditions for three hours (FCN-x x = 0, 200, 400, 500, and 600 °C). Then the FCN-x composites were cooled to ambient temperature in the furnace and used in the experiments given below.

2.3. Materials Characterization and Other Elemental Analysis

An X-ray diffractometer (PANalytical X'Pert Pro, Almelo, The Netherlands) was used to analyze the crystalline structure of the FCN-x composites using 2θ = 5° to 80° scan range (40 kV and 30 mA applied current). Thermal field emission scanning electron microscopy (FE-SEM; SU-8020-Hitachi, Tokyo, Japan) and energy dispersive X-ray spectrometer (EDS) were used to study the morphology, microstructure, and elemental composition of the FCN-x composites. X-ray photoelectron spectroscopy (XPS; model number Shimadzu, Kyoto, Japan) was used to determine the atomic composition and valence states of Fe, C, N, O, and Cr of the samples after Cr(VI) adsorption by FCN-500. All XPS data analyses were carried out with dedicated data fitting code for processing code (Peak Fit V4.12, Systat Software, Inc., Chicago, IL, USA). The FCN-x composites were examined under Fourier transform infrared spectrometer to probe surface reactivities (FT-IR; Thermo Nicolet, Waltham, MA, USA). The thermogravimetric analysis (TGA) curve of sludge was collected on a TGA8000 analyzer in high purity nitrogen from PE. The BET-specific surface area measurements of FCN-x were carried out by Brunauer-Emmett-Teller (BET; Autosorb—IQ3, Houston, TX, USA) method under liquid nitrogen temperature. The magnetic properties of the FCN-x composites were measured by a vibrating sample magnetometer (VSM; 7307, Lake Shore Cryotronics, Inc., Westerville, OH, USA). The content of Cr(VI) in solution was analyzed by laser-plasma mass spectrometry (LA-ICP-MS, Coherent Inc., Santa Clara, CA, USA). An X-ray fluorescence spectrometer (XRF; model number Shimadzu, Kyoto, Japan) was used to determine the chemical composition of the FCN-x.

2.4. Batch Adsorption Experiments

A series of batch experiments were carried out to determine the efficiency of FCN-x and the effects of pH, contact time, and adsorbate/adsorbent loading on Cr(VI) adsorption by FCN-x. A typical batch solution was prepared as outlined below: A 0.04 g FCN-x (0.4 g/L FCN-x) was mixed with 50 mL of a 20 mg/L Cr(VI) solution, and the final volume was adjusted to 100 mL with distilled water. The effect of Cr(VI) adsorption by FCN-x as a function of solution pH (pH 2 to 11), equilibration time (t 10 to 120 min), adsorbate dosage (0.2 to 0.5 g/L), and adsorbent loading (10 mg/L to 50 mg/L) was examined using the batch solution. At the end of a specific time of the reaction, the suspensions were filtered using 0.45 μm membrane filters, and the supernatant Cr(VI) concentration was determined UV spectroscopically at 540 nm (Cary-5000, Agilent, Santa Clara, CA, USA). The adsorption efficiency (H%) and amount of adsorbate are calculated as:

$$H = \frac{(c_0 - c_t)}{c_0} \times 100\% \quad (1)$$

$$q_t = \frac{(c_0 - c_t) \times v}{m} \quad (2)$$

where c_0 (mg/L) and c_t (mg/L) are the concentrations of Cr(VI) before and after the reaction, respectively.

To investigate the mechanism of the adsorption/desorption processes, kinetic adsorption experiments were performed with 40 mg/L of Cr(VI) at 35 °C. Samples were taken at 20 min intervals to determine the adsorption capacity.

2.5. Adsorption Regeneration Experiments

The iron-based carbon material in the recovered solution was mixed with an excess of 1 M NaOH and stirred for 24 h, then filtered and adjusted to pH~7.00 with the distilled water. Then the substrate was dried in an oven at 80 °C for 4 h. The dried material was used to carry out the next Cr(VI) adsorption experiment, as detailed in Section 2.4. The same process was repeated five times.

2.6. Treatment of Wastewater in the Electroplating Industry

The wastewater received from an electroplating wastewater plant was filtered to remove suspended solids. The pH of the waste solution is then adjusted to about 2–3. The Cr(VI) adsorption experiment at a specific pH value was carried out, as discussed in Section 2.4.

3. Results and Discussion

3.1. Characteristics of the Materials

Figure 1 shows the scanning electron microscopic images (SEM) of the FCN-0 and the FCN-500 composites. The two materials are morphologically flocculent aggregates with non-uniform particle sizes. The pores of the FCN-0 are small in number and large in size. The FCN-500 shown in Figure 1b is tightly aggregated and clustered; the number of pores is significantly increased, and the particle size is smaller. The EDS data (Figure 1c) confirms the four elements of C, Fe, N, and O with a weight ratio of 2.5:7.7:0.4:2.5 (right top inset of Figure 1c) are uniformly distributed on the FCN-500. The Fe:O weight ratio of the FCN-500 (2.264) is higher than that of the FCN-0 (2.125, Figure S1), which is mainly caused by the decomposition of crystal water and some organic components during the calcination process.

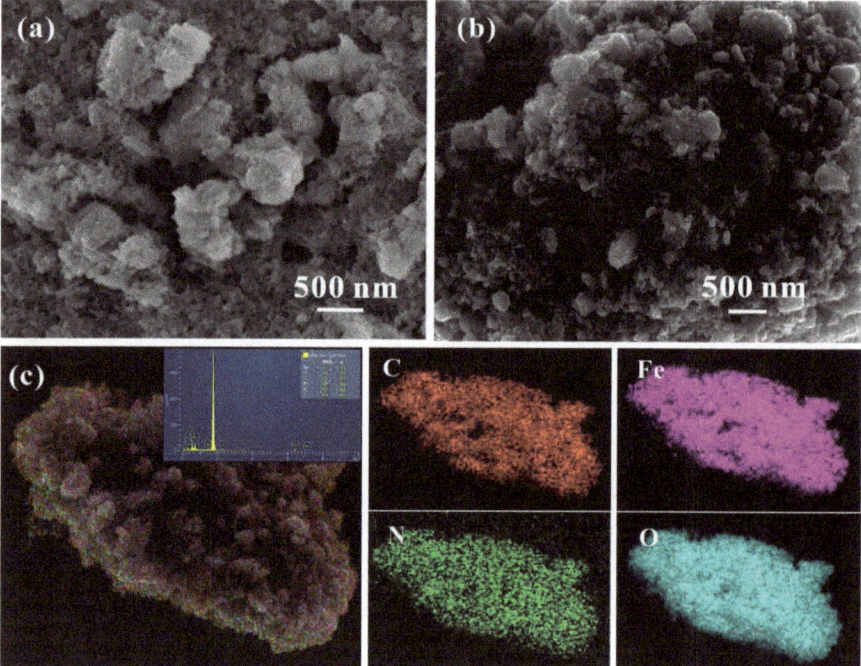

Figure 1. SEM images of the FCN-0 (**a**), FCN-500 (**b**), and elemental mapping images (**c**) of the FCN-500.

Figure 2a shows the X-ray diffraction patterns of the FCN-0, FCN-400, FCN-500, and FCN-600. The diffraction peaks of FCN-0 at 2θ = 28.52°, 36.38°, and 52.81° correspond to the (040), (031), and (151) planes of FeOOH (JCPDS NO.74-1877), respectively. The peaks of FCN-400 at 2θ = 33.2°, 49.5°, and 54.1° can be attributed to the (104), (024), and (116) planes of Fe_2O_3 (JCPDS 79-0007), respectively. While the other three diffraction peaks at 2θ = 36.1°, 41.9°, and 60.7° corresponding to the (111), (200), and (220) planes of FeO (JCPDS 77-2355), respectively. These results indicate that as the temperature rises, the FeOOH in the raw sludge dehydrates to form Fe_2O_3, then reduced to FeO by the carbon in the sludge. The intense peaks of FCN-500 at 2θ = 30.2°, 35.5°, 43.1°, 53.5°, 57.1°, and 62.6° correspond to the (220), (311), (400), (422), (511), and (440) planes of Fe_3O_4 (JCPDS NO.19-0629), respectively. Therefore, it can be speculated that the sludge gradually agglomerated to form a Fe_3O_4 nanoparticle structure during the calcination process [33].

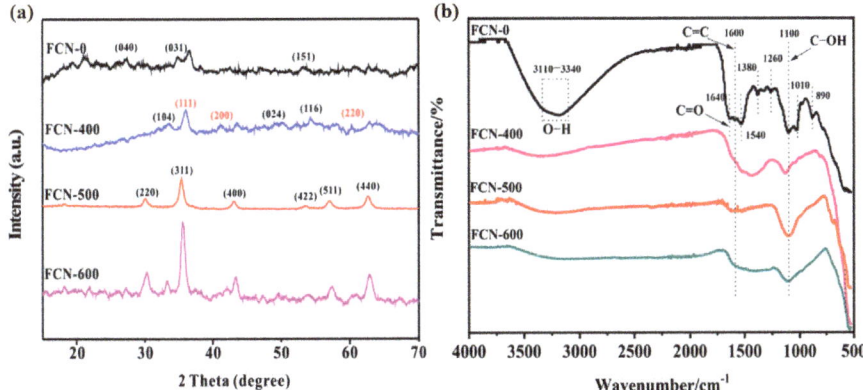

Figure 2. XRD patterns (**a**), FT-IR spectra (**b**) of the FCN-0, FCN-400, FCN-500, and FCN-600.

Figure 2b shows the FT-IR spectra of the FCN-0, FCN-400, FCN-500, and FCN-600. The broad band of FCN-0 in the range of 3110–3340 cm^{-1} is attributed to the stretching vibration of O-H, which is caused by the adsorption of water molecules on the sludge surface. However, the O-H absorption band in FCN-400, FCN-500, and FCN-600 became flat, indicating that the pyrolysis process led to the disappearance of crystal water in the sludge. The two bands at 1600 cm^{-1} and 1540 cm^{-1} belong to the C=C stretching vibrations, and the IR band at 890 cm^{-1} confirms the aromatic ring structure in the FCN-0 composite [34–36]. The wide absorption band at 1110 cm^{-1} indicates the presence of a C-OH bond. The band at 554 cm^{-1} of FCN-0 originated from the vibration of Fe-O, which transformed into the Fe-O-Fe bond (476 cm^{-1} of FCN-500) after calcination at 500 °C, suggesting that 500 °C is the optimal temperature for the crystallization of iron oxides.

Figure 3 shows the magnetization curves of the FCN-0, FCN-400, FCN-500, and FCN-600 composites. The FCN-500 and FCN-600 exhibit typical ferromagnetic behavior, and the highest saturation magnetization value of the FCN-500 is about 54.3 emu/g. This is mainly because the organic components in the sludge are carbonized at high temperatures, and the Fe-containing substances in it are converted into magnetic Fe_3O_4. The high saturated magnetization of the FCN-500 facilitates its separation from the solution under an external magnetic field.

The thermal stability and composition of iron-containing coagulated sludge were studied in the temperature range of 30–1000 °C (Figure 4). The DTG curve showed three obvious weight loss stages during the heating process. The first stage occurred in the temperature range of 30–150 °C, which was mainly due to the loss of crystal water. In the second stage, in the temperature range of 150–350 °C, the weight decreased rapidly, mainly due to the thermal decomposition of organic matter containing nitrogen, sulfur, oxygen, and other elements in the coagulated sludge. The third stage of weight loss occurred at 350–500 °C, during which organics were further carbonized. Combining the XRD results and the magnetization curves, it can be inferred that FeOOH in iron-containing coagulated sludge was dehydrated and transformed into Fe_2O_3 in the temperature range of 30–350 °C. When the temperature was higher than 350 °C, Fe_2O_3 was reduced to FeO by C in the sludge, and gradually converted to Fe_3O_4.

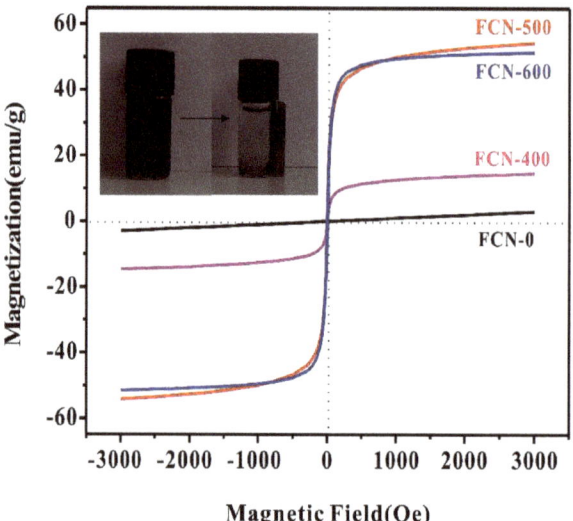

Figure 3. The magnetization curves of the FCN-0, FCN-400, FCN-500, and FCN-600.

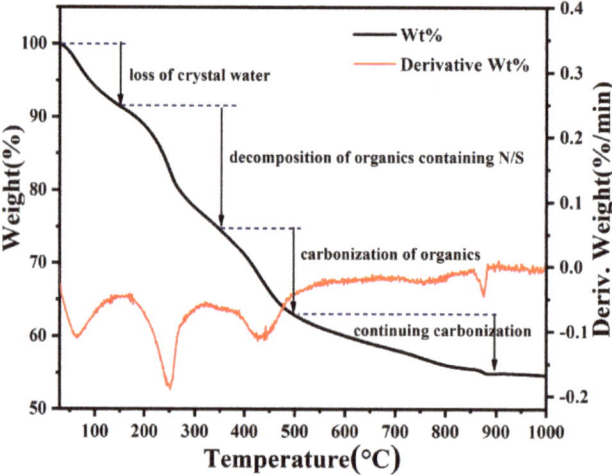

Figure 4. DTG curve of the coagulated sludge.

The N_2 adsorption-desorption isotherms of the FCN-0, FCN-400, FCN-500, and FCN-600 are used to estimate specific surface area and pore size distribution, as displayed in Figure S2. Table S1 shows the pore structure parameters (specific surface area, pore volume, and average pore diameter) of the FCN-0, FCN-400, FCN-500, and FCN-600. It can be seen that calcination has little effect on the specific surface area and pore size of materials.

Table S2 shows the chemical properties of the FCN-0, FCN-400, FCN-500, and FCN-600. According to the XRF results, as shown in the table below, the major metal element in the FCN-x is iron, which is approximately 70%, and the content of each other metal element is less than 1%.

3.2. Parameters Optimization for Cr(VI) Adsorption

The parameters, viz., composite type, pH, reaction time, temperature, adsorbate, and adsorbent loading on Cr(VI) adsorption, were optimized. The removal of Cr(VI) by different composites was first investigated. As shown in Figures 5a and S3a, after 2 h of reaction, the adsorption capacities of the FCN-0, FCN-400, FCN-500 and FCN-600 for Cr(VI) were 2.5 mg/g, 15 mg/g, 37.4 mg/g and 37.5 mg/g, respectively. With the increase in pyrolysis temperature, the adsorption performance of the prepared materials gradually improved. Combined with the results of magnetization curve analysis, FCN-500 showed the most efficient Cr(VI) adsorption capacity and good magnetic separation. This happened because the complex organic matter in the sludge was not completely decomposed at lower pyrolysis temperatures, while at a higher temperature, some functional groups on the sludge surface disappeared, thereby reducing the adsorption effect of Cr(VI), which was consistent with the FT-IR characterization results. Therefore, FCN-500 was selected for further research.

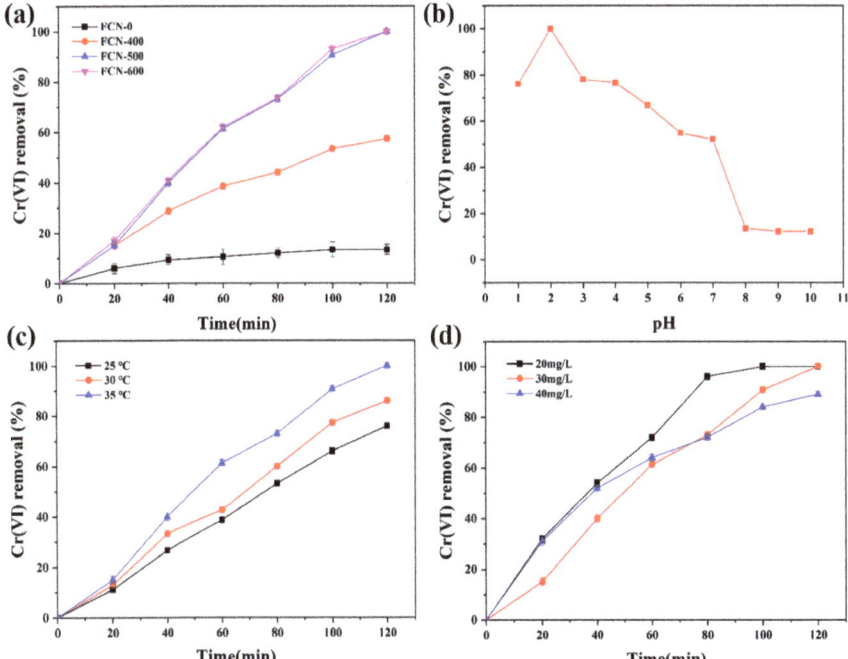

Figure 5. The removal of Cr(VI) by different adsorbents. (**a**) Effect of different adsorbents (T = 35 °C, pH = 2, [Cr(VI)] = 30 mg/L, [Adsorbents] = 0.8 g/L). (**b**) Effect of the pH value (T = 35 °C, [Cr(VI)] = 30 mg/L, [FCN-500] = 0.8 g/L). (**c**) Effect of the temperature (pH = 2, [Cr(VI)] = 30 mg/L, [FCN-500] = 0.8 g/L). (**d**) Effect of the initial Cr(VI) concentration (T = 35 °C, pH = 2, [FCN-500] = 0.8 g/L).

The pH of the solution affects the adsorption of Cr(VI) mainly by changing the surface charge and the degree of ionization of the material, as well as the morphology of Cr(VI) in solution [37]. The adsorption of Cr(VI) by FCN-500 was investigated when the pH value changed from 1.0 to 10.0. It can be observed from Figures 5b and S3b that the adsorption of Cr(VI) reached its best effect at pH = 2, and the removal of Cr(VI) was 99.9%. When the pH value increased, the removal of Cr(VI) started to decrease. The adsorption amount of Cr(VI) by the FCN-500 after 2 h was 20.4 mg/g when the solution pH was 7, and the removal rate dropped from 99.9% to 52%; further increased the pH to 8–10, the adsorption of Cr(VI) dropped sharply to 5 mg/g, and the removal rate decreased to 12%. We supposed the reason for the change was as follows: Cr(VI) in solution mainly existed in the form of $HCrO_4^-$ and $Cr_2O_7^{2-}$ at low pH, while the nitrogen-containing species contained on the surface of the FCN-500 would be protonated under acidic conditions and converted to positively charged nitrogen, thereby adsorbing negatively charged $HCrO_4^-$ and $Cr_2O_7^{2-}$ through electrostatic attraction and reducing Cr(VI) to Cr(III) by a redox reaction. With the pH increasing, the high concentration of OH^- ions would compete with Cr(VI), resulting in a decrease in the adsorption performance of the material. In addition, when the pH was 1, Cr(VI) mainly existed in the form of the H_2CrO_4 compound, which was not conducive to the adsorption of the material [38,39].

The effect of temperature on Cr(VI) adsorption by FCN-500 was studied and shown in Figures 5c and S3c. The removal of Cr(VI) was 76%, 89.5%, and 99.9% when the reaction was carried out at 25 °C 30 °C, and 35 °C, respectively. The increase in reaction temperature is beneficial for accelerating the diffusion rate of pollutant separation, resulting in a higher adsorption capacity and the removal of Cr(VI).

The adsorption performance of FCN-500 at different initial concentrations of Cr(VI) was investigated, as shown in Figures 5d and S3d. When the initial concentration of Cr(VI) was 20 mg/L, the adsorption amount of FCN-500 was 25 mg/g at 2 h of reaction, and the removal rate was 100%. When the concentration of Cr(VI) increased to 30 and 40 mg/L, the adsorption capacity of FCN-500 also increased to 37.4 mg/g and 44.5 mg/g, respectively. However, the removal rate began to decrease gradually. When the initial concentration of Cr(VI) is low, there are many adsorption sites in FCN-500 that can adsorb and bind Cr(VI) in a large amount; when the concentration of pollutants is too high, the material is close to adsorption saturation, and the adsorption process tends to be stable [40].

3.3. Adsorption Isotherms

The adsorption isotherms are used to describe the distributed adsorbed molecules between the liquid and solid phases under equilibrium conditions [41]. The Langmuir model and Freundlich model were adopted to analyze the adsorption performance of 0.8 g/L FCN-500 at a reaction temperature of 35 °C, solution pH = 2.0. The two model equations are shown as follows:

Langmuir model equation:

$$\frac{c_e}{q_e} = \frac{1}{q_m b} + \frac{c_e}{q_m} \tag{3}$$

Freundlich model equation:

$$\ln q_e = \ln K_f + \frac{1}{n}\ln c_e \tag{4}$$

where C_e (mg/L) is the concentration of Cr(VI) in solution at the adsorption equilibrium; q_m (mg/g) is the maximum adsorption capacity; b is the Langmuir constant; K_f and n are the Freundlich constants.

The influence of the initial concentration of Cr(VI) on the equilibrium amount of Cr(VI) adsorbed by FCN-500 was evaluated under the same experimental conditions. The isotherms of Cr(VI) adsorption by FCN-500 were further fitted by Langmuir and Freundlich models, and the results are shown in Figure S4. The correlation coefficient (R^2) of the Langmuir model and Freundlich model was 0.993 and 0.976, respectively. Therefore, the Langmuir model was most suitable for describing Cr(VI) adsorption by FCN-500. The Cr(VI) adsorption capacity with various allied substrates with the FCN 500 is shown in Table 1. The initial conditions, such as pH value, adsorbate, and adsorbent loading, were different. The FCN-500 showed the highest normalized Cr(VI) adsorption compared to other substrates.

Table 1. Comparison of Cr(VI) adsorption capacity using FCN-500 with other reported adsorbents.

Sorbent	pH	q_m (mg/g)	Ref.
Oxidized multiwalled carbon nanotubes	pH = 2.6	2.88	[42]
Activated carbon	pH = 4.0	15.47	[43]
Fe_3O_4/GO	pH = 4.5	32.33	[44]
Magnetic cyclodextrin/chitosan/GO	pH = 3.0	21.6	[45]
FCN-500	pH = 2.0	52.63	This work

3.4. XPS Spectra Analysis

XPS is a characterization analysis method to confirm Cr(VI) reduction on FCN-500 by measuring element species. As shown in Figure 6 of the scan spectra, the distinct peaks observed correspond to Fe 2p, C 1s, N 1s, O 1s, and Cr 2p. Cr(VI) adsorption on FCN-500 is confirmed by the presence of the Cr 2p peak (Figure 6a). In Cr 2p, the peaks at 576.6 and 586.5 eV were marked as Cr 2p3/2 and Cr 2p1/2 of Cr(III) (Figure 6b) [46,47]. The presence of Cr(III) indicates subsequent Cr(VI) reduction. The Fe2p peak is mainly divided into Fe2p3/2 and Fe2p1/2, which were located at 711.7 eV and 724.6 eV, respectively (Figure 6c). The Fe 2p3/2 peak is located at 711.7 eV, indicating Fe_3O_4 in FCN-500, which is consistent with the results of the XRD analysis [48]. The peak corresponding to the C 1s peak shows indifferent behavior in all samples. In the binding spectrum of C1s, the peaks at 284.8 eV, 285.8 eV and 288.3 eV correspond to the C-C/CHx, -C-OR and N-C=N bonds, respectively (Figure 6d) [49,50]. In the O 1s peak (Figure 6e), before Cr(VI) adsorption, O1s has three XPS peaks with binding energies at 530.2 eV, 531.9 eV, and 532.8 eV, which are assigned to Fe-O, C-O, and C=O groups [51]. After Cr(VI) adsorption, the O1s spectrum has fitted into three peaks with the binding energies at 530.2 eV, 531.5 eV, and 532.5 eV, which are assigned to Fe-O, C-O, and C=O groups. After adsorption, the peak intensity of C=O is decreased. The results showed that some of the C=O in FCN-500 had been oxidized to C-O after the adsorption of Cr(VI), indicating that a redox reaction occurred during the adsorption process [27]. The high-resolution spectrum at N 1s shows three peaks at 398.7, 399.8, and 400.7 eV, attributed to =N–, –NH–, and N^+ (=N^{-+} and –NH^{-+}), respectively. After Cr(VI) adsorption, the N^+ content increased significantly, whereas the total amount of =N– and –NH– decreased. The results show that part of nitrogen on FCN-500 is further protonated under acidic conditions, and the non-protonated nitrogen is significantly reduced (Figure 6f) [27].

3.5. Adsorption Mechanism

As shown in half cell reactions (5) and (6), when the pH value of the solution increases from 0 to 14, the oxidation potential of Cr(VI) decreases from 1.33 V to −0.13 V [52]. A strongly acidic environment is conducive to the Cr(VI) species in solution. In the present study, we used pH 2 to endow Cr(VI) in solution.

$$Cr_2O_7^{2-} + 6e^- + 14H^+ \rightarrow 2Cr^{3+} + 7H_2O, E^0 = 1.33 \text{ V (pH = 0)} \quad (5)$$

$$CrO_4^{2-} + 4H_2O + 3e^- \rightarrow Cr^{3+} + 8OH^-, E^0 = -0.13 \text{ V (pH = 14)} \quad (6)$$

The reduction product was confirmed to be Cr(III) by XPS analysis. ICP-MS analysis of chromium species noted a negligible proportion of Cr(III) in solution after adsorption-reduction, indicating that the adsorbed Cr(III) was not released into the solution, but was instead immobilized on the surface of the FCN-500. The nitrogen atoms in the PDA play a key role in Cr(III) chelation [53]. However, protonated nitrogen atoms are not suitable for chelating Cr(III) due to electrostatic repulsion. Therefore, non-protonated nitrogen atoms may be the main reason for the fixation of Cr(III) after the redox reaction [27].

In summary, the reduction reaction is accompanied by the adsorption process, and the reducing agent is the nitrogen atomic group on the FCN-500. After reduction, the produced Cr(III) was immobilized on the adsorbent. It should be noted that the adsorption mechanism for Cr(VI) involves several steps. (1) The adsorption of Cr(VI) occurred in an aqueous solution, resulting in the protonated nitrogen groups of the active site on the FCN-500 at pH = 2; (2) The adsorption of Cr(VI) on the active sites was reduced to Cr(III) by nitrogen groups on the FCN-500; (3) The reduced Cr(III) was immobilized in situ on the adsorbent surface by chelating with aprotic nitrogen atoms.

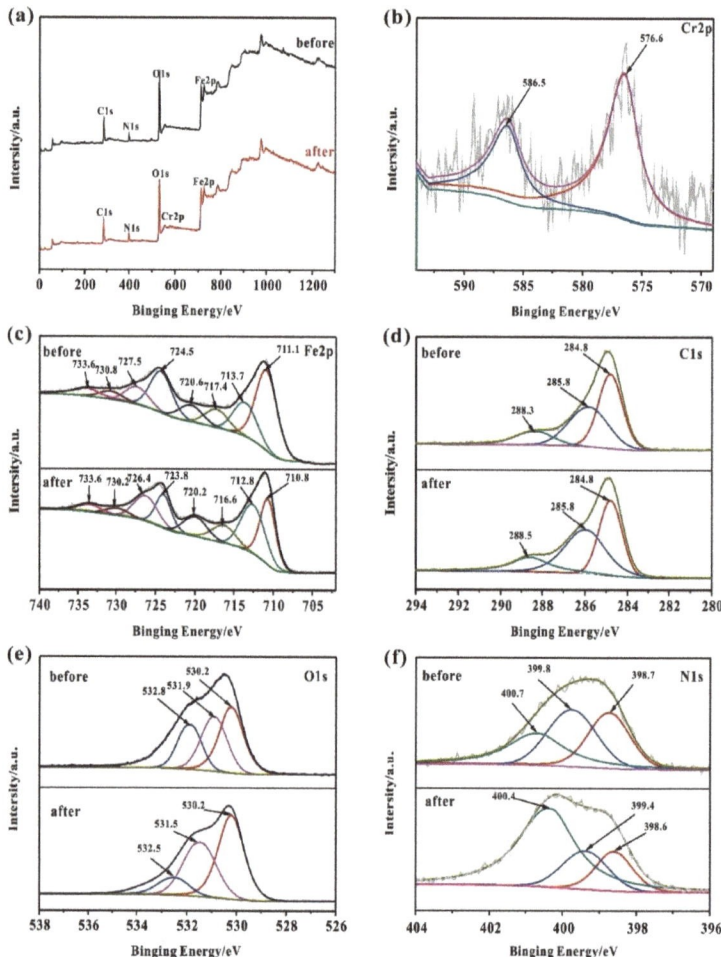

Figure 6. XPS spectra of FCN-500 before and after Cr(VI) removal: wide scan (**a**) and high-resolution spectra of (**b**) Cr2p, (**c**) Fe2p, (**d**) C1s, (**e**) O1s and (**f**) N1s.

3.6. Adsorption Regeneration Experiments

After adsorption-reduction, chromium occurs on FCN-500 as Cr(III). The regeneration of reactive sites was carried out using 1 M NaOH for 12 h (each cycle). After five adsorption-desorption cycles (Figure 7a), the adsorption capacity of FCN-500 decreased slightly, reaching about 80% optimal capacity. As the number of regenerations increased, the adsorption capacity declined after the third regeneration step. The adsorption capacity of the regenerated FCN-500 remained unchanged after five cycles. The XRD patterns of fresh and used FCN-500 were also collected as shown in Figure 7b. Results suggested that after five cycles of adsorption-regeneration, the basic structure of FCN-500 was well maintained, proving its good regenerative adsorption performance of FCN-500. In addition, a new peak near 42° was observed for the used FCN-500, which can be associated to the crystal plane of some Cr-N-Fe species, indicating that Cr interacted with nitrogen groups on FCN-500 [54]. This result also proved the proposed mechanism, that nitrogen-containing groups were the active sites for Cr adsorption.

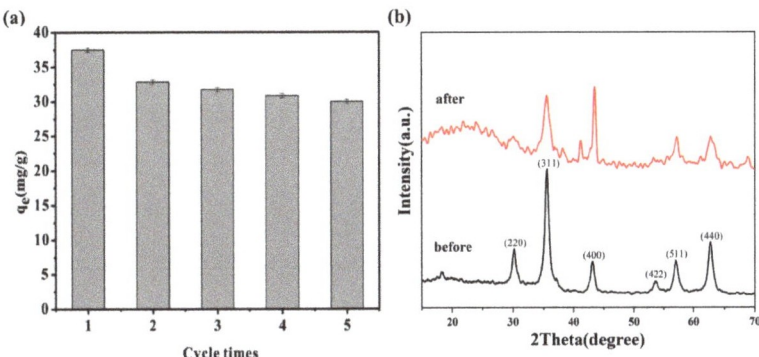

Figure 7. (**a**) Experimental diagram of adsorption and regeneration. (**b**) XRD patterns of fresh and used FCN-500.

3.7. Treatment of Wastewater in the Electroplating Industry

Electroplating wastewater discharged from electroplating production contains a large amount of chromium, heavy metals, and organic matter. The pH of electroplating wastewater is around 2–3, so the FCN-500 can be used to remove Cr(VI) without further adjustment. As shown in Figure S5, the removal of total chromium by FCN-500 reached 99.93% after 2 h reaction under the conditions of T = 25 °C, pH = 3, [Cr] = 117 mg/L, and [FCN-500] = 2.4 g/L. The results showed that the FCN-500 magnetic material also had a good adsorption effect on Cr in the actual wastewater.

4. Conclusions

In this study, the magnetic Fe-C-N composite was successfully prepared from the coagulation sludge produced from dye chemical wastewater as a starting material to develop a Cr(VI) removal substrate designated as FCN. The material obtained when the pyrolysis temperature was 500 °C (FCN-500) performed the best adsorption performance for Cr(VI). The maximum adsorption capacity for Cr(VI) by FCN-500 was 52.63 mg/g at 35 °C and pH = 2. The Cr(VI) removal from the solution by FCN 500 occurs due to the adsorptive–reduction process. The Cr(VI) → Cr(III) reduction preferentially occurred at hetero N sites, followed by Cr(III) adsorption on aprotic N on FCN-500. Besides, FCN-500 also showed good reusability, with approximately 80% removal of Cr(VI) after 5 cycles. This work provides a new method for recycling chemical solid waste sustainably as a value-added product.

Supplementary Materials: The following supporting information can be downloaded at: https://www.mdpi.com/article/10.3390/w15122290/s1.

Author Contributions: Conceptualization, X.C.; Methodology, X.L., H.L., Z.D. and B.W.; Formal analysis, X.L.; Investigation, X.L., H.L., Z.D. and B.W.; Writing—original draft, X.L. and H.L.; Writing—review & editing, R.W. and X.C.; Supervision, K.C. and X.C.; Funding acquisition, K.C. and X.C. All authors have read and agreed to the published version of the manuscript.

Funding: The authors acknowledge the financial support from the National Key R&D Program of China (2019YFC0408500), and the Key Science and Technology Projects of Anhui Province (202003a07020004).

Data Availability Statement: Not applicable.

Acknowledgments: R.W. acknowledges the Program of Distinguished Professor in B&R Countries (Grant No. G20200012010).

Conflicts of Interest: The authors declare no conflict of interest.

References

1. Kumar, A.S.K.; Jiang, S.-J.; Tseng, W.L. Effective adsorption of chromium(VI)/Cr(III) from aqueous solution using ionic liquid functionalized multiwalled carbon nanotubes as a super sorbent. *J. Mater. Chem. A* **2015**, *3*, 7044–7057. [CrossRef]
2. Al-Othman, Z.A.; Naushad, M.; Inamuddin. Organic-inorganic type composite cation exchanger poly-o-toluidine Zr(IV) tungstate: Preparation, physicochemical characterization and its analytical application in separation of heavy metals. *Chem. Eng. J.* **2011**, *172*, 369–375. [CrossRef]
3. Albadarin, A.B.; Mangwandi, C.; Al-Muhtaseb, A.a.H.; Walker, G.M.; Allen, S.J.; Ahmad, M.N.M. Kinetic and thermodynamics of chromium ions adsorption onto low-cost dolomite adsorbent. *Chem. Eng. J.* **2012**, *179*, 193–202. [CrossRef]
4. Mohammed, Y.A.Y.A.; Abdel-Mohsen, A.M.; Zhang, Q.J.; Younas, M.; Zhong, L.B.; Yang, J.C.E.; Zheng, Y.M. Facile synthesis of ZIF-8 incorporated electrospun PAN/PEI nanofibrous composite membrane for efficient Cr (VI) adsorption from water. *Chem. Eng. J.* **2023**, *461*, 141972. [CrossRef]
5. Zhang, Y.; Xu, M.; Li, H.; Ge, H.; Bian, Z. The enhanced photoreduction of Cr(VI) to Cr(III) using carbon dots coupled TiO_2 mesocrystals. *Appl. Catal. B-Environ.* **2018**, *226*, 213–219. [CrossRef]
6. Putz, A.M.; Ciopec, M.; Negrea, A.; Grad, O.; Ianasi, C.; Ivankov, O.I.; Milanovic, M.; Stijepovic, I.; Almasy, L. Comparison of structure and adsorption properties of mesoporous silica functionalized with aminopropyl groups by the co-condensation and the post grafting methods. *Materials* **2021**, *14*, 628. [CrossRef]
7. Dehghani, M.H.; Taher, M.M.; Bajpai, A.; Heibati, B.; Tyagi, I.; Asif, M.; Agarwal, S.; Gupta, V.K. Removal of noxious Cr (VI) ions using single-walled carbon nanotubes and multi-walled carbon nanotubes. *Chem. Eng. J.* **2015**, *279*, 344–352. [CrossRef]
8. Liang, H.; Sun, R.; Song, B.; Sun, Q.; Peng, P.; She, D.J. Preparation of nitrogen-doped porous carbon material by a hydrothermal-activation two-step method and its high-efficiency adsorption of Cr(VI). *Hazard. Mater.* **2020**, *387*, 121987. [CrossRef]
9. Ianăși, C.; Ianăși, P.; Negrea, A.; Ciopec, M.; Ivankov, O.I.; Kuklin, A.I.; Almásy, L.; Putz, A.M. Effects of catalysts on structural and adsorptive properties of iron oxide-silica nanocomposites. *Korean J. Chem. Eng.* **2021**, *38*, 292–305. [CrossRef]
10. Lyu, W.; Wu, J.; Zhang, W.; Liu, Y.; Yu, M.; Zhao, Y.; Feng, J.; Yan, W. Easy separated 3D hierarchical coral-like magnetic polyaniline adsorbent with enhanced performance in adsorption and reduction of Cr(VI) and immobilization of Cr(III). *Chem. Eng. J.* **2019**, *363*, 107–119. [CrossRef]
11. Zhang, Y.; Li, M.; Li, J.; Yang, Y.; Liu, X. Surface modified leaves with high efficiency for the removal of aqueous Cr (VI). *Appl. Surf. Sci.* **2019**, *484*, 189–196. [CrossRef]
12. Norseth, T. The carcinogenicity of chromium. The carcinogenicity of chromium. *Environ. Health Perspect.* **1981**, *40*, 121–130. [CrossRef] [PubMed]
13. Pakade, V.E.; Tavengwa, N.T.; Madikizela, L.M. Recent advances in hexavalent chromium removal from aqueous solutions by adsorptive methods. *Rsc Adv.* **2019**, *9*, 26142–26164. [CrossRef] [PubMed]
14. Tadeusz, P.; Paweł, M.; Małgorzata, M.; Monika, B.; Małgorzata, K.; Izabella, J. Adsorption and degradation of phenoxyalkanoic acid herbicides in soils: A review. *Environ. Toxicol. Chem.* **2016**, *35*, 271–286.
15. Fu, H.R.; Wang, N.; Qin, J.H.; Han, M.L.; Ma, L.F.; Wang, F. Spatial confinement of a cationic MOF: A SC-SC approach for high capacity Cr(VI)-oxyanion capture in aqueous solution. *Chem. Commun.* **2018**, *54*, 11645–11648. [CrossRef] [PubMed]
16. Jamshidifard, S.; Koushkbaghi, S.; Hosseini, S.; Rezaei, S.; Karamipour, A.J.; Rad, A.; Irani, M.J. Incorporation of UiO-66-NH_2 MOF into the PAN/chitosan nanofibers for adsorption and membrane filtration of Pb(II), Cd(II) and Cr(VI) ions from aqueous solutions. *Hazard. Mater.* **2019**, *368*, 10–20. [CrossRef]
17. Kong, Q.; Wei, J.; Hu, Y.; Wei, C.J. Fabrication of terminal amino hyperbranched polymer modified graphene oxide and its prominent adsorption performance towards Cr(VI). *Hazard. Mater.* **2019**, *363*, 161–169. [CrossRef]
18. Abdelhamid, H.N.; Georgouvelas, D.; Edlund, U.; Mathew, A.P. Cello ZIF Paper: Cellulose-ZIF hybrid paper for heavy metal removal and electrochemical sensing. *Chem. Eng. J.* **2022**, *446*, 136614. [CrossRef]
19. Gao, Y.; Wei, Z.; Li, F.; Yang, Z.M.; Chen, Y.M.; Zrinyi, M.; Osada, Y. Synthesis of a morphology controllable Fe_3O_4 nanoparticle/hydrogel magnetic nanocomposite inspired by magnetotactic bacteria and its application in H_2O_2 detection. *Green Chem.* **2014**, *16*, 1255–1261. [CrossRef]
20. Zhang, F.; Tang, X.; Huang, Y.; Keller, A.A.; Lan, J. Competitive removal of Pb^{2+} and malachite green from water by magnetic phosphate nanocomposites. *Water Res.* **2019**, *150*, 442–451. [CrossRef]
21. Meng, Y.; Chen, D.; Sun, Y.; Jiao, D.; Zeng, D.; Liu, Z. Adsorption of Cu^{2+} ions using chitosan-modified magnetic Mn ferrite nanoparticles synthesized by microwave-assisted hydrothermal method. *Appl. Surf. Sci.* **2015**, *324*, 745–750. [CrossRef]
22. Feng, Y.; Du, Y.; Chen, Z.; Du, M.; Yang, K.; Lv, X.; Li, Z.J. Synthesis of Fe_3O_4 nanoparticles with tunable sizes for the removal of Cr(VI) from aqueous solution. *Coat. Technol. Res.* **2018**, *15*, 1145–1155. [CrossRef]
23. Raj, A.; Bauman, M.; Lakic, M.; Dimitrusev, N.; Lobnik, A.; Kosak, A. Removal of Pb^{2+}, CrT, and Hg^{2+} Ions from Aqueous Solutions Using Amino-Functionalized Magnetic Nanoparticles. *Int. J. Mol. Sci.* **2022**, *23*, 16186.

24. Wang, L.; Wang, J.M.; Zhang, R.; Liu, X.G.; Song, G.X.; Chen, X.F.; Wang, Y.; Kong, J.-L. Highly efficient As(V)/Sb(V) removal by magnetic sludge composite: Synthesis, characterization, equilibrium, and mechanism studies. *Rsc Adv.* **2016**, *6*, 42876–42884. [CrossRef]
25. He, Y.T.; Traina, S.J. Cr(VI) reduction and immobilization by magnetite under alkaline pH conditions: The role of passivation. *Environ. Sci. Technol.* **2005**, *39*, 4499–4504. [CrossRef]
26. Liu, Y.; Zhang, Z.; Sun, X.; Wang, T.J. Design of three-dimensional macroporous reduced graphene oxide-Fe_3O_4 nanocomposites for the removal of Cr(VI) from wastewater. *Porous Mater.* **2019**, *26*, 109–119. [CrossRef]
27. Wan, C.; Zhang, R.; Wang, L.; Liu, X.; Bao, D.; Song, G. Enhanced reduction and in-situ stabilization of Cr(VI) by Fe_3O_4@polydopamine magnetic microspheres embedded in sludge-based carbonaceous matrix. *Appl. Surf. Sci.* **2021**, *536*, 147980. [CrossRef]
28. Zhao, D.; Gao, X.; Wu, C.; Xie, R.; Feng, S.; Chen, C. Facile preparation of amino functionalized graphene oxide decorated with Fe_3O_4 nanoparticles for the adsorption of Cr(VI). *Appl. Surf. Sci.* **2016**, *384*, 1–9. [CrossRef]
29. Islam, M.S.; Ahmed, M.K.; Raknuzzaman, M.; Habibullah-Al-Mamun, M.; Kundu, G.K. Heavy metals in the industrial sludge and their ecological risk: A case study for a developing country. *J. Geochem. Explor.* **2017**, *172*, 41–49. [CrossRef]
30. Liu, X.; Li, X.M.; Yang, Q.; Yue, X.; Shen, T.T.; Zheng, W.; Luo, K.; Sun, Y.H.; Zeng, G.M. Landfill leachate pretreatment by coagulation–flocculation process using iron-based coagulants: Optimization by response surface methodology. *Chem. Eng. J.* **2012**, *200–202*, 39–51. [CrossRef]
31. Chen, Y.D.; Ho, S.H.; Wang, D.W.; Wei, Z.S.; Chang, J.S.; Ren, N.Q. Lead removal by a magnetic biochar derived from persulfate-ZVI treated sludge together with one-pot pyrolysis. *Bioresour. Technol.* **2018**, *247*, 463–470. [CrossRef] [PubMed]
32. Leite, A.J.B.; Lima, E.C.; dos Reis, G.S.; Thue, P.S.; Saucier, C.; Rodembusch, F.S.; Dias, S.L.P.; Umpierres, C.S.; Dotto, G.L. Hybrid adsorbents of tannin and APTES (3-aminopropyltriethoxysilane) and their application for the highly efficient removal of acid red 1 dye from aqueous solutions. *J. Environ. Chem. Eng.* **2017**, *5*, 4307–4318. [CrossRef]
33. Liu, X.; Tian, J.; Li, Y.; Sun, N.; Mi, S.; Xie, Y.; Chen, Z.J. Enhanced dyes adsorption from wastewater via Fe_3O_4 nanoparticles functionalized activated carbon. *Hazard. Mater.* **2019**, *373*, 397–407. [CrossRef] [PubMed]
34. Zhu, W.; Sun, F.; Goei, R.; Zhou, Y. Construction of WO_3-g-C_3N_4 composites as efficient photocatalysts for pharmaceutical degradation under visible light. *Catal. Sci. Technol.* **2017**, *7*, 2591–2600. [CrossRef]
35. Proniewicz, L.M.; Paluszkiewicz, C.; Weselucha-Birczynska, A.; Baranski, A.; Dutka, D.J. FT-IR and FT-Raman study of hydrothermally degraded groundwood containing paper. *Mol. Struct.* **2002**, *614*, 345–353. [CrossRef]
36. Barroso-Bogeat, A.; Alexandre-Franco, M.; Fernandez-Gonzalez, C.; Gomez-Serrano, V. FT-IR analysis of pyrone and chromene structures in activated carbon. *Energy Fuels* **2014**, *28*, 4096–4103. [CrossRef]
37. Wu, C.; Tu, J.; Liu, W.; Zhang, J.; Chu, S.; Lu, G.; Lin, Z.; Dang, Z. The double influence mechanism of pH on arsenic removal by nano zero valent iron: Electrostatic interactions and the corrosion of Fe-0. *Environ. Sci. Nano* **2017**, *4*, 1544–1552. [CrossRef]
38. Gu, H.; Rapole, S.B.; Huang, Y.; Cao, D.; Luo, Z.; Wei, S.; Guo, Z.J. Synergistic interactions between multi-walled carbon nanotubes and toxic hexavalent chromium. *Mater. Chem. A* **2013**, *1*, 2011–2021. [CrossRef]
39. Wu, Y.; Luo, H.; Wang, H.; Wang, C.; Zhang, J.; Zhang, Z.J. Adsorption of hexavalent chromium from aqueous solutions by graphene modified with cetyltrimethylammonium bromide. *Colloid Interface Sci.* **2013**, *394*, 183–191. [CrossRef]
40. Xu, H.; Liu, Y.; Liang, H.; Gao, C.; Qin, J.; You, L.; Wang, R.; Li, J.; Yang, S. Adsorption of Cr(VI) from aqueous solutions using novel activated carbon spheres derived from glucose and sodium dodecylbenzene sulfonate. *Sci. Total Environ.* **2021**, *759*, 143457. [CrossRef]
41. Liang, Q.; Luo, H.; Geng, J.; Chen, J. Facile one-pot preparation of nitrogen-doped ultra-light graphene oxide aerogel and its prominent adsorption performance of Cr(VI). *Chem. Eng. J.* **2018**, *338*, 62–71. [CrossRef]
42. Hu, J.; Chen, C.; Zhu, X.; Wang, X.J. Removal of chromium from aqueous solution by using oxidized multiwalled carbon nanotubes. *Hazard. Mater.* **2009**, *162*, 1542–1550. [CrossRef] [PubMed]
43. Babel, S.; Kurniawan, T.A. Cr (VI) removal from synthetic wastewater using coconut shell charcoal and commercial activated carbon modified with oxidizing agents and/or chitosan. *Chemosphere* **2004**, *54*, 951–967. [CrossRef] [PubMed]
44. Liu, M.; Wen, T.; Wu, X.; Chen, C.; Hu, J.; Li, J.; Wang, X. Synthesis of porous Fe_3O_4 hollow microspheres/graphene oxide composite for Cr(VI) removal. *Dalton Trans.* **2013**, *42*, 14710–14717. [CrossRef]
45. Li, L.; Fan, L.; Sun, M.; Qiu, H.; Li, X.; Duan, H.; Luo, C. Adsorbent for chromium removal based on graphene oxide functionalized with magnetic cyclodextrin-chitosan. *Colloids Surf. B-Biointerfaces* **2013**, *107*, 76–83. [CrossRef]
46. Lei, Y.; Chen, F.; Luo, Y.; Zhang, L.J. Three-dimensional magnetic graphene oxide foam/Fe_3O_4 nanocomposite as an efficient absorbent for Cr(VI) removal. *Mater. Sci.* **2014**, *49*, 4236–4245. [CrossRef]
47. Hu, J.; Chen, G.H.; Lo, I.M.C. Removal and recovery of Cr(VI) from wastewater by maghemite nanoparticles. *Water Res.* **2005**, *39*, 4528–4536. [CrossRef]
48. Guo, X.; Du, B.; Wei, Q.; Yang, J.; Hu, L.; Yan, L.; Xu, W.J. Synthesis of amino functionalized magnetic graphenes composite material and its application to remove Cr(VI), Pb(II), Hg(II), Cd(II) and Ni(II) from contaminated water. *Hazard. Mater.* **2014**, *278*, 211–220. [CrossRef]
49. An, T.D.; Nguyen, P.C.T.; Nguyen, N.T.; Nguyen, P.H.; Hoai, T.Q.T.; Huong, T.D.; Van Thang, N.; Hoang, L.H.; Le Tuan, N.; Nhiem, D.N.; et al. One-step synthesis of oxygen doped g-C_3N_4 for enhanced visible-light photodegradation of Rhodamine B. *Phys. Chem. Solids* **2021**, *151*, 109900.

50. Sevilla, M.; Fuertes, A.B. Chemical and Structural Properties of Carbonaceous Products Obtained by Hydrothermal Carbonization of Saccharides. *Chem.-A Eur. J.* **2009**, *15*, 4195–4203. [CrossRef]
51. Zhang, C.; Su, J.; Zhu, H.; Xiong, J.; Liu, X.; Li, D.; Chen, Y.; Li, Y. The removal of heavy metal ions from aqueous solutions by amine functionalized cellulose pretreated with microwave-H_2O_2. *RSC Adv.* **2017**, *7*, 34182–34191. [CrossRef]
52. Ding, J.; Pu, L.; Wang, Y.; Wu, B.; Yu, A.; Zhang, X.; Pan, B.; Zhang, Q.; Gao, G. Adsorption and Reduction of Cr(VI) Together with Cr(III) Sequestration by Polyaniline Confined in Pores of Polystyrene Beads. *Environ. Sci. Technol.* **2018**, *52*, 12602–12611. [CrossRef] [PubMed]
53. Zhu, K.; Chen, C.; Xu, H.; Gao, Y.; Tan, X.; Alsaedi, A.; Hayat, T. Cr(VI) Reduction and Immobilization by Core-Double-Shell Structured Magnetic Polydopamine@Zeolitic Idazolate Frameworks-8 Microspheres. *ACS Sustain. Chem. Eng.* **2017**, *5*, 6795–6802. [CrossRef]
54. Peng, D.L.; Sumiyama, K.; Oku, M.; Konno, T.J.; Wagatsuma, K.; Suzuki, K.J. X-ray diffraction and X-ray photoelectron spectra of Fe-Cr-N films deposited by DC reactive sputtering. *Mater. Sci.* **1999**, *34*, 4623–4628. [CrossRef]

Disclaimer/Publisher's Note: The statements, opinions and data contained in all publications are solely those of the individual author(s) and contributor(s) and not of MDPI and/or the editor(s). MDPI and/or the editor(s) disclaim responsibility for any injury to people or property resulting from any ideas, methods, instructions or products referred to in the content.

MDPI
St. Alban-Anlage 66
4052 Basel
Switzerland
www.mdpi.com

Water Editorial Office
E-mail: water@mdpi.com
www.mdpi.com/journal/water

Disclaimer/Publisher's Note: The statements, opinions and data contained in all publications are solely those of the individual author(s) and contributor(s) and not of MDPI and/or the editor(s). MDPI and/or the editor(s) disclaim responsibility for any injury to people or property resulting from any ideas, methods, instructions or products referred to in the content.

www.ingramcontent.com/pod-product-compliance
Lightning Source LLC
LaVergne TN
LVHW070425100526
838202LV00014B/1532